CONCEPTS AND METHODS IN MODERN THEORETICAL CHEMISTRY

STATISTICAL MECHANICS

Atoms, Molecules, and Clusters
Structure, Reactivity, and Dynamics
Series Editor: Pratim Kumar Chattaraj

Aromaticity and Metal Clusters
Edited by Pratim Kumar Chattaraj

Concepts and Methods in Modern Theoretical Chemistry: Electronic Structure and Reactivity
Edited by Swapan Kumar Ghosh and Pratim Kumar Chattaraj

Concepts and Methods in Modern Theoretical Chemistry: Statistical Mechanics
Edited by Swapan Kumar Ghosh and Pratim Kumar Chattaraj

Quantum Trajectories
Edited by Pratim Kumar Chattaraj

ATOMS, MOLECULES, AND CLUSTERS

CONCEPTS AND METHODS IN MODERN THEORETICAL CHEMISTRY

STATISTICAL MECHANICS

EDITED BY

SWAPAN KUMAR GHOSH

PRATIM KUMAR CHATTARAJ

CRC Press
Taylor & Francis Group
Boca Raton London New York

CRC Press is an imprint of the
Taylor & Francis Group, an **informa** business

CRC Press
Taylor & Francis Group
6000 Broken Sound Parkway NW, Suite 300
Boca Raton, FL 33487-2742

Printed on acid-free paper
Version Date: 20121214

International Standard Book Number: 978-1-4665-0620-6 (Hardback)

Library of Congress Cataloging-in-Publication Data

Concepts and methods in modern theoretical chemistry : statistical mechanics / edited by Swapan Kumar Ghosh and Pratim Kumar Chattaraj.
pages cm. -- (Atoms, molecules, and clusters)
Includes bibliographical references and index.
ISBN 978-1-4665-0620-6 (hardback)
1. Chemistry, Physical and theoretical. I. Ghosh, Swapan Kumar.

QD453.3.C64 2013
541'.2--dc23 2012044530

Visit the Taylor & Francis Web site at
http://www.taylorandfrancis.com

and the CRC Press Web site at
http://www.crcpress.com

Contents

Series Preface

ATOMS, MOLECULES, AND CLUSTERS: STRUCTURE, REACTIVITY, AND DYNAMICS

While atoms and molecules constitute the fundamental building blocks of matter, atomic and molecular clusters lie somewhere between actual atoms and molecules and extended solids. Helping to elucidate our understanding of this unique area with its abundance of valuable applications, this series includes volumes that investigate the structure, property, reactivity, and dynamics of atoms, molecules, and clusters.

The scope of the series encompasses all things related to atoms, molecules, and clusters including both experimental and theoretical aspects. The major emphasis of the series is to analyze these aspects under two broad categories: approaches and applications. The *approaches* category includes different levels of quantum mechanical theory with various computational tools augmented by available interpretive methods, as well as state-of-the-art experimental techniques for unraveling the characteristics of these systems including ultrafast dynamics strategies. Various simulation and quantitative structure–activity relationship (QSAR) protocols will also be included in the area of approaches.

The *applications* category includes topics like membranes, proteins, enzymes, drugs, biological systems, atmospheric and interstellar chemistry, solutions, zeolites, catalysis, aromatic systems, materials, and weakly bonded systems. Various devices exploiting electrical, mechanical, optical, electronic, thermal, piezoelectric, and magnetic properties of those systems also come under this purview.

The first two books in the series are (a) *Aromaticity and Metal Clusters* and (b) *Quantum Trajectories*. A two-book set on *Concepts and Methods in Modern Theoretical Chemistry*, edited by Swapan Kumar Ghosh and Pratim Kumar Chattaraj, is the new addition to this series. The first book focuses on the electronic structure and reactivity of many-electron systems and the second book deals with the statistical mechanical treatment of collections of such systems.

Pratim Kumar Chattaraj
Series Editor

Foreword

A certain age comes when it is no longer unseemly to reflect on one's contribution to the world and, in the case of a scientist, the mark one has left on one's career. Professor B. M. Deb has reached such an age and can look back with considerable satisfaction on his scientific legacy. I knew him long ago, when his career was still to come, when he was at Oxford and was forming his aspirations and skills. Now, long after, in these volumes, we are seeing where those aspirations and skills in due course led.

One of the principal contributions of theoretical chemistry to what might be called "everyday" chemistry is its development of powerful computational techniques. Once such techniques were regarded with suspicion and of little relevance. But in those days the techniques were primitive, and the hardware was barely adequate for the enormous computations that even the simplest molecules require. Then, over the decades, techniques of considerable sophistication emerged, and the hardware evolved in unimaginable ways to accommodate and inspire even more imagination and effort. Now, the computations give great insight and sometimes surpass even actual measurements.

Of these new techniques, the most intriguing, and currently one in high fashion, has been the density functional theory. That Professor Deb has contributed so much in this field is demonstrated by the number of contributions in these volumes that spring from his work. Fashions, of course, come and go, but these techniques are currently having a considerable impact on so many branches of chemistry that they are undoubtedly a good reason for Professor Deb to reflect, with characteristic but misplaced modesty, on what he has done to promote and advance the technique.

It was for me a great pleasure to know the young Professor Deb and to discern promise and to know that the contributions to these volumes show that that promise has been more than amply fulfilled in a lifetime of contributions to theoretical chemistry. Professor Deb must be enormously proud of having inspired these volumes, and justly so.

Peter Atkins
Oxford

Preface

This collection presents a glimpse of selected topics in theoretical chemistry by leading experts in the field as a tribute to Professor Bidyendu Mohan Deb in celebration of his seventieth birthday.

The research of Professor Deb has always reflected his desire to have an understanding and rationalization of the observed chemical phenomena as well as to predict new phenomena by developing concepts or performing computations with the help of available theoretical, modeling, or simulation techniques. Formulation of new and more powerful theoretical tools and modeling strategies has always formed an ongoing and integral part of his research activities. Proposing new experiments, guided by theoretical insights, has also constituted a valuable component of his research that has a fairly interdisciplinary flavor, having close interconnections with areas like physics and biology.

The concept of single-particle density has always fascinated him, perhaps starting with his work on force concept in chemistry, where the density is sufficient to obtain Hellmann–Feynman forces on the nuclei in molecules. His two reviews on "Force Concept in Chemistry" and "Role of Single Particle Density in Chemistry," published in *Reviews of Modern Physics*, have provided a scholarly exposition of the intricate concepts, inspiring tremendous interest and growth in this field. These have culminated in two edited books. The force concept provided the vehicle to go to new ways of looking at molecular shapes, the HOMO postulate being an example of his imaginative skills. The concept of forces on the nuclei was soon generalized to the concept of stress tensor within the electron cloud in molecules, the role of which in determining chemical binding and stability of molecules was also explored. Various aspects of the density functional theory (DFT) were investigated. The static aspects were soon viewed as only a special case of the corresponding dynamical theory, the so-called quantum fluid dynamics (QFD), which was developed in 3-D space and applied to study collision phenomena, response to external fields, and other related problems.

His mind has always opened new windows to bring in the fresh flavor of novel concepts for interpreting the "observed," predicting the "not yet observed," and also created tools and strategies to conquer unknown territory in the world of molecules, materials, and phenomena. "Concepts are the fragrance of science," he always emphasizes. His research has often seemed to be somewhat unconventional in the sense that he has always stressed conceptual developments that are often equally suited for practical applications as well. He has a thirst for looking into the secret of "why things are the way they are" and the mystery behind "being to becoming," focusing on the structure and dynamics of systems and phenomena, both of which have been enriched immensely by his contributions. Aptly, we have the two present books covering structure and dynamics, respectively.

The topics in *Concepts and Methods in Modern Theoretical Chemistry: Electronic Structure and Reactivity* include articles on DFT, particularly the functional and conceptual aspects, excited states, molecular electrostatic potentials, intermolecular interactions, general theoretical aspects, application to molecules, clusters and solids, electronic stress, the information theory, the virial theorem, new periodic tables, the role of the ionization potential and electron affinity difference, etc. The majority of the chapters in *Concepts and Methods in Modern Theoretical Chemistry: Statistical Mechanics* include time-dependent DFT, QFD, photodynamic control, nonlinear dynamics, molecules in laser field, charge carrier mobility, excitation energy transfer, chemical reactions, quantum Brownian motion, the third law of thermodynamics, transport properties, nucleation, etc.

In the Indian context, theoretical chemistry has experienced significant growth over the years. Professor Deb has been instrumental in catalyzing this growth by providing the seed and nurturing young talents. It is the vision and effort of Professor Deb that made it possible to inspire the younger generation to learn, teach, and practice theoretical chemistry as a discipline. In this context, it is no exaggeration to describe him as the doyen of modern theoretical chemistry in India.

Professor Deb earned a PhD with Professor Charles Coulson at the University of Oxford and then started his professional career at the Indian Institute of Technology, Bombay, in 1971. Being a scientist–humanist of the highest order, he has always demanded a high sense of integrity and a deep involvement from his research group and other students. He has never sacrificed his own human qualities and never allowed other matters to overtake the human aspects of life.

While his research has focused on conceptual simplicity, computational economy, and sound interpretive aspects, his approach to other areas of life reflects the same. We have often wondered at the expanse of his creativity, which is not restricted to science but also covers art, literature, and life in general. His passion for work has, of course, never overshadowed his warmth, affection, and helpfulness to others. He has an extraordinary ability to act as a creative and caring mentor. His vast knowledge in science, art, literature, and many other of the finer aspects of life in general, together with his boundless sources of enthusiasm, creativity, and imagination, has often made him somewhat unconventional in his thinking, research, and teaching. Designing new experiments in class and introducing new methods in teaching have also been his passion. His erudition and versatility are also reflected in his writings on diverse topics like the cinema of Satyajit Ray and lectures on this as well as various aspects of art.

We are privileged to serve as editors of these two books on Concepts and Methods in Modern Theoretical Chemistry and offer the garland of scholarly essays written by experts as a dedication to this great scientist–humanist of recent times with affection and a deep sense of respect and appreciation for all that he has done for many of us and continues to do so. We also gratefully acknowledge the overwhelming and hearty response received from the contributors, to whom we express our indebtedness.

We are grateful to all the students, associates, and collaborators of Professor B. M. Deb who spontaneously contributed to the write-up of the "Reminiscences" and, in particular, Dr. Amlan K. Roy for compiling it in a coherent manner to the present form. Finally, we are deeply indebted to Professor B. M. Deb for his kind help, guidance, and encouragement throughout our association with him.

Swapan Kumar Ghosh
Pratim Kumar Chattaraj

Reminiscences

It is indeed a great pleasure to pen this note in celebration of Professor B. M. Deb's seventieth birthday. For many of us, he is a mentor, confidante, and adviser. Many others look at him as an extraordinary teacher; a patient, encouraging, and motivating guide; a warm and caring human being; and a connoisseur of literature, art, and so on. His dedication and passion for science is infectious.

Many of us have been fortunate to attend his lectures on quantum chemistry, structure, bonding, symmetry, and group theory, which were all about the interlinking of abstract concepts that are often sparsely scattered. After trudging along a series of lectures, one is rewarded with the eventual conclusion that all chemical bonds are mere manifestations of a single phenomenon, namely, the redistribution of electron density. Often, he would explain physics from real-life analogies rather than try to baffle and intimidate audiences with lots of mathematics—a popular trick often used in the community. Just paying attention in his class gives one enough confidence to tackle the most challenging problems in quantum chemistry. His recent endeavor to initiate a course on Indian heritage has been highly appreciated. It is not a history class, as the title may imply to some people, but rather a scientific evaluation of the Indian past. Taking examples from our glorious past, the course differentiates between easy and right about scientific ethics and logically establishes the path one should follow for uplifting individual souls and society as a whole. Although a theoretician, his enthusiasm and excitement for practical applications of science is no less. The experiments on beating hearts and chemical oscillations are among the most popular in the class.

His books *The Force Concept in Chemistry* and *The Single-Particle Density in Physics and Chemistry* were hugely influential among those who sought, in quantum chemistry, not just a computational tool for the calculation of molecular properties, but a fundamental understanding of the physics of chemical bonding and molecular reactivity. The application of the Hellmann–Feynman theorem to provide qualitative insights into chemical binding in molecules as well as molecular shapes caught the interest of even R. P. Feynman. As a research student, his communication with Professor Feynman was a matter of great amazement, motivation, and pride for many of his early PhD students, as Dr. Anjuli S. Bamzai recalls. Despite his considerable work in density functional theory (DFT), he held an agnostic attitude toward it, in the sense that he did not regard the search for a functional as the holy grail of DFT or see DFT as being somehow in opposition to wave function–based theories. He was also not against approximations and freely employed them wherever useful. But he was convinced that the electron density held the key to a deeper understanding of the chemical phenomena. Thus, in a way, he was willing to entertain the need for considering the phase in addition to density to achieve a consistent treatment of excited states and time-dependent phenomena.

To have worked with him has been a major turning point in our lives. We discover him as a scientist with high morality and professional ethics. It is not only learning the concepts in theoretical chemistry but also a more holistic approach toward research, learning, and science itself. While scrupulously fair, he expected his students to be conscientious. He gave his all to his students and to his research. Reasonably enough, he expected no less from his students and from his colleagues, a favorite expression being that he wanted the students "to go flat out" on their prospective research problems. The amount of hard work that he put, propelled by tiny seeds of imagination and analytical logic, always inspired us. But while the force of his scientific conviction was strong, he was always open to arguments and discussion. Even in turbulent times and under less-than-ideal conditions, he was not willing to compromise on his scientific standards or integrity. He had a knack for choosing and working on problems that were emerging frontiers of theoretical chemistry. That was because of his intuition to choose research projects for us so that we could contribute to the field effectively, despite the fact that all his research works were done in India in relative isolation. Although much of his research career spanned the overlap between physics and chemistry, he had no sympathy for those who would regard chemistry as inferior to physics. When a physicist, after hearing Professor Deb speak about his current research, praised him with the words, "You are almost doing physics," he rejoined with a wry smile, "No, I am doing good chemistry." With this statement, even his detractors would agree!

It feels amazing that we have learned as much from anecdotal informal interaction with him as from the research experience. What added to the pleasure of working with him were discussions about science and nonscientific matters. It was fascinating to listen to him talk about poetry, literature, movies, food, art, and cultures across the world. We would occasionally visit his residence and spend time with him at the dining table discussing the progress of our projects while partaking of delicious snacks and meals prepared by Mrs. Deb. For many of us, it was something like a home away from home, and we soon learned that a combination of food and food for thought goes well together. The amazement of such an experience is narrated here by Dr. Bamzai. Their home was decorated with the works of some of the greatest artists of all time. Often one would come across a discussion about Leonardo da Vinci's *The Last Supper* or Picasso's *Guernica* and how the artist, through his work, had conveyed the tragedies of war and its horrific impact on innocent civilians. At other times, he would discuss how M. C. Escher's art effectively conveys important concepts such as symmetry and transformations in crystallography. He has serious concern also about science, culture, and heritage. He constantly engages into the popularization of science as well as the improvement of the education system in India. It is surprising how he was able to impart knowledge on such a diverse array of topics.

Given his varied interests and the positive energy that he imbues into his surroundings, we know that he will never stop being an academic. Despite his own and Mrs. Deb's deteriorating health, they have stood beside their students and colleagues with constant support and encouragement. Many of us remember the act of good Samaritan-ship by Professor Deb and his family toward his colleagues. One such act is vividly recollected here by Professor Harjinder Singh, whose daughter was struggling in an intensive care unit at that time. They needed to stay at a place close to the hospital. Deb's family extended their wholehearted support during that crisis, not minding any inconvenience caused to them, especially when the city of Chandigarh was going through the political turmoil of a full-blown secessionist movement, regular terrorist threats, shootings, bus bombings, and assassinations.

A lesson we learned from Professor Deb that we have carried throughout our life was his admonition: "Beware of the fourth rater who calls the third rater good." It was a call and a challenge to aspire to the highest standards of excellence in life, and it is the pursuit of this gold standard that he strived to inculcate in us, despite potential temptations to discard it so often! We consider ourselves very fortunate to have Professor Deb as our teacher, philosopher, and guide. His work and work ethic will continue to influence and nurture future generations via many students and postdocs he has taught and guided. He remains a source of inspiration to all who wish to be an ideal teacher, a thorough researcher, and, above all, a decent human being. We feel privileged to be a part of his extended family and take this opportunity to express our sincere gratitude to him for his support, kindness, and patience. We are indebted to him and send our best wishes to his family.

Anjuli S. Bamzai
Pratim K. Chattaraj
Mukunda Prasad Das
Swapan K. Ghosh
Neetu Gupta
Geeta Mahajan
Smita Rani Mishra
Amitabh Mukherjee
Aniket Patra
Amlan K. Roy
Mainak Sadhukhan
R. P. Semwal
Harjinder Singh
Ranbir Singh
Nagamani Sukumar
Vikas
Amita Wadehra

Editors

Swapan Kumar Ghosh earned a BSc (Honors) and an MSc from the University of Burdwan, Bardhaman, India, and a PhD from the Indian Institute of Technology, Bombay, India. He did postdoctoral research at the University of North Carolina, Chapel Hill. He is currently a senior scientist with the Bhabha Atomic Research Centre (BARC), Mumbai, India, and head of its theoretical chemistry section. He is also a senior professor and dean-academic (Chemical Sciences, BARC) of the Homi Bhabha National Institute, Department of Atomic Energy (DAE), India, and an adjunct professor with the University of Mumbai–DAE Centre of Excellence in Basic Sciences, India.

He is a fellow of the Indian Academy of Sciences, Bangalore; Indian National Science Academy, New Delhi; National Academy of Sciences, India, Allahabad; Third World Academy of Sciences (TWAS), Trieste, Italy (currently known as the Academy of Sciences for the Developing World); and Maharashtra Academy of Sciences. He is a recipient of the TWAS prize in chemistry; silver medal of the Chemical Research Society of India (CRSI); the Jagdish Shankar Memorial Lecture Award of the Indian National Science Academy; the A. V. Rama Rao Prize of Jawarharlal Nehru Centre for Advanced Scientific Research, Bangalore, India; and the J. C. Bose Fellowship of the Department of Science and Technology, India. He is currently also one of the vice presidents of CRSI.

His research interests are theoretical chemistry, computational materials science, and soft condensed matter physics. He has been involved in teaching and other educational activities including the Chemistry Olympiad Program. He has twice been the mentor and delegation leader of the Indian National Chemistry Olympiad Team participating in the International Chemistry Olympiad at Athens (Greece) and Kiel (Germany).

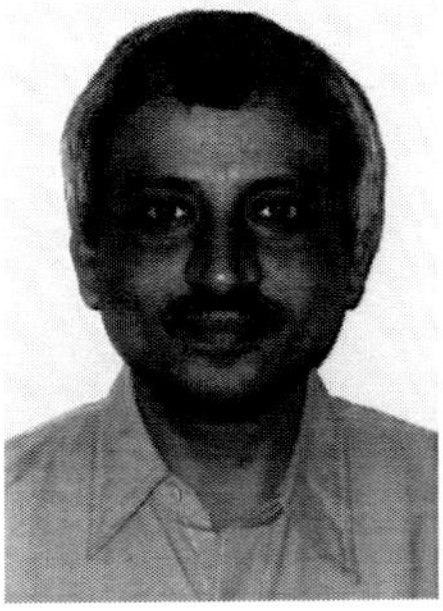

Pratim Kumar Chattaraj earned a BSc (Honors) and an MSc from Burdwan University and a PhD from the Indian Institute of Technology (IIT), Bombay, India, and then joined the faculty of the IIT, Kharagpur, India. He is now a professor with the Department of Chemistry and also the convener of the Center for Theoretical Studies there. In the meantime, he visited the University of North Carolina, Chapel Hill, as a postdoctoral research associate and several other universities throughout the world as a visiting professor. Apart from teaching, Professor Chattaraj is involved in research on density functional theory, the theory of chemical reactivity, aromaticity in metal clusters, *ab initio* calculations, quantum trajectories, and nonlinear dynamics. He has been invited to deliver special lectures at several international conferences and to contribute chapters to many edited volumes.

Professor Chattaraj is a member of the editorial board of *J. Mol. Struct. Theochem* (currently *Comp. Theo. Chem.*), *J. Chem. Sci., Ind. J. Chem.-A*, *Nature Collections Chemistry*, among others. He has edited three books and special issues of different journals. He was the head of the Department of Chemistry, IIT, Kharagpur, and a council member of the Chemical Research Society of India. He is a recipient of the University Gold Medal, Bardhaman Sammilani Medal, INSA Young Scientist Medal, B. C. Deb Memorial Award, B. M. Birla Science Prize, and CRSI Medal. He was an associate of the Indian Academy of Sciences. He is a fellow of the Indian Academy of Sciences (Bangalore), the Indian National Science Academy (New Delhi), the National Academy of Sciences, India (Allahabad), and the West Bengal Academy of Science and Technology. He is also a J. C. Bose National Fellow and a member of the Fonds Wetenschappelijk Onderzoek (FWO), Belgium.

Contributors

Satrajit Adhikari
Department of Physical Chemistry
Indian Association for the Cultivation of Science
Kolkata, India

Bidhan Chandra Bag
Department of Chemistry
Siksha Bhavana
Visva Bharati
Santiniketan, India

Biman Bagchi
Solid State and Structural Chemistry Unit
Indian Institute of Science
Bangalore, India

Ranjit Biswas
Department of Chemical, Biological and Macromolecular Sciences
S. N. Bose National Centre for Basic Sciences
Kolkata, India

María Luisa Cerón
Laboratorio de Química Teórica Computacional (QTC)
Facultad de Química
Pontificia Universidad Católica de Chile
Santiago, Chile

Amalendu Chandra
Department of Chemistry
Indian Institute of Technology
Kanpur, India

V. K. Chandrasekar
Centre for Nonlinear Dynamics
School of Physics
Bharathidasan University
Tiruchirappalli, India

Jun Cheng
Department of Chemistry
University of Cambridge
Cambridge, United Kingdom

Shih-I Chu
Center for Quantum Science and Engineering and Department of Physics
National Taiwan University
Taipei, Taiwan

and

Department of Chemistry
University of Kansas
Lawrence, Kansas

Snehasis Daschakraborty
Department of Chemical, Biological and Macromolecular Sciences
S. N. Bose National Centre for Basic Sciences
Kolkata, India

Ayan Datta
Department of Spectroscopy
Indian Association for the Cultivation of Science
Jadavpur, Kolkata
West Bengal, India

Sushanta Dattagupta
Visva–Bharati
Bolpur, West Bengal, India

and

Indian Institute of Science Education and Research, Kolkata
Mohanpur Campus
Nadia, West Bengal, India

Arnab Ghosh
Indian Association for the Cultivation of Science
Jadavpur, Kolkata, India

Soledad Gutiérrez-Oliva
Laboratorio de Química Teórica Computacional (QTC)
Facultad de Química
Pontificia Universidad Católica de Chile
Santiago, Chile

Bárbara Herrera
Laboratorio de Química Teórica Computacional (QTC)
Facultad de Química
Pontificia Universidad Católica de Chile
Santiago, Chile

John T. Heslar
Center for Quantum Science and Engineering and Department of Physics
National Taiwan University
Taipei, Taiwan

Peter Holland
Green Templeton College
University of Oxford
Oxford, United Kingdom

Vandana Kurkal-Siebert
BASF-SE
Ludwigshafen, Germany

M. Lakshmanan
Centre for Nonlinear Dynamics
School of Physics
Bharathidasan University
Tiruchirappalli, India

Deepak Mathur
Tata Institute of Fundamental Research
Mumbai, India

Manoj K. Mishra
Department of Chemistry
Indian Institute of Technology Bombay
Mumbai, India

and

University of Lucknow
Lucknow, India

S. Mohakud
Theoretical Sciences Unit
Jawaharlal Nehru Center for Advanced Scientific Research
Jakkur Campus
Bangalore, India

S. K. Pati
Theoretical Sciences Unit
and
New Chemistry Unit
Jawaharlal Nehru Center for Advanced Scientific Research
Jakkur Campus
Bangalore, India

Aniket Patra
Indian Institute of Science Education and Research, Kolkata
Mohanpur Campus
Nadia, West Bengal, India

Deb Shankar Ray
Indian Association for the Cultivation of Science
Jadavpur, Kolkata, India

Mantu Santra
Solid State and Structural Chemistry Unit
Indian Institute of Science
Bangalore, India

Manabendra Sarma
Department of Chemistry
Indian Institute of Technology Guwahati
Guwahati, India

K. L. Sebastian
Department of Inorganic and Physical Chemistry
Indian Institute of Science
Bangalore, India

Bhavesh K. Shandilya
Department of Chemistry
Indian Institute of Technology Bombay
Mumbai, India

Jane H. Sheeba
Centre for Nonlinear Dynamics
School of Physics
Bharathidasan University
Tiruchirappalli, India

Rakesh S. Singh
Solid State and Structural Chemistry Unit
Indian Institute of Science
Bangalore, India

Sudarson Sekhar Sinha
Indian Association for the Cultivation of Science
Jadavpur, Kolkata, India

Michiel Sprik
Department of Chemistry
University of Cambridge
Cambridge, United Kingdom

R. S. Swathi
School of Chemistry
Indian Institute of Science Education and Research
Thiruvananthapuram, India

Dmitry A. Telnov
Department of Physics
St. Petersburg State University
St. Petersburg, Russia

Ashwani K. Tiwari
Indian Institute of Science Education and Research
Kolkata, India

Alejandro Toro-Labbé
Laboratorio de Química Teórica Computacional (QTC)
Facultad de Química
Pontificia Universidad Católica de Chile
Santiago, Chile

An Interview with B. M. Deb

(This interview was conducted by Richa Malhotra for the journal *Current Science*. An edited version of the interview was published in *Current Science* on January 25, 2012, and an expanded version appears in the present book. Courtesy of *Current Science*.)

- **How has the field of theoretical and computational chemistry evolved over the years?**

  One has to write a book to answer this! It is similar to answering how science has evolved in the last century and the present century so far. Let me try to explain how the broad contours of theoretical and computational chemistry have developed over many years.

  Theoretical chemistry has been operating at the interface between chemistry, physics, biology, mathematics, and computational science. It deals with *systems* and *phenomena* concerning these large subjects. The *systems* are microscopic, mesoscopic, and macroscopic, *viz.*, atoms, molecules, clusters, and other nanosystems, soft and hard condensed matter. The *phenomena* involve a holistic combination of *structure*, *dynamics*, and *function*. *Structure* concerns itself with geometry, where both Group Theory and Topology (especially, Graph Theory) come in. *Dynamics* deal with evolution in time; structure is a consequence of dynamics, and *vice versa*. *Function* implies all kinds of properties, *viz.*, electrical, magnetic, optical, chemical, biological, and even mechanical properties. Let me show this by a triangular SDF figure, which is actually *valid for every field of human endeavor,* including literature and arts.

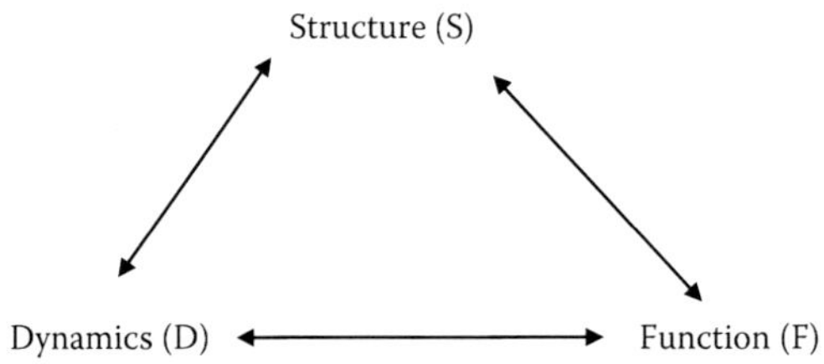

The disciplines which study all of these are quantum chemistry (both nonrelativistic and relativistic), quantum biology and biochemistry, quantum pharmacology, spectroscopy, molecular reaction dynamics, equilibrium and nonequilibrium statistical mechanics, equilibrium and nonequilibrium thermodynamics, nonlinear dynamics, mathematical methods of chemistry, etc., with various subdisciplines. It is interesting to note that Graph Theory, which is a branch of mathematics and is also used by chemists, physicists, biologists, sociologists, electrical engineers, neural and other network scientists, had drawn primary inspiration in the 1870s from structural formulas in chemistry which denote *connectivity*. As you see, at this level, it is really not possible to distinguish between theoretical chemistry and theoretical physics or, for that matter, theoretical biology. Atomic and molecular physics, polymer and condensed matter physics—bringing in materials science—and even certain issues of structure and interaction in nuclear physics are of interest to theoretical chemists. Present-day mathematical chemistry, which uses topology—though not necessarily in conjunction with quantum mechanics—to develop quantitative structure–activity relations for drug design, hazard chemicals assessment, etc., is another aspect of theoretical chemistry.

Computational chemistry has been primarily concerned with the development and application of computer software, using theoretical chemistry methodologies, utilizing numerical methods and computer programming in a significant way. Nowadays, not all theoretical chemists and computational chemists develop their own codes. Only some do, if necessary, whereas others employ standard and/or commercially available software packages for performing computations on electronic structure; geometry; various chemical, physical, and biological properties; as well as various kinds of classical and semiclassical simulations of structures and dynamics of large molecular systems such as proteins, polymers, and liquids. Since the 1930s, theoretical and computational chemists have been a major driving force behind developments in computational sciences, including both computer hardware and software development, especially number crunching and graphics. Graphics are particularly important because chemists find it difficult to work without detailed visualization. Also, representing millions of computed numbers in terms of colorful pictures, which could undulate in time as well, greatly enhances our insight into the phenomenon being studied. Presently, we feel that any equation which cannot be solved analytically but the solution exists can be solved numerically with an accuracy which goes beyond experimental accuracy as long as the variables are not too many in number.

Because of their subject's complex multidisciplinary nature, theoretical chemists have been somewhat like orphans! You can find them everywhere, in chemistry, biochemistry, physics, mathematics, computer science, chemical engineering, materials science, as well as in industries across the world, though I believe very few, if any at all, in Indian industries. Since earth scientists are currently using theoretical chemistry computations for interpretations of their data, perhaps we will soon have a theoretical chemist in an earth science department!

Historically, mathematicians, physicists, chemists, computer scientists, and even economists have contributed to theoretical chemistry. Apart from the mathematician J. Sylvester's realization in the 1870s that structural formulas in chemistry have a hidden algebraic structure, I think Gibbs's development of thermodynamics, followed by the axiomatic development of the subject by Carathéodory and Born, Debye–Hückel–Onsager theory of strong electrolytes, Lewis's electronic theory of valence, the vector atom model, etc., are some of the important landmarks in the early days of theoretical chemistry. Once quantum mechanics came into being, there resulted an explosive growth in the areas that I have mentioned earlier.

- **What have been your key contributions and areas of interest in chemistry?**

It is embarrassing to talk about "key contributions" of a mediocre scientist. I always believed that (1) theory should not only explain current experiment but should also make predictions for future experiment and that (2) concepts are the fragrance of science. Therefore, along with my research students, I have been struggling to develop rigorous concepts in chemistry which can lead to deep insights, as well as accurate results which are amenable to physical and pictorial interpretation. Whatever we have been able to do has been possible only because of my courageous students.

Because of my persistent interest in geometry, our first work in India was to develop a purely qualitative and general molecular–orbital approach (without computation and by using group theory extensively), leading to a force model for explaining and predicting various features of molecular geometry of small- and medium-sized molecules based on the electron density in the highest occupied molecular orbital. This was followed by semi-empirical computations of electronic structure and geometry of quite a few unknown molecules, predicting that they are capable of independent existence. Along with these, we wrote an article entitled "the force concept in chemistry." The responses to this article changed the course of our research, especially Fano's suggestion that we think of how the electron density can be calculated without the wave function and Feynman's suggestion that we look into internal stresses in molecules. Even though we did not know then of Hohenberg–Kohn theorems and Kohn–Sham equations, we were already completely convinced of the fundamental significance of electron density in three-dimensional space and strongly felt that, through the electron density, nonrelativistic quantum phenomena can have "classical" interpretations, which are necessary for pictorial understanding. Based on Feynman's suggestion, we defined an electrostatic stress tensor using the electron density and showed that this has the same form as Maxwell's stress tensor for classical electromagnetic fields. Furthermore, along the bond direction in a diatomic molecule, the appropriate component of the stress tensor shows an extremum at the equilibrium internuclear distance. In trying to understand why stress tensor should be such an important entity, we realized that we have to go to classical fluid dynamics. The fluid dynamical interpretation of one-electron systems was already in existence but taking it to many-electron systems was rather difficult. We therefore developed what we called a quantum fluid dynamical interpretation of many-electron systems in terms of the electron density and defined a comprehensive stress tensor for such a system in terms of the full nonrelativistic Hamiltonian, i.e., by incorporating kinetic, electrostatic, exchange, and correlation terms. This still had the same form as Maxwell's tensor. We then defined a general criterion for the stability of matter, *viz.,* the force density obtained from this stress tensor must vanish at every point in three-dimensional space. The stress tensor, however, did not yield a deterministic equation for the electron density which has to incorporate both *space and time.*

Time was of the essence in our struggle. The two interlinked bottlenecks in the electron density approach were time dependence and excited states. We first developed a rigorous time-dependent density functional theory for a certain class of potentials by utilizing QFD. Since this version of density functional theory was not exact for all potentials, we also developed a similar approach in terms of natural orbitals which are exact in principle. This approach yielded an equation for the ground state density whose accuracy was very good. Using this, we calculated the frequency-dependent multipole (2^l-pole, $l = 1, 2, 3, 4$) polarizabilities of atoms. Some of these computed numbers still await experimental verification.

Efforts continued to generate more accurate equations for directly determining density by a single equation no matter how many electrons are there in the system. One such effort yielded a quadratic—rather than a differential—time-independent equation whose density yielded many interesting results. This equation led to an effort to justify the existence of empirically finite atomic, ionic, and Van der Waals radii—even though quantum mechanically these radii are infinite—by adopting a conjecture that such finite radii are decided by a single universal value of the electron density in space. Finally, we were able to obtain a fascinating nonrelativistic nonlinear single partial differential equation for the direct determination of electron density and properties of many-electron systems. Besides applying this equation to the ground state and time-dependent situations, application was also made to proton-atom high-energy scattering; it was possible to identify *approach, encounter, and departure regimes,* which should be helpful in chemical reactions. This equation has a number of interesting mathematical properties, some of which we have examined while others remain unexplored. A relativistic quantum fluid dynamical density approach was also developed. Additionally, we have written quite a bit in trying to emphasize the fundamental significance of electron density in understanding structures, dynamics, and properties.

An important job of theory is to explain and predict phenomena. Two decades ago, we became interested in atoms and molecules under extreme conditions such as intense laser and strong magnetic fields. We pushed our time-dependent density equation into these difficult situations. With lasers, quite exciting results and insights were obtained into various multiphoton processes such as spatial shifting of density in both femtoseconds and attoseconds, photoionization spectra, high-order harmonics generation (its implication is the creation of attosecond and X-ray lasers), suppression of ionization under a superintense laser, Coulomb explosion in molecular dissociation, etc. The mechanism of shortening of bond length in a diatomic molecule under strong magnetic fields was also studied. We have predicted that, if an oscillating strong magnetic field of an appropriate frequency interacts with a hydrogen atom, coherent radiation should be emitted. This remains to be experimentally verified. We have also proposed a new dynamical signature of quantum chaos and demonstrated it with strong magnetic fields.

We now come to excited states. Using a hybrid wave function-density approach and an interpretation of exchange proposed by other workers, we have been able to calculate excited state energies and densities of several hundred excited states of various atoms. These were singly, doubly, and triply excited states, autoionizing states, satellite states, etc., and involved both small and large energy differences.

- **You have worked mainly outside the boundaries of "chemistry." What are the interdisciplinary areas that you have worked on?**

I do not think I have worked outside the boundaries of chemistry, which are actually limitless. An interdisciplinary research area that I have pursued is the quantum theory of structures, dynamics, and properties of atoms and molecules. I have had great pleasure in devoting some time over the years for designing exciting and colorful chemical experiments, based on research literature, for undergraduate and postgraduate teaching laboratories. Each of these brings as many sciences as possible on the common platform of one single experiment. This was to partly satisfy my hunger for experimental chemistry! Also, writing on integrated learning in sciences, designing new curricula and developing new courses have been something of a passion. I have had a life-long interest in science, mathematics, literature, and art in Ancient and Medieval India, on which I am writing a book for the last five years. The idea of holism of Ancient Indians that everything is connected to everything else has always fascinated me because this is the essence of multidisciplinarity.

- **What do you see as things that have changed in the field of chemistry, especially theoretical chemistry and computational chemistry?**

I see considerable development in the interfaces between chemistry and biology, as well as between chemistry and materials science. Some development has also occurred in the interface between chemistry and earth science, as well as between chemistry and archaeology (e.g., archaeometry). With the advent of improved computer hardware and software, the way chemistry used to be done has changed, in the way data are recorded and analyzed. Computational chemistry software are being used almost routinely by many experimental chemists. Computational chemists are themselves using standard software packages to tackle more and more exciting and challenging problems. Two- and three-dimensional visualizations (graphics) are increasingly being employed. Experimentally, attempts are being made to probe single molecules rather than molecules in an assembly. Combinatorial chemistry, as well as green chemistry, has been in existence for some time. A synthesis protocol using artificial intelligence also exists. Attosecond (10^{-18} s) phenomena, concerned with electronic motions, have emerged very recently. Overall, I sense a great churning taking place in chemistry.

- **What do you think lies in the future of theoretical chemistry and computational chemistry?**

  If I am not wrong, of the total global population of theoretical and computational chemists, 90% or even more are computational chemists. Two things ought to be noted here. Software packages represent the technology of theoretical chemistry, and they employ existing theories which cannot be regarded as "perfect." Everybody knows that "all exact sciences are dominated by approximations." Chemical systems being highly complex, it would be rather unrealistic to play with toy models which admit analytical solutions. Therefore, the need for developing new concepts for improving existing theories would remain strong because this is an open-ended quest. Needless to say, software packages should not be used as "black boxes."

  I have a feeling that the number of theoretical chemists who can traverse the whole gamut of theoretical chemistry, *viz.*, generation of concepts, formalisms, algorithms, computer codes, and new ways of interpreting computed numbers, is decreasing steadily all over the world. Urgent replenishments are needed through the induction of bright, imaginative, and capable young chemists. In a way, theoretical chemists are akin to poets, admittedly with a practical bent of mind. We need to ensure that poetry, imagination, and the fun of making predictions do not disappear from chemistry.

- **Where do you think physical chemistry stands relative to other areas like inorganic and organic chemistry? (in terms of number of researchers, publications, Nobel Prizes, etc.)**

  Since my undergraduate days, I have been acutely uncomfortable with the attitude that chemistry can be completely classified into inorganic, organic, and physical chemistry. These are artificial intellectual barriers. The numbers of researchers and publications in certain areas of chemistry have been steadily increasing and will continue to do so. In terms of the number of researchers in various areas, there has been a seriously lopsided development in some countries because of the tripartite classification. One even hears of cases where there is a large number of Ph.D. students with just one Supervisor. I hope the situation will improve and a balanced development will take place. Until 1960s, successive Nobel Committees apparently did not find theoretical chemists worthy of the Nobel Prize, although the latter had enormous impact on the whole of chemistry. That also changed from the 1960s. Of late, even a theoretical condensed matter physicist has received the Nobel Prize in chemistry. So, the earlier we teach ourselves to climb over these barriers, the better for the growth of chemistry.

- **How are Physical Chemistry and Chemical Physics different from each other?**

  Since both the terms refer to the interface between chemistry and physics, they should have the same meaning. However, in usage, this is not so. The term "chemical physics" was coined in the postquantum mechanical era and popularized by journals in chemical physics. One might simplistically say that, if, in the chemistry–physics interface, one is veering more toward chemistry, then one is doing physical chemistry, whereas, if one veers more toward physics, one is doing chemical physics. Alternatively, since science develops by progressive quantification, one might say that chemical physics is the modern more quantified version of physical chemistry. But I think all such distinctions are somewhat contrived. However, chemical physics has certainly been enriched by contributions from many physicists who probably felt more comfortable with this term than "physical chemistry." It may be worth noting here that an "overzealous" scientist had once defined physical chemistry as "the study of everything that is interesting"!

- **How has computation changed the way research in chemistry is carried out?**

  Over the years, there has been a sea of change in the attitude of chemists. Earlier, any theoretical method and the numbers computed from it had to be justified by comparing with experimental results. This has drastically changed because of two reasons: First, the sophistication in theory, algorithms, and computer codes is now so good that these frequently deliver computed numbers much beyond present-day experimental accuracy. Second, there are many situations in which it is extremely difficult to perform an experiment, e.g., to study a very short-lived molecule or study a phenomenon like the folding of a protein in a biological environment. Theoretical and computational chemistry could be the only route to take in such cases.

  Let me give you an example of accuracy of a theoretical method. In the last five years, it has been possible to numerically solve the Schrödinger equation for some systems with a precision of forty significant figures! While I do not understand the experimental significance of numbers beyond a certain significant figure or what we can do with such precise numbers, the fact remains that such computational accuracy is now deliverable and it challenges experiment. This is definitely good for overall development. Another recent development is the experimental tomographical picture of the highest occupied molecular orbital of the N_2 molecule, which proved the physical existence of a wave function.

With the availability of dependable and commercially available software packages developed by theoretical and computational chemists, in collaboration with experts in numerical methods, an interesting situation has come about. The synthesis and structure of a new molecule discovered in the chemical laboratory is nowadays justified by experimentalists by doing a geometry optimization according to a good software package. On the other hand, the experimental determination of structure generally resorts to a combination of methods.

However, we should never forget that experiment and theory are the two wings of a bird named science. It cannot fly on only one wing.

- **On one hand, the boundaries between chemistry and other sciences are blurring, and on the other, chemistry is branching out into specialized courses/fields. How do you think this is making a difference? Is the effect getting balanced out in some way?**

I look at it differently, instead of a balance or an imbalance. I like to spell "Chemistry" as "Chemistree." The Tree of Chemistry is large. It has deep roots and spreads in all directions. It has many branches and subbranches. New branches, subbranches, and leaves sprout in the course of time. As chemists, we are like birds living on this tree. A group of birds might nest in a small subbranch. There is no harm in that as long as the birds leave their nest once in a while and become aware of the large tree.

I believe there is a network of sciences, humanities, and social sciences with chemistry as a central science. I think future teaching and research in chemistry might develop along such a network.

- **What significance did the International Year of Chemistry (2011) have for you? What would you like to see changing in the future about research in chemistry?**

Let me answer the first question first. Chemistry has always been a deeply humanistic subject. For six thousand years, chemistry has worked for the benefit of humankind. Therefore, IYC did not remind me anew of chemistry in the service of humankind. Instead, it reminded me of two individuals whom I greatly admire: Madame Marie Sklodowska Curie and Acharya (Sir) Prafulla Chandra Rây. Besides being the centenary of Madame Curie's Nobel Prize in chemistry, 2011 was also the 150th birth anniversary of Acharya Rây, the first modern chemist of India and, along with Acharya (Sir) J. C. Bose, the founder of modern scientific research in India. As a teacher, Acharya Rây had inspired Meghnad Saha (the founder of modern astrophysics), Satyendra Nath Bose (the founder of quantum statistics), Jnan Chandra Ghosh (pre-Debye–Hückel theory of strong electrolytes), and many others. Besides his own well-known researches in chemistry, he was the founder of the chemical and pharmaceutical industries in India and an indefatigable social reformer. He was one of the greatest sons and builders of modern India, greatly admired by Mahatma Gandhi and Rabindranath Tagore, as well as numerous other persons, because of his asceticism, scientific modernism, deep knowledge about classical Indian culture, and a life totally dedicated to others.

There are striking parallels in the lives of Madame Curie and Acharya Rây which set guidelines for other human beings: poverty, suffering, indomitable spirit which does not recognize any obstacle, pioneering works in spite of continued ill health, tremendous leadership, building of multiple institutions, as well as an ascetic life totally devoid of self and completely dedicated to the welfare of others.

Coming to the second question. Within the global scenario, I believe we are not too bad in dealing with problems of fundamental importance in chemistry. However, I would like to see much greater intensity here, in terms of issues which were not tackled before. Where I would like to see extensive leapfrogging is in the development of new and sophisticated technologies born in chemical research laboratories, in collaboration with other scientists and engineers, wherever necessary. Some examples would be attosecond and X-ray lasers, a working quantum computer, new drug molecules by drastically cutting down the laboratory-to-market time schedule through a clever but absolutely safe multidisciplinary approach, etc. The list is actually quite long. Increasingly sophisticated chemical technologies which would be inexpensive and eco-friendly and can improve the lives of common people, especially those in rural and impoverished areas all over the world, need to be developed as quickly as possible.

- **You were involved in the development of chemistry curriculum for Indian universities. What are the key aspects of a good chemistry curriculum according to you?**

A curriculum involves a combination of teaching, learning, and assessment. Irrespective of what an individual chemist may practice in his/her own research, a chemistry curriculum must not split chemistry into inorganic, organic, and physical chemistry, and there should be no specialization in any of these three up to the graduation level (pre-Ph.D.). I strongly believe that this tripartite splitting has done enormous damage to the free development of chemistry in certain countries. Throughout the undergraduate years, there should be self-exploration by the student through as many small and medium projects as possible. Science can be learnt only through a dialog with Nature, through experiments in the laboratory, and in natural environments outside the laboratory. Laboratory programs in certain countries are not in a good state. We must bring back imagination, excitement, and wonder into the laboratory courses in chemistry. This is easier said than done. Here, theoretical and computational chemists should join hands with their experimental colleagues in devising concept-oriented, technique-intensive, and generally fun experiments for students. We must also bring back experimental demonstrations during classroom lectures. Let us not forget that chemistry is a subject combining magic, logic, and aesthetics.

The life-blood of any educational program is a dedicated and conscientious band of teachers. I would request the teachers concerned that, in formulating any chemistry curriculum, they should keep in mind that chemistry is a *central science*, overlapping with practically any subject under the sun and even the processes occurring in the sun, in the earth, and elsewhere. Here, I would like to draw a connectivity network which depicts chemistry as a central science, and teachers as well as students may keep this in mind. Note that seven lines, some of which are coincident, radiate out from every subject toward other subjects. Both teaching and research in chemistry might develop in the future along this network.

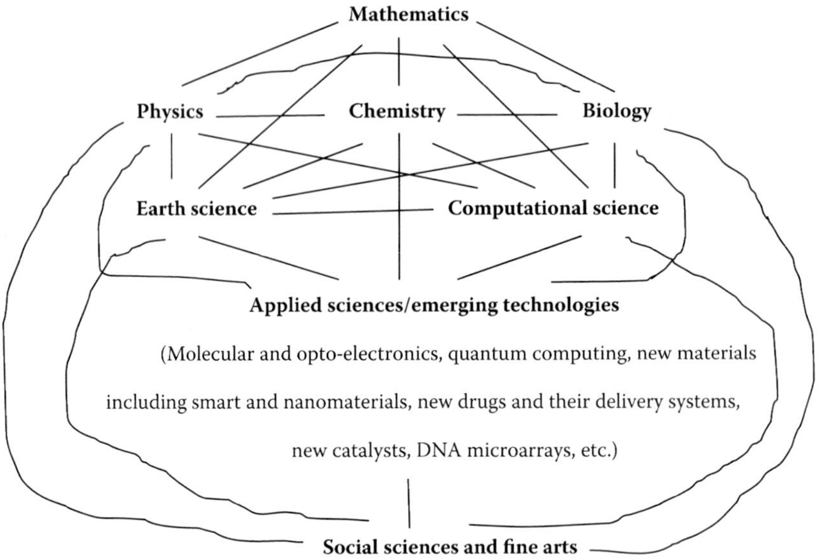

Within this pattern, a chemistry curriculum must impart both *intellectual and manual skills* to the student and try to *integrate both skills*. This is the essence of creativity.

- **What kind of prospects do young theoretical and computational chemists have?**

  I strongly feel that every chemistry department in colleges and universities should appoint at least one theoretical and computational chemist. In universities, the critical number would be three. Because of the multidisciplinary nature of the subject, a theoretical chemist can teach quite a few areas and would therefore lend strength to the teaching programs. Secondly, industries in a number of countries do not seem to have felt the need to appoint theoretical and computational chemists. All these have drastically reduced the employability of young theoretical and computational chemists, who show enormous personal courage to go into these areas. As a result, theoretical and computational chemists have found employment only in a limited number of institutions. I find this overall situation fraught with danger for the future development of chemistry.

- **What sparked your interest in Chemistry? (You had done a Ph.D. in mathematics.)**

  I drifted into chemistry. I could see myself also in literature or medicine or biology or physics. However, even though my father was a legendary teacher of mathematics, I never saw myself as a mathematician. An encounter with a highly charismatic teacher of chemistry put me into chemistry at Presidency College, Kolkata. I began to love the subject because of its all-encompassing nature. I was seriously thinking about going into experimental chemistry. It was a distinguished polymer chemist who advised me to pursue my doctoral studies with Professor C. A. Coulson. Since Professor Coulson was the Director of the Mathematical Institute of Oxford University, my D.Phil. degree was under the Mathematics Faculty at Oxford. Still, it took me six months at Oxford to firmly conclude that theoretical chemistry with its unlimited expanse will be my life, mainly because I knew that I would never be able to come to grips with it.

Looking back, I am convinced that it was my teachers right from high school to the doctoral level who were instrumental in charting my professional life.

- **Is there a particular incident from your research career, an anecdote, that you would like to share with the readers?**

I recall an incident which helped to redefine the course of my research. Around 1970, I had written an article on what I called "the force concept in chemistry" for students and teachers of chemistry. The chemistry journals I sent it to declined to consider it for publication, saying this would be beyond their readership. Exasperated, I sent it to Professor Coulson for his critical comments. Professor Coulson decided to communicate it himself to Reviews of Modern Physics (RMP). It was highly interesting that, while the referee(s) accepted the article, Professor U. Fano, the Editor of RMP, wanted me to rewrite parts of it, making an extremely important point that I comment on how the electron density can be calculated directly. I wrote whatever I could, and the article was published. I was rather unnerved but elated when I received many letters from people belonging to various disciplines, including several highly respected scientists. One of them was Professor R. P. Feynman who, besides telling me that he liked the article, suggested that I look into stresses in molecules, which he himself was interested in at one time but never published anything on it. Enclosed with his letter came the Xerox copies of relevant pages on stresses from his B.S. dissertation (under J. C. Slater) at MIT, which contained his famous work on the Hellmann–Feynman theorem. These two suggestions changed my research.

1 Theoretical Studies of Nucleation and Growth

Rakesh S. Singh, Mantu Santra, and Biman Bagchi

CONTENTS

1.1 INTRODUCTION

Nucleation and growth of a new (daughter) phase from an old (parent) phase are two related topics of tremendous current interest, with wide-ranging practical value and with applications ranging from atmospheric research to materials science. First-order phase transitions usually occur via nucleation and subsequent growth of the postcritical nucleus. The last stage is again divided into two parts: growth and aging. The latter is also referred to as ripening. The formation of a droplet of the stable phase within the metastable bulk phase through an activated process is called nucleation. Growth follows nucleation and leads to phase transition. Aging occurs in the late stage of first-order phase transition and takes place when the system is closed for mass exchange. After nucleation and growth, minimization of the total interfacial energy drives competitive late-stage growth. The mechanism recognized for the aging process is Ostwald ripening, which is the phenomenon in which the smaller clusters lose monomers and decay and larger clusters capture monomers and grow.

The classical nucleation theory (CNT) (of Becker–Döring–Zeldovich) provides a simple yet elegant description of homogeneous nucleation in terms of free energy barrier, with the size of the cluster as the sole order parameter describing nucleation

[1–6]. CNT assumes that a spherical droplet of the new stable phase grows in a sea of parent metastable bulk phase by addition of single molecules and that this droplet has to grow beyond a certain "critical size" (R^*) to compensate for the energy required to form the surface between the two phases. The free energy of formation of a droplet of radius R is

$$\Delta G(R) = -\frac{4\pi}{3} R^3 \left|\Delta G_V\right| + 4\pi R^2 \gamma, \tag{1.1}$$

where $\left|\Delta G_V\right|$ is the free energy difference per unit volume between the daughter and the parent phases, and γ is the surface tension of the interface between them. The preceding relation gives the following well-known expressions for the size of the critical nucleus (R^*) and the free energy barrier ($\Delta G(R^*)$):

$$R^* = \frac{2\gamma}{\Delta G_V}$$

$$\Delta G(R^*) = \frac{16\pi\gamma^3}{3(\Delta G_V)^2}. \tag{1.2}$$

Note that the free energy barrier depends more strongly on surface tension than on the free energy difference. From the preceding discussion, we get the expected result that both the critical cluster size (R^*) and the free energy barrier ($\Delta G(R^*)$) decrease with increase in supersaturation or supercooling and the rate of nucleation increases following the Arrhenius rate expression.

However, the Becker–Döring theory is known to become unreliable at large supersaturation or supercooling. Recent studies have shown that the temperature dependence of rate undergoes a crossover from Arrhenius-type temperature dependence at low supersaturation to weaker non-Arrhenius type temperature dependence at large supersaturation [7–11]. Quantitative estimates of an experimentally observed rate are also known to be greatly at variance with the predictions of CNT [12]. Although several aspects of this CNT have recently been analyzed critically in both two- and three-dimensional systems [13–17] and different lacunae have been removed, the important problem of the mechanism of nucleation at large metastability remains unresolved and somewhat controversial.

In fact, it is not surprising that Becker–Döring–like simple theories face difficulties. These are based on many simplifying assumptions, such as (1) growth by single particle addition, (2) capillary approximation to evaluate the free energy surface, and (3) only a single large growing cluster dominating nucleation. It can be shown that each of these approximations becomes invalid at high supersaturation. Recently, a new formulation has been developed that avoids making all these approximations by directly evaluating the free energy surface by using modern methods of constrained variation (umbrella sampling, transition matrix Monte Carlo, and metadynamics) and using a larger set of order parameters than a single cluster size. In Section 1.2, we shall discuss gas–liquid nucleation at large metastability in detail.

Crystallization is a more complex process than condensation, and one order parameter description often fails as density and order both change on crystallization. It has been long known that, when a solid crystallizes out from a solution or melt, it is not the thermodynamically stable phase that forms; rather, it is the thermodynamically *least* stable (metastable) phase that separates out first, which progressively transforms to the stable phase, given adequate condition to attain equilibrium. In zeolite systems, one often observes that the less stable faujasite phase precipitates first. The most stable alpha-quartz phase forms either at high temperature or after a long time [18]. More recently, Chung et al. have also observed the same scenario in the crystallization of metal phosphate [19]. This observation was made of Wilhelm Ostwald long ago and has been generalized as the Ostwald step rule of successive crystallization [20,21]. Despite enormous interest in the Ostwald step rule, many aspects of the selection of solid polymorphs from solution have remained unclear. According to the original statement of Ostwald, the phase that precipitates out first is not the most stable phase but the one closest in likeness to the parent sol phase. The fact that the most stable phase precipitates out at high temperatures (or after a long time) is explained by appealing to CNT. CNT expressions show that, even when ΔG_V is the largest for the most stable phase, it might not form if the surface tension is large. In such cases, the phase of intermediate or least stability can win on kinetic grounds if the surface tension term is less. This is what Ostwald probably meant by the statement that the phase that precipitates out at first stage is closest in nature to the sol phase.

The issue of transformation of the least stable phase to the most stable phase, however, remains unsettled. For example, what is the nucleation scenario in the case of such transformations? Simulations have shown that, for Lennard–Jones (LJ) fluid, it is always the metastable body centered cubic (bcc) phase that forms first and then transforms into the stable face centered cubic (fcc) phase [22]. Alexander and McTague have also suggested that a metastable bcc phase can easily be formed from supercooled liquid [23]. This is a good example of "disappearing polymorph." Although many theoretical approaches have been developed to understand this phenomenon [24–26], an elegant quantitative explanation is still lacking. In Section 1.3, we shall discuss the energy landscape view of crystallization and the Ostwald rule of stages. In the last section (Section 1.4), we shall describe Ostwald ripening, which is late-stage growth in a system undergoing phase transition.

1.2 GAS–LIQUID NUCLEATION AT LARGE METASTABILITY

There are several issues regarding nucleation at large metastability. First, of course, is the issue of the quantitative accuracy of CNT. It is now well established that the use of CNT with surface tension obtained from the equilibrium gas–liquid coexistence with a planar interface does not provide quantitative agreement. The second issue concerns the validity of the free energy decomposition embodied in Equation 1.1. While such decomposition can be valid at low supersaturation when the size of the critical cluster is large, it becomes questionable when the embryo size becomes small at large supersaturation. Density functional theoretical approaches suggest that the critical nucleus at large metastability has different characteristics from the critical nucleus at low supersaturation [13].

Recently, Parrinello and coworkers have studied the freezing of LJ fluid as a function of the degree of supercooling and found that the nucleation acquires a spinodal-like character at large supercooling. Crystallization proceeds in collective fashion and becomes spatially more diffuse as the spinodal is approached [17]; that is, several large clusters grow at large metastability (contrary to the small supersaturation where only one cluster grows), and the ultimate fate of these clusters is stochastic. This has also been verified in the case of condensation and is discussed in detail in the next section.

1.2.1 Two-Dimensional Free Energy Surface: Elucidation of Nucleation Near Kinetic Spinodal

To study the behavior of the entire system on changing supersaturation in the gas–liquid nucleation, a two-dimensional free energy surface as a function of two order parameters, i.e., the total number of particles present in the system (N_{tot}) and the "liquidness" (analogous to the magnetization in the Ising model) of the system (N_{liq}), has been constructed. The latter is given by the total number of liquidlike particles (N_{liq}) identified by its local density. A particle is considered to be liquidlike if it has more than four nearest neighbors within a cutoff distance ($R_c = 1.5\sigma$, corresponding to the first shell of a particle in liquid phase). Liquidlike particles that are connected by a neighborhood (within the cutoff distance of 1.5σ) form liquidlike clusters.

Figures 1.1a and 1.1b show the free energy surfaces of formation of liquidlike clusters of spherical particles interacting with the LJ potential at two different supersaturations (S). S is defined as P/P_0, where P is the pressure of the system and P_0 is the pressure at coexistence at the same temperature. The figures show that, at intermediate supersaturation ($S \sim 1.8$), both the activation barrier and the number of liquidlike particles (~50) at the barrier are large (9.5 k_BT). On the other hand, at large supersaturation ($S \sim 2.4$), the free energy surface near the saddle is very flat. Here, the number of liquidlike particles at the barrier is just about 35, and the free energy barrier from the minimum is even less than 4 k_BT. Importantly, these liquidlike particles are dispersed among several intermediate-sized clusters. The disappearance of the free energy barrier at large metastability (from Figure 1.1 and expected to be around $S \sim 2.6$) makes the system unstable. The flatness of the free energy surface and the collective growth at large metastability demand an alternate theoretical formalism that can describe the simultaneous growth of several clusters. One then needs to introduce a set of order parameters that allows to order the clusters according to their sizes. However, in order to develop any such formalism, we first need to find out the metastability (supersaturation) where a transition from single to collective growth occurs.

Unlike chemical reactions, in nucleation, the activation of the largest cluster controls the phase transition of the entire system. Once the largest cluster crosses the critical size, it grows rapidly and engulfs the entire system, leading to the phase transition. At low supersaturation, there is a separation of time scale between nucleation and growth of the largest cluster and other clusters present in the system. However, this separation of time-scale assumption breaks down at large metastability (or high

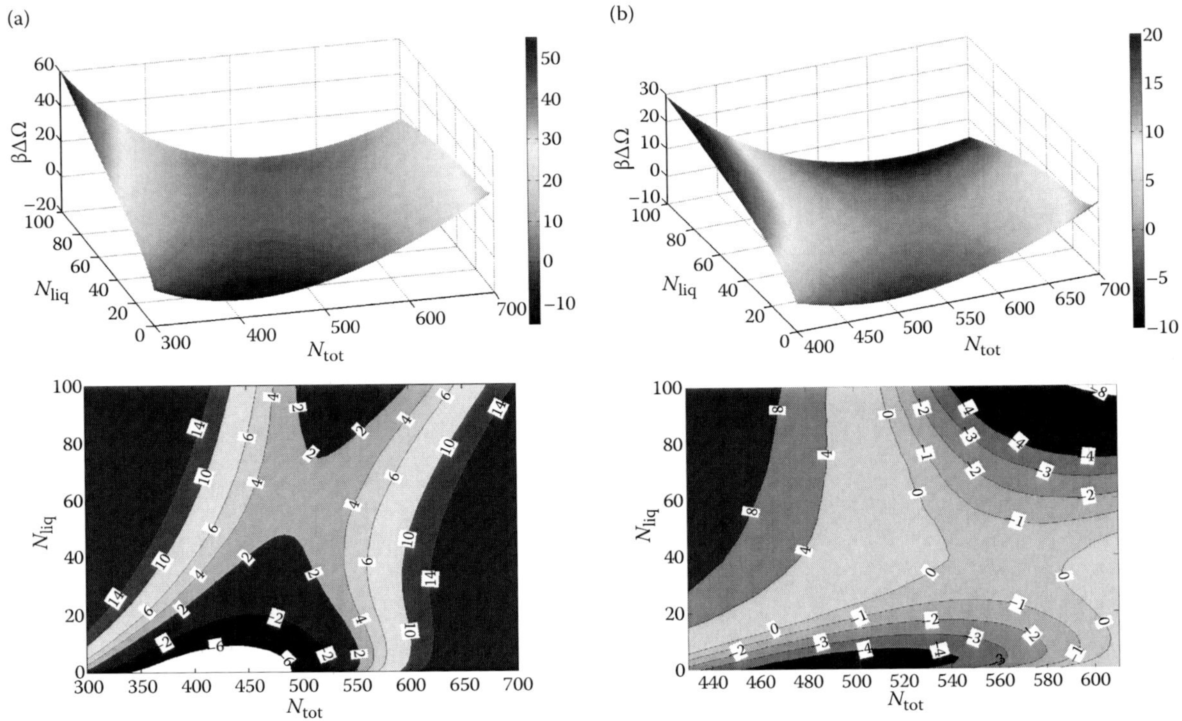

FIGURE 1.1 Computed three-dimensional free energy surfaces and contour plots computed in grand canonical ensemble at reduced temperature $T^* = 0.741$ and volume $V = (25\sigma)^3$ for (a) activity $\xi = 0.018$ (corresponding to $S \sim 1.8$ in the *NPT* ensemble) and (b) activity $\xi = 0.020$ (corresponding to $S \sim 2.4$ in the *NPT* ensemble). Note that the energy scales are given on the right-hand side of the free energy surfaces. In the case of contour diagrams, the free energy values are indicated by respective numbers. (Reproduced from Bhimalapuram, P. et al., *Phys. Rev. Lett.*, 98, 206104, 2007. With permission.)

supersaturation). Thus, it is more appropriate to consider the largest cluster as the reaction coordinate for quantifying the supersaturation where a transition from single nucleus growth to collective growth occurs.

Figure 1.2 shows the computed free energy surfaces of the largest liquidlike cluster ($F^L(n)$) for a wide range of supersaturation. As the supersaturation is increased, the free energy barrier for the largest cluster ($\Delta F^L(n^*)$) becomes lower and vanishes entirely. At the same time, the minimum at the intermediate cluster size also becomes deeper and shifts toward a larger size. The size of the critical cluster also becomes progressively smaller on increasing supersaturation. The disappearance of $\Delta F^L(n^*)$ signifies the onset of the *kinetic spinodal* point, which is the actual limit of metastability of the system [7–11]. On the other hand, the thermodynamic spinodal point (or spinodal) corresponds to the disappearance of the free energy barrier for all clusters. Note that the thermodynamic spinodal lies at much deeper supersaturation than the kinetic spinodal. At large metastability, the spatially diffused nature of growth arises due to the appearance of the free energy minimum at a small cluster size (or abundance of intermediate-sized clusters) in the system.

In simulations, it has been observed that $\Delta F^L(n^*)$ is system size dependent (note that thermodynamic spinodal is system size independent) and given approximately by $\beta\Delta F^L(n^*) = \beta\Delta F(n^*) - \ln V$ [11]. On increasing system size, the location of minima shifts toward the larger cluster size, and $\Delta F^L(n^*)$ decreases. For a sufficiently large system size, $\Delta F^L(n^*)$ vanishes entirely, and the system becomes unstable. Although $\Delta F^L(n^*)$ vanishes for a sufficiently large system size (or large supersaturation), there is still a considerably large barrier for the growth of a cluster of size n ($F(n)$). Therefore, the barrier for $F(n)$, which is still present in the system, is no longer the rate-determining factor for the nucleation. The rate is determined by the (barrierless) growth of the largest cluster present in the system. The system size dependence can be explained by the following theoretical analysis:

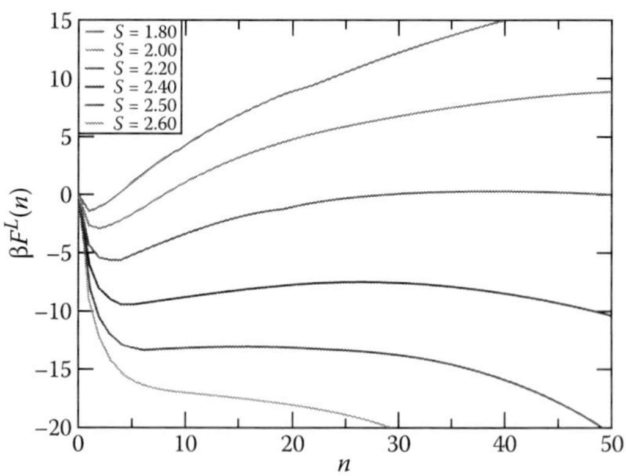

FIGURE 1.2 Free energy versus size of the largest liquidlike cluster in the *NPT* ensemble (T^* = 0.741) for different supersaturations (S) in the LJ system. (Reproduced from Bhimalapuram, P. et al., *Phys. Rev. Lett.*, 98, 206104, 2007. With permission.)

Let us assume that there are M clusters present in the system at a particular supersaturation S, with their sizes X_i in increasing order. The probability of the largest cluster having size n is given by

$$P^L(X_M = n) = P(X_M \leq n) - P(X_M \leq n - 1). \tag{1.3}$$

Now, if we assume that the clusters are noninteracting, then

$$P(X_M \leq n) = P(X \leq n)^M. \tag{1.4}$$

Therefore, one can write

$$P^L(X_M = n) = P(X \leq n)^M - P(X \leq n - 1)^M. \tag{1.5}$$

Here, $P(X \leq n)$ is the probability of any cluster having size $\leq n$ and is given by

$$P(X \leq n) = \sum_{i=0}^{n} P(X = i) \tag{1.6}$$

where $P(X = i)$ is the probability of any cluster having size i, as given by CNT. Thus, we obtain the free energy of the largest cluster as

$$\begin{aligned}
\beta F^L(n) &= -\ln P^L(X_M = n) = -\ln\left[P(X \leq n)^M - P(X \leq n-1)^M\right] \\
&= -\ln\left[\{P(X \leq n) - P(X \leq n-1)\} \times \sum_{i=1}^{M} P(X \leq n)^{M-i} P(X \leq n-1)^{i-1}\right] \\
&= -\ln\left[P(X = n) \times \sum_{i=1}^{M} P(X \leq n)^{M-i} P(X \leq n-1)^{i-1}\right] \\
&= \beta F(n) - \ln\left\{\sum_{i=1}^{M} P(X \leq n)^{M-i} P(X \leq n-1)^{i-1}\right\}.
\end{aligned} \tag{1.7}$$

This derivation is for the largest cluster. In general, following a similar approach, an expression for the free energy of the kth largest cluster can also be obtained and is discussed later in this section.

Now, $M = \sum_{n=0}^{n^*} M(n)$, where $M(n)$ is the number of clusters of size n and n^* is the size of the critical cluster. Supersaturation S can be approximated as $S = \rho/\rho^o$, where ρ^o is the density of the vapor phase at coexistence and ρ is the density of the supersaturated vapor phase. According to CNT, the population of clusters of size n can be given by

$$M(n) = M(0)\exp[-\beta F(n)]. \tag{1.8}$$

For a system of volume V and density ρ,

$$M(n) = V\rho \exp[-\beta F(n)], \tag{1.9}$$

and hence,

$$M = V\rho \sum_{n=0}^{n^*} \exp[-\beta F(n)]. \tag{1.10}$$

Let us now analyze the reason for the system-size dependence of $F^L(n)$ from Equation 1.7. As the theoretical analysis shown above can calculate $F^L(n)$ from $F(n)$ itself, it is possible to find out the kinetic spinodal limit for a given system size where $F(n)$ is implied to be known. $F^L(n)$ is limited by the inclusion of integral probability of all smaller clusters present in the system. Therefore, as the system size is increased, the probability of a particular size of the largest cluster will involve a larger number of small cluster probability integral. As a result, the precritical minimum of $F^L(n)$ will move toward a larger value, and the free energy barrier $\Delta F^L(n^*)$ will decrease.

In a more physical way, one can understand the system-size dependence as follows: The observed minimum in $F^L(n)$ corresponds to the average size of the largest cluster in the system at the metastable state. The average number of clusters of size $n(M(n))$ is an extensive quantity, and on increasing system size, the distribution of $M(n)$ broadens and shifts toward a larger cluster size. This implies that the average number of larger clusters keeps increasing with system size, and the observed disappearance of the minimum in $F^L(n)$ above a certain system size originates from this extensive nature of $M(n)$. As the average size of the largest cluster progressively increases with system size, it has to climb a shorter barrier thereafter. Thus, the lifetime of the metastable state is dependent on the size of the system.

1.2.2 Dynamical Crossover at Large Metastability

A theoretical formalism based on an extended set of order parameters (which can describe the simultaneous growth of several clusters beyond kinetic spinodal) has also been developed [9]. This is a generalization of the theoretical formalism developed considering the largest cluster as an order parameter (see the preceding discussion). The expression for the free energy of the kth largest cluster ($k = 1$ stands for the largest cluster in the system, $k = 2$ is the second largest cluster, $k = 3$ is the third largest cluster, and so on) of size n is given by [9]

$$\begin{aligned}\beta F_k^L(n) = \beta F(n) - \ln\left\{\frac{M!}{(k-1)!(M-k)!}\right\} - (M-k)\ln\{G(n)\} \\ - (k-1)\ln\{1-G(n)\} + \ln(M(n)),\end{aligned} \tag{1.11}$$

where $\beta F(n) = -\ln P(n)$ is the free energy of a cluster of size n and $P(n)$ is the probability density function of n. $M(n)$ is the average number of clusters of size n. M is the average number of total clusters (including both liquidlike clusters and single gas particles) present in the system and is the summation of $M(n)$ over n. $G(n)$ is the cumulative probability distribution function of $P(n)$. The only approximation made in deriving Equation 1.11 is that the clusters are noninteracting. The preceding expression establishes a relation between the free energy of formation of the kth largest cluster of size n, $F_k^L(n)$, and the CNT free energy, $F(n)$.

The change in mechanism from activated to barrierless dynamics is reflected in the plot of variation of the rate with density (equivalent to supersaturation), as shown in Figure 1.3. In this figure, the change in slope at density ρ_{ks} signifies the crossover from activated to barrierless growth and is the onset density for the kinetic spinodal point. The calculated rate of nucleation scales differently with supersaturation below and above the kinetic spinodal. At supersaturations between the kinetic and thermodynamic spinodal points, the clusters grow through *barrierless diffusion*. Just beyond the kinetic spinodal, only a few clusters grow, whereas at the thermodynamic spinodal point, all the clusters grow spontaneously. Kinetic spinodal is also sensitive to the range of interaction potential [27]. This sensitivity arises due to the sensitivity of surface tension on the range of interaction.

At two different supersaturations, the snapshots of the clusters when the largest cluster crosses critical size are shown in Figure 1.4. Figure 1.4a depicts the situation at low supersaturation (below the kinetic spinodal point), and Figure 1.4b depicts the situation at large supersaturation (above the kinetic spinodal point). From these snapshots, it is quite evident that, above the kinetic spinodal, nucleation and growth become more collective (many clusters cross the critical size through barrierless diffusion). An earlier theoretical analysis of nucleation in Ising model systems found a similar supersaturation dependence of the crossover of the nucleation mechanism [28].

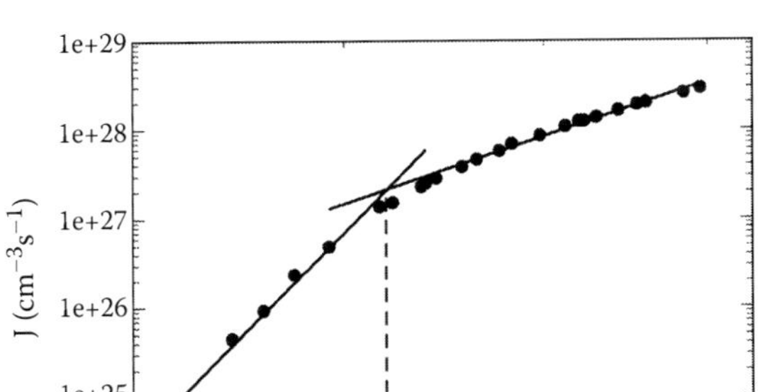

FIGURE 1.3 Rate of nucleation at different densities (or supersaturation) at $T^* = 0.741$ and $N = 500$ in the *NVT* system. The figure clearly shows the crossover of nucleation from activated to barrierless diffusion at density ρ_{ks}, indicated by a dashed bar. The supersaturation corresponding to the dashed line is the kinetic spinodal point. (Reproduced from Santra, M. and Bagchi, B., *Phys. Rev. E*, 83, 031602, 2011. With permission.)

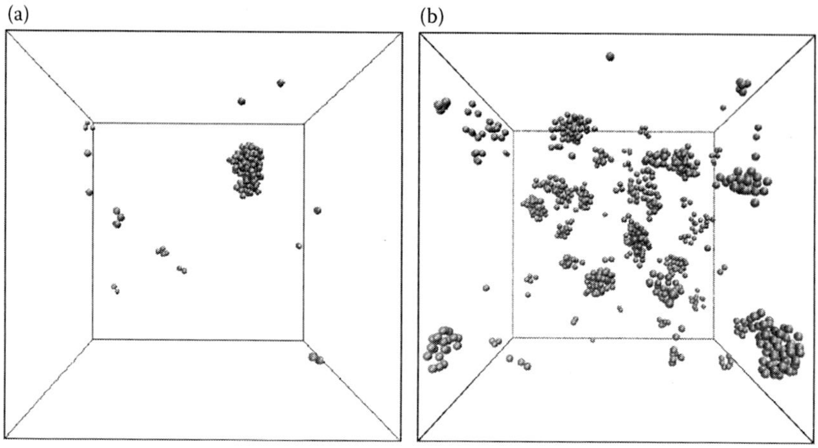

FIGURE 1.4 (a) Snapshot of liquidlike particles in the system at low supersaturation ($\rho^* = 0.0334$) when the largest cluster crosses the critical size. From the snapshot, it is clear that only the largest cluster undergoes nucleation and growth. Compared to the largest cluster, all the other clusters are very small in size and are engulfed by the largest cluster. (b) Snapshot of all the liquidlike clusters present in the system when the largest cluster crosses the critical size at high supersaturation ($\rho^* = 0.0554$). Many liquidlike clusters are present in the system. The snapshot shows that, at high supersaturation, more than one cluster start growing, and they are of comparable sizes. (Reproduced from Santra, M. and Bagchi, B., *Phys. Rev. E*, 83, 031602, 2011. With permission.)

1.3 ENERGY LANDSCAPE VIEW OF OSTWALD STEP RULE

Protein folding and crystallization, two phenomena of current interest in biology and materials science, share a common physical chemistry basis. Both have similar rugged energy landscape arising from definite entropy–enthalpy relationships. Recently, Chung et al. [19] have also confirmed that, for crystallization of metal phosphate, the energy landscape is rugged with multiple minima. A schematic representation of rugged free energy landscape is shown in Figure 1.5a.

However, characterization of these multiple minima and maxima poses a serious problem because the order parameters are not always well defined. In the case of protein folding, the radius of the protein and the number of native contacts serve as two good order parameters. Further quantification can be accomplished by considering quantities such as the number of hydrophobic pair contacts.

In the case of crystallization of a solid from liquid or melt, we need order parameters that uniquely identify different structures; that is, ideally, each minimum should be characterized by a set of values of the order parameters and should correspond to a unique structure. In the Ramakrishnan–Yusouff density functional theory of freezing [29], the order parameters are the density components evaluated at the reciprocal lattice vectors of the solid, along with the fractional density change. It is relatively easy to make either equilibrium or a dynamic calculation of freezing to

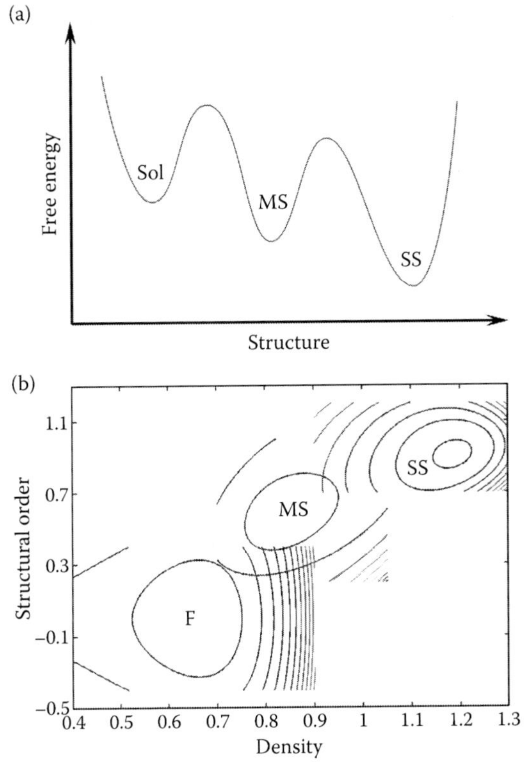

FIGURE 1.5 (a) Schematic rugged free energy landscape of crystallization. Sol stands for the sol phase, MS for metastable solid, and SS for stable solid. (b) Contour diagram of two-dimensional free energy surfaces given by Equations 1.13 and 1.14 is shown. F stands for fluid phase, MS for metastable solid, and SS for stable solid.

different lattice types; that is, one can explain why argon freezes into an fcc lattice whereas liquid sodium freezes into a bcc solid. However, the situation is far more difficult in the case of complex solids like zeolites. Here, we do not have information about the liquid structure necessary for a microscopic theory. We therefore discuss an approach based on a coarse-grained approach developed recently by Oxtoby and coworkers [30], which has been rather successful in capturing some aspects of complex systems.

In the zeolite system, faujasite has the lowest value of the order parameter, and in the quartz system, it has the highest [18]. In the energy landscape picture, the surface tension and difference between the minimum of free energy corresponding to two structures play an important role in determining nucleation barrier between two metastable phases or one metastable phase and one stable solid (SS) phase. However, it is very hard to obtain surface tension between two metastable solid (MS) phases.

In the energy landscape view, the polymorphs are the inherent structures of the sol phase and shall be obtained when the vibrational degrees of freedom and the kinetic energy are removed from the molecules. Thus, the polymorphs form a rugged landscape with the most stable structure at the bottom of the energy ladder, just like in the folding funnel of a protein.

When we consider the formation of the MS (which we assume to be the closest to the sol phase), then the free energy gap ΔG_V is lower than the most stable phase, referred to as SS. Thus, according to CNT (Equation 1.2), the only way that the phase MS can precipitate out at any temperature is to have such a lower surface tension that the nucleation barrier is lowest for MS. Because of lower solid–liquid surface tension leading to a lower nucleation barrier, MSs are kinetically favored. Thus, kinetics seems to play a very dominant role.

At high temperature, the following proposed scenario holds: Since the energy of the system is high, it can probe all the minima of the system. Even if it gets trapped in a low-lying minimum, like in the MS1 or MS2 phase, it can escape from the minimum before the phase grows to macroscopic size. In other words, when the nucleus forms, it can melt within a time comparable to the relaxation time of the system. It, of course, gets trapped many times in the low-lying minima and gets out again and again. When it gets trapped in the deep minimum of the most stable phase, it grows. However, when the temperature is low, it gets trapped in the closest minimum, as envisaged by Ostwald.

In the next section, we construct a density functional theory, which provides a quantitative explanation of the sequential formation of metastable states before transforming to the most stable phase.

1.3.1 Free Energy Functional

For simplicity, we consider one fluid (F) (the sol) phase and two solid phases that compete with each other. The two solid phases are referred to as MS and SS, as shown in Figure 1.5a. Thus, the free energy exhibits only three minima (states). The proposed free energy functionals for the three phases, i.e., F, MS, and SS, are

$$\Omega_i[\rho(\mathbf{r}),m(\mathbf{r})] = \int d\mathbf{r}[f_i(\rho(\mathbf{r}),m(\mathbf{r})) - \mu\rho(\mathbf{r})] + \frac{1}{2}\int d\mathbf{r}\left[K_{\rho i}(\nabla\rho(\mathbf{r}))^2\right] + \frac{1}{2}\int d\mathbf{r}\left[K_{mi}\rho_S^2(\nabla m(\mathbf{r}))^2\right], \tag{1.12}$$

where f_i is a local Helmholtz free energy density function of the average number density $\rho(\mathbf{r})$ and structural order parameter $m(\mathbf{r})$, and μ is the chemical potential. Here, "i" indicates respective phases. ρ_s is the average density of bulk solid phase. The last two terms (square gradient terms) account for the nonlocal effects in the system due to inhomogeneity in density and structural order parameters. $K_{\rho i}$ and K_{mi} are related to the correlation lengths for ρ and m. Following Talanquar and Oxtoby [30], the Helmholtz free energy density for homogeneous fluid is

$$f_f(\rho,m) = k_B T\rho[\ln \rho - 1 - \ln(1 - b\rho)] - a\rho^2 + k_B T\alpha_l m^2. \tag{1.13}$$

The preceding free energy functional is a trivial generalization of van der Waal's free energy functional. In a similar spirit, one can also write the Helmholtz free energy functional for solids (metastable and stable) as

$$f_{js}(\rho,m) = k_B T\rho[\ln \rho - 1 - \ln(1 - b\bar{\rho})] - a\rho^2 + k_B T\left[\alpha_{1js} m^2 + \alpha_{2js}\right] \tag{1.14}$$

where $j = m$ for metastable solid and s for stable solid. Here, k_B is the Boltzmann's constant, T is the absolute temperature, and a and b are van der Waal's parameters and account for the effect of interactions between dissolved molecules. $\bar{\rho}$ is the weighted average density and is given as

$$\bar{\rho} = \rho[1 - \alpha_{3js} m(\alpha_{4js} - m)]. \tag{1.15}$$

The values of the parameters are $a = 1.0$, $b = 1.0$, $\alpha_l = \alpha_{1ms} = \alpha_{1ss} = 0.25$, $\alpha_{2ms} = 1.5$, $\alpha_{3ms} = 0.22$, $\alpha_{4ms} = 1.85$, $\alpha_{2ss} = 2.0$ $\alpha_{3ss} = 0.30$, $\alpha_{4ms} = 2.0$, $K_{\rho i} = a/2$, and $K_{mi} = a/8$. A contour plot of free energy functions given by Equations 1.13 and 1.14 is shown in Figure 1.5b.

1.3.2 Phase Diagram

For the homogeneous system, the densities and structural orders of coexisting phases can be determined by equating the chemical potentials and thermodynamic grand potential densities (pressures) of respective phases:

$$\mu_\alpha(\rho_\alpha) = \mu_\beta(\rho_\beta) \text{ and } \omega_\alpha(\rho_\alpha) = \omega_\beta(\rho_\beta), \tag{1.16}$$

where $\mu_i = \left(\frac{\partial f_i(\rho,m)}{\partial \rho}\right)_T$ and $\omega_i = f_i - \mu_i\rho_i$.

The preceding two conditions ensure that the system is in both thermodynamic and mechanical equilibrium. Figure 1.6 shows the results of quantitative calculation of the coexistence of all the phases. Here, we show four phases in all: gas, liquid, MS, and SS phases. From the phase diagram, it is quite evident that, at a fixed temperature, the density (or order parameter) difference between F and SS is larger than that between F and MS. As we know from the extensive study of the gas–liquid system showing that surface tension between two phases depends strongly on the order parameter difference, one can qualitatively conclude that the surface tension between the F and SS phases is larger than that between the F and MS phases. More accurate quantitative results for surface tension are discussed in the next section.

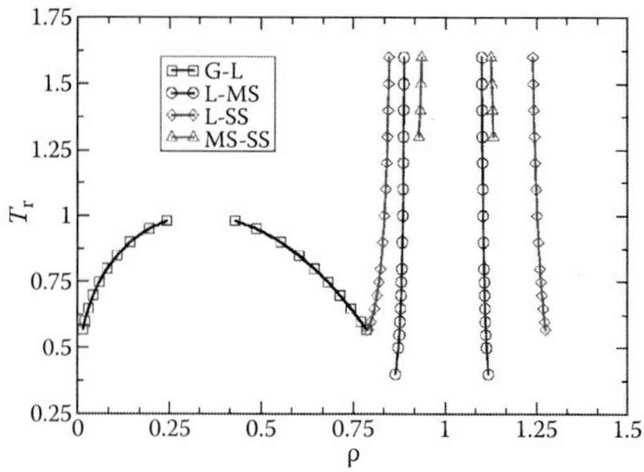

FIGURE 1.6 Phase diagram for the proposed free energy surfaces is shown. Reduced temperature T_r is defined as T/T_c. The lines with square symbols indicate the gas–liquid coexistence, the lines with diamond symbol indicate the coexistence between fluid and stable solid, the lines with circles indicate the coexistence between fluid and metastable solid and, the lines with upright triangles indicate the coexistence between metastable solid and stable solid.

1.3.3 Surface Tension

For the phase diagram depicted in Figure 1.6, we can evaluate the values of the surface tension between the coexisting phases for a planar interface along the z-axis by solving the Euler–Lagrange equations associated with the following equilibrium conditions:

$$\frac{\delta\Omega}{\delta\rho(z)} = 0 \quad \text{and} \quad \frac{\delta\Omega}{\delta m(z)} = 0, \tag{1.17}$$

where $\Omega(\rho(z), m(z))$ is the grand canonical free energy functional corresponding to the inhomogeneous system with density profile $\rho(z)$ and order profile $m(z)$:

$$\Omega[\rho(z), m(z)] = \int dz[f(\rho(z), m(z)) - \mu\rho(z)] + \frac{1}{2}\int dz\left[K_\rho(\nabla\rho(z))^2\right] + \frac{1}{2}\int dz\left[K_m\rho_S^2(\nabla m(z))^2\right]. \tag{1.18}$$

Here, $f = \min\{f_f, f_{js}\}$.

The density and order profiles shown in Figure 1.7 are obtained by solving the corresponding Euler–Lagrange equations under appropriate boundary conditions. The surface tension is extra energy cost for the formation of an interface (density and order profile shown in Figure 1.7) and is defined as

$$\gamma = \frac{(\Omega(\rho(z), m(z)) - \Omega_{hl/hs})}{A}, \tag{1.19}$$

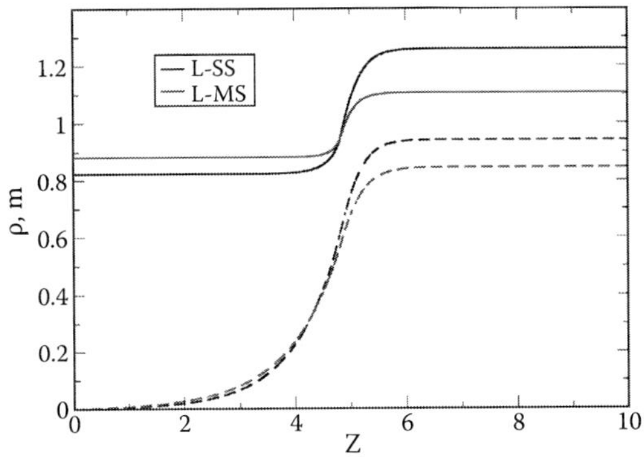

FIGURE 1.7 Density and order profile for the proposed free energy functional for planar interface along the z-axis at $T_r = 0.80$. Solid lines indicate the density profiles along the L-SS and L-MS interfaces, respectively. Dotted lines indicate the order profiles along the L-SS interface and the L-MS interface, respectively.

where $\Omega_{hl/hs}$ is the free energy of the coexisting homogeneous liquid or solid phase and A is the area of the interface. After inserting the equilibrium density and order profile in Equation 1.18, the calculated surface tension using Equation 1.19 for the liquid–MS (L-MS) interface is $\gamma_{L/MS} = 6.8 \times 10^{-2}$ and that for liquid–SS (L-SS) interface is $\gamma_{L/MS} = 14.7 \times 10^{-2}$ (in units of $a/b^{5/3}$).

If we consider Figure 1.5a as representative of metastable crystallization, the free energy difference between different phases will follow the trend $\Delta G_{LMS_1} < \Delta G_{LMS_2} < \cdots < \Delta G_{LSS}$, in general, and the aforementioned theoretical analysis suggests that surface tension would follow $\gamma_{LMS_1} < \gamma_{LMS_2} < \cdots < \gamma_{LSS}$. Physically, it can be understood that less energy is required for the formation of disordered or open structured solids from the melt or sol.

If we assume that the energy landscape is invariant to the temperature, then the only way in which the system explores the free energy landscape can change with temperature and also with time. At low temperature or less reaction time, only those states will be accessible, which have a low free energy barrier or surface tension, because the system does not have sufficient energy (at low temperature) or sufficient time (for structural organization in the case of short reaction time) to overcome the free energy barrier of nucleation. However, at high temperature or for long reaction time, the system has sufficient energy and time to overcome multitudes of maxima and minima on the free energy landscape and reach/find the deepest minimum of the most stable state.

Thus, the preceding theoretical analysis provides a molecular view of "disappearing polymorphs" common in crystal engineering. However, many experimental factors such as temperature and rates of precipitation may control the ultimate formation of the polymorphs. From the discussion, it is also quite evident that there is a close resemblance between protein folding and crystal engineering, and both share a common physical chemistry basis.

1.4 OSTWALD RIPENING

Ostwald ripening is a commonly observed phenomenon during the late stage of first-order phase transformations in solid or liquid solutions. It describes the change in an inhomogeneous structure over time and describes a process where, over time, small crystals or sol particles dissolve and redeposit onto larger crystals. The minimization of the total interfacial energy drives the competitive growth between clusters of various sizes. Thus, the smaller particles disappear, and the larger particles grow. The Ostwald ripening phenomenon is observed in both inorganic and macromolecular systems, including proteins. The phenomenon was first described by Wilhelm Ostwald in 1896.

In 1961, Lifshitz, Slyozov, and Wagner [31] (LSW) performed a mathematical investigation of Ostwald ripening in the case where diffusion of solute particles is the slowest rate-determining process (famously known as the LSW theory). The derivation first states how a cluster grows in a supersaturated solution. This theory assumes the coarsening phase as noninteracting, spherical, and fixed in space.

LSW defined the number distribution of sizes of dispersed clusters ($n(R,t)$) to describe the morphology of the system. The flux of clusters passing through size range R and $R + dR$ is $J = Vn(R,t)$, where $V = dR/dt$ is the rate of change of size of a cluster by solute transport to or from the cluster. Under the assumption that $n(R,t)$ is a conserved quantity (i.e., clusters are not disappearing or forming in the system and their sizes are only changing with time), the time evolution of distribution function $n(R,t)$ can be described by continuity equation

$$\frac{\partial n(R,t)}{\partial t} = -\frac{\partial}{\partial R}[Vn(R,t)]. \tag{1.20}$$

In order to find V, LSW assumed a local equilibrium between the cluster of size R and solutes in solution. This assumption allows obtaining the concentration of molecules at the cluster–solution interface by using the Gibbs–Thomson (GT) relation, $C_{eq}(R) = C_\infty(1 + \alpha/R)$, where C_∞ is the bulk solubility and $\alpha = 2\gamma v_m/k_BT$ is the capillary length (γ is the surface tension, v_m is the molar volume of solute, and T is the absolute temperature). The GT relation provides the curvature-induced solubility of a cluster, and the capillary length α defines a length scale for curvature-induced solubility of clusters. As discussed earlier, the driving force for Ostwald ripening or aging of the clusters is the curvature-induced solubility of clusters and the fact that small particles are in equilibrium with higher solute concentrations (given by the GT relation) than larger clusters; the larger clusters grow and derive the system to decrease its surface volume ratio.

By using Fick's law, $J = D\partial C/\partial R$ (D is the diffusion constant for solute or monomers) and the GT relation, LSW derived the expression for V [31]:

$$V = Dv_m\left[\frac{C(t) - C_{eq}(R)}{R}\right] = Dv_m\left[\frac{C(t) - C_\infty(1+\alpha/R)}{R}\right]. \tag{1.21}$$

In the preceding expression, D is the solute diffusion constant and v_m is the solute molar volume. From Equation 1.21 we can obtain the critical cluster size, $R_C(t) = C_\infty\alpha/(C(t) - C_\infty)$. Clusters having sizes less than the critical cluster size (R_C) have negative V and will dissolve, and clusters having sizes $R > R_C$ will grow with time. Note that R_C is time dependent, and this dependency arises due to the time-dependent supersaturation $S(t) = C(t) - C_\infty$. This means that the cluster that grows at time t may dissolve at some later time t', depending on the supersaturation. During nucleation and early stage of growth, the supersaturation of system does not change (significantly) with time.

Using Equations 1.20 and 1.21, along with the condition of mass conservation, given as

$$C(t) = C_0 - \frac{4\pi}{3v_m}\int R^3 n(R,t)dR, \tag{1.22}$$

where C_0 is the initial concentration of solutes, LSW concluded that, in an asymptotic limit ($t \to \infty$), the average of the cube of the radius of the clusters or the third moment of the $n(R,t)$, $\langle R^3 \rangle$, grows as

$$\langle R^3 \rangle = \frac{4}{9} D\alpha C_\infty t, \tag{1.23}$$

where the average is taken over the cube of the radius of all the particles and D is the diffusion constant of the solute (atom/molecule). C_∞ is the bulk solubility, and α is the capillary length. Note that Equation 1.23 determines only the asymptotic growth of clusters.

Some recent experimental studies in both protein crystallization and inorganic systems have observed deviations from the prediction of the LSW theory [32]. Deviations arise mainly due to a few assumptions involved in the LSW theory: (1) surface energy of a cluster can be approximated as that of the spherical cluster, (2) surface tension is independent of the size of the cluster (that is, the effect of curvature on surface tension is ignored), and (3) solute deposition on the cluster is a diffusion-controlled phenomena. On the contrary, in many cases of crystal growth, the solute deposition is not diffusion limited but depends on the availability of the incorporation sites at the surface of the cluster. Kinetic barrier for incorporation is also equally important. The shape of the cluster is also highly irregular due to the slow relaxation of highly dense solidlike clusters. In the case of protein crystallization, the deviation mainly arises due to the two-step process of crystallization. Due to the submerged vapor–liquid critical point, a highly dense metastable amorphous phase forms first (due to large-scale fluctuations in density), and then, the nucleation and growth of the stable crystal phase occur within the dense metastable high-density amorphous phase [32(b),33].

1.5 CONCLUSION

In this article, we have discussed a newly developed theory of gas–liquid nucleation at large metastability. The theory is based on an extended set of order parameters (unlike CNT, which has only one order parameter). These order parameters are ordered cluster sizes. These clusters show remarkable changes in their free energy profile at large metastability and can describe the experimentally observed crossover to weaker-than-activated dynamics at large metastability. The main reason is that, at large metastability, many clusters grow simultaneously, and nucleation becomes a collective diffusion process spread over the whole system. The crossover temperature/density is the onset of kinetic spinodal point beyond which nucleation occurs spontaneously.

We extend the discussion to the well-known problem of "disappearing polymorphs." We have also presented an energy landscape view (based on the two order parameter density functional theory) of the problem of "disappearing polymorphs." Surface tension between the liquid and the MS phase could be much lower than the liquid and the SS phase, leading to a kinetic preference of the system to initially go over to the MS phase. The final fate of the MS phase depends on several other factors such as the free energy barrier between the MS and SS phases and the temperature and solution conditions. Lastly, we describe a theoretical approach to treat Ostwald ripening, which is the late-stage growth in a system undergoing phase transition.

ACKNOWLEDGMENTS

This work was supported in part by grants from DST and CSIR, India. BB acknowledges DST for further support through a J. C. Bose Fellowship.

REFERENCES

1. (a) Debenedetti, P. G. *Metastable Liquids: Concepts and Principles* (Princeton University Press, **1996**). (b) Frenkel, J. *Kinetic Theory of Liquids* (Dover, New York, **1955**). (c) Zettlemoyer, A. C. *Nucleation* (Dekker, New York, **1969**).
2. Becker, R., and W. Döring. *Ann. Phys.* **1935**, *24*, 719.
3. Debenedetti, P. G. *Nature* **2006**, *441*, 168.
4. Langer, J. S. *Ann. Phys.* **1967**, *41*, 108.
5. Oxtoby, D. W. *Nature*, **2001**, *413*, 694.
6. Binder, K. *Ann. Phys.* **1976**, *98*, 390.
7. Bhimalapuram, P., S. Chakrabarty, and B. Bagchi. *Phys. Rev. Lett.* **2007**, *98*, 206104.
8. Wedekind, J., G. Chkonia, J. Wölk, R. Strey, and D. Reguera. *J. Chem. Phys.*, **2009**, *131*, 114506.
9. Santra, M. and B. Bagchi. *Phys. Rev. E*, **2011**, *83*, 031602.
10. Santra, M., R. S. Singh, and B. Bagchi. *J. Stat. Mech.* **2011**, *P03017*.
11. (a) Chakrabarty, S., M. Santra, and B. Bagchi. *Phys. Rev. Lett.* **2008**, *101*, 019602. (b) Maibaum, L. *Phys. Rev. Lett.* **2008**, *101*, 256102.
12. Sinha, S., A. Bhabhe, H. Laksmono, J. Wölk, and R. Strey. *J. Chem. Phys.* **2010**, *132*, 064304
13. Oxtoby, D. W. *J. Chem. Phys.* **1988**, *89*, 7521.
14. Pan, A. C., and D. Chandler. *J. Phys. Chem. B* **2004**, *108*, 19681.
15. ten Wolde, P. R., and D. Frenkel. *J. Chem. Phys.* **1998**, *109*, 9901.
16. Santra, M., S. Chakrabarty, and B. Bagchi. *J. Chem. Phys.* **2008**, *129*, 234704.
17. Trudo, F., D. Donadio, and M. Parrinello. *Phys. Rev. Lett.* **2006**, *97*, 105701.

18. (a) Davis, M. E., and R. F. Lobo. *Chem. Mater.* **1992**, *4*, 756. (b) Petrovic, I., A. Navrotsky, M. E. Davis, and S. I. Zones. *Chem. Mater.* **1993**, *5*, 1805. (c) Henson, N. J., A. K. Cheetham, and D. Gale. *J. Chem. Mater.* **1994**, *3*, 27.
19. Chung, S. Y., Y. M. Kim, J. G. Kim, and Y. Kim. *J. Nat. Phys.* **2009**, *5*, 68.
20. Ostwald, W. *Z. Phys. Chem.* **1897**, *22*, 289.
21. Stranski, I. N., and D. Totomanow. *Z. Phys. Chem.* **1933**, *163*, 399.
22. (a) ten Wolde, P. R., M. J. Ruiz-Montero, and D. Frenkel. *Phys. Rev. Lett.* **1995**, *75*, 2714. (b) ten Wolde, P. R., and D. Frenkel. *Phys. Chem. Chem. Phys.* **1999**, *1*, 2191.
23. Alexander, S., and J. P. McTague. *Phys. Rev. Lett.* **1978**, *41*, 702.
24. Shen, Y. C., and D. W. Oxtoby. *Phys. Rev. Lett.* **1996**, *77*, 3585.
25. Toth, G. I., J. R. Morris, and L. Granasy. *Phys. Rev. Lett.* **2011**, *106*, 045701.
26. Oxtoby, D. W. *Annu. Rev. Mater. Res.* **2002**, *32*, 39.
27. Singh, R. S., M. Santra, and B. Bagchi. *J. Chem. Phys.* **2012**, *136*, 084701.
28. Rikvold, P. A., H. Tomita, S. Miyashita, and S. W. Sides. *Phys. Rev. E*, **1994**, *49*, 5080.
29. Ramakrishnan, T. V., and M. Yussouf. *Phys. Rev. B*, **1977**, *19*, 2775.
30. Talanquer, V., and D. W. Oxtoby. *J. Chem. Phys.* **1998**, *109*, 223.
31. (a) Lifshitz, I. M., and V. V. Slyozov. *J. Phys. Chem. Solids*, **1961**, *19*, 35. (b) Wagner, C. *Z. Electrochem.* **1961**, *65*, 581.
32. (a) Vekilov, P. G. *Cryst. Growth Des.* **2007**, *7*, 2796. (b) Streets, A. M., and S. R. Quake. *Phys. Rev. Lett.* **2010**, *104*, 178102.
33. ten Wolde, P. R., and D. Frenkel. *Science*, **1997**, *277*, 1975.

2 Transport Properties of Binary Mixtures of Asymmetric Particles

A Simulation Study

Snehasis Daschakraborty and Ranjit Biswas

CONTENTS

2.1 INTRODUCTION

Study of the transport properties of asymmetric particles is of fundamental importance because real molecules more often than not are asymmetric in nature [1–8]. The structural aspects and transport properties of fluids and fluid mixtures containing asymmetric particles are significantly different from those made of spherical entity. Asymmetry in particle shape and interaction can lead to microscopic heterogeneity in solution structure, even in model systems under normal conditions. Hard rod and disk model is the simplest example of these types of systems, which have been used by many authors to study different structural and dynamical behaviors [1–7]. These model systems are governed by hard repulsive interactions only, and therefore, attractive interaction among particles finds no role in determining various properties of either neat or mixed systems. Gay–Berne (GB) potential, on the other hand, includes both repulsive and attractive interactions and, thus, somewhat more realistic for studying the properties of liquids made of asymmetric particles. At a very simplistic level, GB interaction has some similarities with that among Lennard–Jones particles [8,9]. Several simulations using GB potential have already explored structural and dynamical aspects of several asymmetric systems [10–12]. The phase behavior of GB fluids is also very interesting because the modified form of GB potential [13] can give rise to liquid crystal. This is an important observation, as optoelectronic industries require materials that could be used intelligently for designing and fabricating liquid-crystal-display devices.

Phase behavior has been extensively studied for GB fluid, and three distinctly different phases have been identified, i.e., the isotropic, nematic, and smectic phases [14–16]. Among these three phases, whereas the smectic phase is the most orientationally ordered, the isotropic phase has no orientational ordering, and the nematic one lies in between these two. The dynamics of pure GB fluid has been seen to be different for different phases. Interestingly, the diffusion coefficient parallel to the molecular axis shows an anomalous increase with density as the system enters from the isotropic to the nematic region. The Debye diffusion model appears to explain the reorientational mechanism for the nematic phase although it fails to explain in the isotropic region [11,12]. Both bulk and shear viscosities have been simulated by several researchers, and a good agreement between simulations and experiments has been observed [17]. Molecular dynamics simulations for molecules represented by GB ellipsoid particles and transverse point dipoles have also been reported [18]. Results for polar GB fluid have been compared with the nonpolar GB fluid, and it has been seen that, for polar ones, the smectic phase is formed at lower density compared to the nonpolar variety.

Several simulation studies have already been carried out on the translational and rotational dynamics of GB fluid near the isotropic–nematic phase transition (I-N) point, as well as in the isotropic phase region [11,12]. These works are more focused toward the verification of hydrodynamic relationships in these regions. Simulations of single particle and collective reorientation correlation functions reveal some interesting results. For example, decay of the second rank ($l = 2$) collective orientational relaxation slows down as the I-N transition point is approached. Moreover, the rank dependence predicted by the Debye law also breaks down in this region. The translational diffusion coefficient (D_T) and reorientational correlation time (τ_l) have also been simulated where the product $D_T \times \tau_l$ remains independent only at higher density and lies between the slip and stick limits of the Stokes–Einstein–Debye relation only for GB particles having lower aspect ratio ($\kappa \leq 1.5$). For higher aspect ratio ($\kappa \sim 3.0$), however, it rarely shows the preceding behavior.

Detailed molecular dynamics simulations have been carried out also for GB particles in the sea of spheres. These studies have indicated anisotropic diffusion for the ellipsoids at higher density. In addition, the ratio between parallel and perpendicular diffusion coefficients rises from unity to the value of aspect ratio as density of the system increases [19–21]. The product of the translational diffusion coefficient and reorientational correlation time behaves in a manner similar to that found for pure GB fluid.

The preceding survey suggests that the binary mixture of GB fluid has not been studied so far by simulation or numerical methods although, as already mentioned, this is important because real systems are more likely to possess either size, shape, or interaction asymmetry, or any combination of them. The verification of hydrodynamic relations is important for uncovering the nature of solute–solvent interactions in these more complex but model systems. This will certainly help to understand the composition dependence of the binary mixture of GB fluids. One expects in these studies a high degree of nonlinearity in composition dependence because asymmetric interaction-induced nonideal solution behavior has been observed for LJ mixtures of size-symmetric particles [22,23].

In this chapter, we have carried out equilibrium molecular dynamics simulations for binary mixtures of GB fluid containing two components of varying aspect ratios near I-N transition in order to study the transport properties of the binary mixture and investigate nonideality in this system. Our objective is to investigate the composition dependence of radial distribution function ($g(r)$), pressure (P), shear viscosity coefficient (η), translational diffusion coefficients (D_T) (overall, self, and mutual), and rotational correlation time constants (τ_l) of rank $l = 1$ and 2. We report the product $D_T \times \tau_l$, which has been found to be nearly independent of mixture composition. The rotational dynamics has been studied, where the Debye diffusion model fails to explain the reorientational mechanism. Nonideality has been observed for pressure, self, and overall diffusion coefficients, even though the extent of nonideality is always less than 10%. Interestingly, nonideality is absent for viscosity and mutual diffusion coefficients. The mutual diffusion coefficient remains nearly independent throughout the mole fraction range, which qualitatively suggests that the mixture is probably homogeneous at all compositions although further analyses are required for a definitive answer [24].

2.2 MODEL AND SIMULATION DETAILS

In this section, we will discuss the model we have used and the details of the simulation method.

Molecular dynamics simulations were carried out for binary mixtures using 500 ellipsoids interacting via the following GB interaction potential [9,13].

$$U_{GB}\left(\hat{u}_i,\hat{u}_j,\hat{r}_{ij}\right) = 4\varepsilon\left(\hat{u}_i,\hat{u}_j,\hat{r}_{ij}\right) \times\left[\left(\frac{d_w\sigma_0}{r_{ij}-\sigma\left(u_i,u_j,r_{ij}\right)+d_w\sigma_0}\right)^{12}-\left(\frac{d_w\sigma_0}{r_{ij}-\sigma\left(u_i,u_j,r_{ij}\right)+d_w\sigma_0}\right)^{6}\right], \tag{2.1}$$

where σ_0 is the diameter of the major axis of the ellipsoid, and $\sigma\left(\hat{r}_{ij},\hat{u}_i,\hat{u}_j\right)$ is given by

$$\sigma\left(\hat{u}_i,\hat{u}_j,\hat{r}_{ij}\right) = \sigma_0\left[1-\left\{\frac{\chi\alpha^2\left(\hat{u}_i\cdot\hat{r}_{ij}\right)+\chi\alpha^{-2}\left(\hat{u}_j\cdot\hat{r}_{ij}\right)-2\chi^2\left(\hat{u}_i\cdot\hat{r}_{ij}\right)\left(\hat{u}_j\cdot\hat{r}_{ij}\right)\left(\hat{u}_i\cdot\hat{u}_j\right)}{1-\chi^2\left(\hat{u}_i\cdot\hat{u}_j\right)}\right\}\right] \tag{2.2}$$

and

$$\chi = \left[\frac{\left(l_i^2-d_i^2\right)\left(l_j^2-d_j^2\right)}{\left(l_j^2-d_i^2\right)\left(l_i^2+d_j^2\right)}\right]^{-1/2} \tag{2.3}$$

$$\alpha^2 = \left[\frac{\left(l_i^2 - d_i^2\right)\left(l_j^2 + d_i^2\right)}{\left(l_j^2 - d_j^2\right)\left(l_i^2 + d_j^2\right)}\right]^{-1/2}, \tag{2.4}$$

where l and d denote the length and breadth, respectively, of each particle.

The total well depth parameter can be computed as follows:

$$\varepsilon\left(\hat{u}_i, \hat{u}_j, \hat{r}_{ij}\right) = \varepsilon_0 \varepsilon_1^{\nu}\left(\hat{u}_i, \hat{u}_j\right) \varepsilon_2^{\mu}\left(\hat{u}_i, \hat{u}_j, \hat{r}_{ij}\right). \tag{2.5}$$

The orientation-dependent strength terms are calculated in the following manner:

$$\varepsilon_1\left(\hat{u}_i, \hat{u}_j\right) = \left[1 - \chi^2\left(\hat{u}_i \cdot \hat{u}_j\right)\right]^{-1/2} \tag{2.6}$$

$$\varepsilon_2\left(\hat{u}_i, \hat{u}_j, \hat{r}_{ij}\right) = 1 - \left\{\frac{\chi'\alpha'^2\left(\hat{u}_i \cdot \hat{r}_{ij}\right) + \chi'\alpha'^{-2}\left(\hat{u}_j \cdot \hat{r}_{ij}\right) - 2\chi'^2\left(\hat{u}_i \cdot \hat{r}_{ij}\right)\left(\hat{u}_j \cdot \hat{r}_{ij}\right)\left(\hat{u}_i \cdot \hat{u}_j\right)}{1 - \chi'^2\left(\hat{u}_i \cdot \hat{u}_j\right)}\right\}, \tag{2.7}$$

where

$$\alpha'^2 = \left[1 + (\varepsilon_E/\varepsilon_S)^{1/\mu}\right]^{-1/2} \text{ and } \chi' = \frac{1 - (\varepsilon_E/\varepsilon_S)^{1/\mu}}{1 + (\varepsilon_E/\varepsilon_S)^{1/\mu}}. \tag{2.8}$$

The total number of particles was kept constant ($N = 500$) across the composition, and NVT ensembles were considered for simulations. A cubic box with conventional periodic boundary conditions was employed for binary mixtures of 500 GB prolate ellipsoids, with components having different aspect ratios. The first component (C1) was of aspect ratio $\kappa_1 = 2.0$, and the second component (C2) was of $\kappa_2 = 1.5$. The mole fraction of C1 was then varied to have binary mixtures at different compositions. All the quantities in the simulation were scaled to appropriate units and the scaled quantities of density, temperature, and time denoted by ρ^*, T^*, and t^*, respectively. Present simulations were carried out at $\rho^* = 0.4$ and $T^* = 1.0$. The time step Δt^* used was 0.001. The system was equilibrated for 2×10^5 time steps, and the production involved 1.3×10^6 steps for all the mixtures. d_w in the potential form was set to 1, and parameters μ and ν were set to their canonical values of 2.0 and 1.0, respectively. The asymmetry in energy $\kappa' = \frac{\varepsilon_s}{\varepsilon_E}$ was set to 5.0 for all the mixtures, with ε_s and ε_E denoting the energy parameters for the ellipsoids having end–end and end–side configurations, respectively.

Translational self-diffusion coefficients (D_T) were calculated from both mean-squared displacements (MSDs, $\langle|\Delta\vec{r}(t)|^2\rangle$) and velocity autocorrelation functions

(VACFs). The MSDs were calculated from the simulated center-of-mass positional vectors ($\vec{r}_i^{\,c}(t)$) [25,26]

$$\left\langle \left|\Delta\vec{r}(t)\right|^2 \right\rangle = \frac{1}{N}\left\langle \sum_{i=1}^{N} \left|\vec{r}_i^{\,c}(t) - \vec{r}_i^{\,c}(0)\right|^2 \right\rangle, \tag{2.9}$$

which produced D_T via the connection

$$D_T = \left[\frac{1}{6t}\left(\left\langle \left|\Delta\vec{r}(t)\right|^2 \right\rangle\right)\right]_{t\to\infty}. \tag{2.10}$$

D_T from the VACF were obtained by the following manner [25,26]:

$$D_T = \frac{1}{3}\int_0^{\infty} dt \left\langle \vec{v}_i(t).\vec{v}_i(0) \right\rangle, \tag{2.11}$$

where $\vec{v}_i$ is the center-of-mass velocity vector associated with the ith particle, and averaging was done over both time and number of particles.

In binary mixtures, mutual diffusion describes the ability of one species diffusing into another. The mutual diffusion coefficient $D_{12}(=D_{21})$ in a binary mixture of species 1 and 2 is defined by Green–Kubo relation as [27–29]

$$D_{12} = \frac{Q}{3Nx_1x_2}\int_0^{\infty} \left\langle \vec{J}_{12}(t)\cdot\vec{J}_{12}(0) \right\rangle dt, \tag{2.12}$$

where relative velocity $\vec{J}_{12}$ is defined as

$$\vec{J}_{12}(t) = x_2\sum_{k=1}^{N_1} \vec{v}_k(t) - x_1\sum_{l=1}^{N_2} \vec{v}_l(t). \tag{2.13}$$

N denotes the total number of particles, and x_1 and x_2 are the mole fractions of species 1 and 2, respectively. $\vec{v}_k(t)$ is the velocity of the kth particle of specie 1 at time t, and $\vec{v}_l(t)$ is the velocity of the lth particle of specie 2 at time t. The thermodynamic factor Q can be expressed as

$$Q = [1 + x_1x_2\rho(G_{11} + G_{22} - 2G_{12})]^{-1} \tag{2.14}$$

with

$$G_{ij} = 4\pi \int_0^\infty r^2 [g_{ij}(r) - 1] dr, \tag{2.15}$$

where ρ is the number density and $g_{ij}(r)$ is the radial distribution function for the pair of species ij.

To study the reorientational motion associated with $l = 1$ and 2, we calculated the single-particle reorientational correlation functions defined by [25]

$$C_l^{(s)}(t) = \frac{\left\langle P_l\left(\hat{e}_i(0)\cdot\hat{e}_i(t)\right)\right\rangle}{\left\langle P_l\left(\hat{e}_i(0)\cdot\hat{e}_i(0)\right)\right\rangle}, \tag{2.16}$$

where $\hat{e}_i(t)$ is the unit vector along the symmetry axis of molecule i, and P_l is the lth-order Legendre polynomial. In the preceding equations, the angular bracket implies an average over the particles and over the time origins.

Shear viscosity coefficient (η) was calculated using the Green–Kubo relation [30,31]

$$\eta = \frac{V}{9k_BT} \int_0^\infty \langle P_{\alpha\beta}(t) P_{\alpha\beta}(0)\rangle \, dt, \tag{2.17}$$

where $\alpha,\beta = x,y,z$, and $P_{\alpha\beta}$ denotes the off-diagonal term of the pressure tensor

$$P_{\alpha\beta} = \frac{1}{V}\left(\sum_i \frac{p_{i\alpha}p_{i\beta}}{m_i} + \sum_i \sum_{j>i} r_{ij\alpha} f_{ij\beta}\right). \tag{2.18}$$

As before, the preceding correlation functions were also averaged over particles and time.

The pressure was then obtained from the simulated diagonal terms of the pressure tensor by employing the following expression:

$$P = \frac{1}{3} Tr[P] = \frac{1}{3}\sum_\alpha P_{\alpha\alpha}. \tag{2.19}$$

2.3 RESULTS AND DISCUSSION

Effects of C1 (component with higher aspect ratio) on the *average* radial distribution function ($g(r)$) have been depicted in Figure 2.1 for four representative compositions.

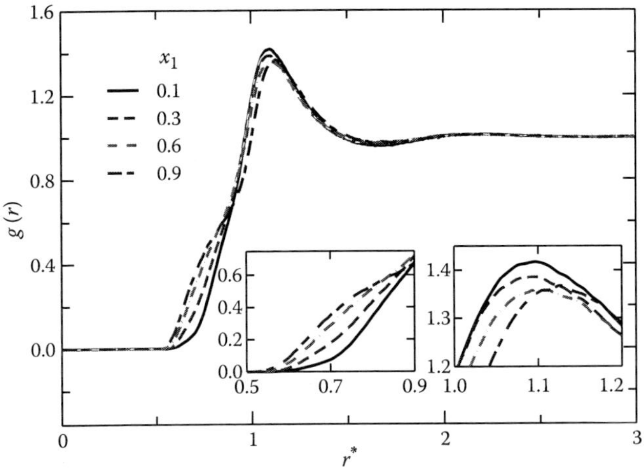

FIGURE 2.1 Plots of simulated overall radial distribution function ($g(r)$) for systems with different compositions. Two insets represent $g(r)$ for $r \sim 0.75$ and 1 for better visualization. Different curves are color coded and explained in the plot. x_1 represents the mole fraction of the first component ($\kappa = 2$) in the binary mixture.

It is evident from this figure that the simulated $g(r)$ undergoes several modifications as mixture composition is altered by changing the mole fraction (x_1) of C1. The peak position of $g(r)$ shifts toward longer distance, along with decrease in peak height as x_1 in the mixture is increased. Interestingly, a hump at $r \sim 0.75$ may be noticed, which becomes more prominent upon increasing x_1. These two regions have been shown separately in the insets for better visualization. This indicates the gradual rise of the probability of cross configuration over end–end and side–side configurations as the binary mixture becomes enriched with particles of higher aspect ratio. This is supported by the potential energy diagram for the GB interaction, where it has been found that the depth of the potential energy well for cross configuration is higher than that for end–end and side–side configurations for both ellipsoids and disks [32]. It should be noted that the formation of hump at $r \sim 0.75$ has been found previously by other researchers as well as for the isotropic and discotic–nematic phases [33].

Figure 2.2 represents the mean square displacement (MSD); normalized VACF as a function of time in the upper and middle panels, respectively; and translational diffusion coefficient ($D^*_{overall}$) as a function of x_1 in the lower panel. Both the MSD and VACF plots suggest weak composition dependence. This is reflected in the lower panel, where $D^*_{overall}$ (calculated from both Equations 2.10 and 2.11) has been plotted as a function of x_1. $D^*_{overall}$ increases to a maximum value at $x_1 \sim 0.5$ and then decreases upon further increase in x_1. Although the magnitude of variation is small (~13%), the systematic decrease probably suggests a kind of structural transition of the binary mixture. As the system passes through the 0.5 mole fraction, the system makes a transition from the C2-dominated to the C1-enriched regime. The higher diffusion coefficient at $x_1 \sim 0.5$ mole fraction may arise due to the least effective packing of the ellipsoids in the system. A previous theoretical investigation involving binary

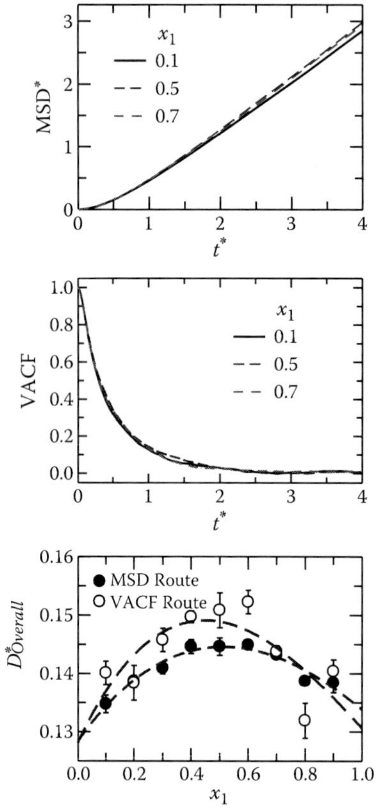

FIGURE 2.2 (Upper panel) Simulated MSDs for the overall systems as a function of time at four representative compositions. (Middle panel) Simulated normalized VACF as a function of time at the same four representative compositions. (Lower panel) Overall translational diffusion coefficient $D^*_{Overall}$ (calculated from the VACF and MSD plot) as a function of the mole fraction of the first specie. Error bars have been computed via block average. Lines going through the data are for visual guide.

mixtures of hard spheres and ellipsoids predicted relatively less compact packing nearly at 50:50 composition [34]. The extrapolated value of diffusion coefficient ($D^*_{overall}$) for pure C1 is ~0.13, which is in close agreement to the earlier simulated value obtained by using pure GB particles having aspect ratio $\kappa = 2$ at comparable density and temperature [12].

Self-diffusion coefficients for the two components have been calculated separately by using Equations 2.10 (MSD route) and 2.11 (VACF route), and finally, the mean values have been plotted in Figure 2.3. The upper and middle panels of Figure 2.3 are the plots for the self-diffusion coefficients of C1 (D_1^*) and C2 (D_2^*), respectively, as a function of x_1. D_1^* decreases almost steadily with x_1, although the extent of decrease is somewhat small (~12%). D_1^* is nearly equal to $D^*_{overall}$ at $x_1 = 0.9$, which is expected because self-diffusion of any species in a binary mixture dominated by that

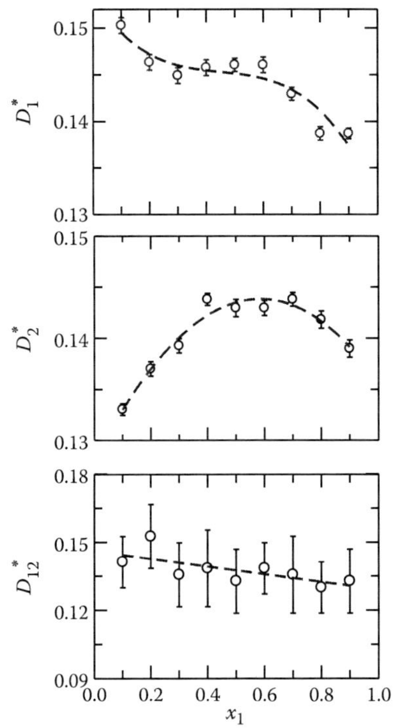

FIGURE 2.3 Composition dependence of the simulated self-diffusion coefficient of the (upper panel) first component having aspect ratio κ = 2.0, (middle panel) second component having κ = 1.5, and (lower panel) the mutual diffusion coefficient. Error bars have been computed via block average. Lines going through the data are for visual guide.

specie should be nearly equal to the overall diffusion of the system. Unlike D_1^*, D_2^* exhibits a nonmonotonic composition dependence with a peak at $x_1 \approx 0.6$, although the overall change is only ~11% of the initial value of D_2^*. As expected, D_2^* is also nearly equal to $D_{overall}^*$ at $x_1 = 0.1$. A closer inspection of these two plots in Figure 2.3 reveals that the self-diffusion coefficients of two species are nearly the same at $x_1 =$ 0.5. This probably signals a structural transition occurring at this composition. This structural transition has not been seen earlier for asymmetric binary fluid mixtures. The lower panel of Figure 2.3 represents the mutual diffusion coefficient of the species as a function of x_1. In binary liquid mixtures, mutual diffusion is related to the ability of one specie diffusing into the other. This is different from self-diffusion, which is a measure of mobility of each component in the absence of any external force that means the diffusion of a given species in an environment created only by that species. Therefore, mutual diffusion involves collective motion of many particles of different species together in the mixture and arises due to the gradient of the composition (or chemical potential). Mutual diffusion can be expressed in terms of velocity correlation functions of the collective motion of the system or in terms of MSD of the center of mass of the particles of either of the two components. The mutual diffusion coefficient $D_{12}(= D_{21})$ in a binary mixture of species 1 and 2 has been obtained by using the Green–Kubo relation [27–29] expressed in Equation 2.12. The lower panel of Figure 2.3 shows that the mutual diffusion coefficient is nearly constant to the variation in composition within the uncertainty limits. The statistical error in mutual diffusivity is of great concern and could be reduced up to some extent by averaging over more simulation runs. This insensitivity of mutual diffusion coefficient to mixture composition may arise from strong miscibility of components in the binary mixture, but a more precise study warrants simulations using the thermodynamic integration method [24].

Figure 2.4 represents the plot for viscosity coefficient (η^*) as a function of x_1. η^* has been calculated from the integration of stress autocorrelation function [30,31] by using Equation 2.17. The figure shows that η^* remains almost constant in the entire mole fraction range and, more interestingly, does not show any type of nonideal behavior, which is not expected from diffusion behavior (Figure 2.3). The absence of the nonideality of η^* may be due to the large estimation error, which is very much clear from the plot. We would like to mention here that, with η^* being more collective in nature than the diffusion coefficient, simulations of the former are less trivial than the latter. Thus, the collectiveness and large error of estimation of viscosity coefficient have made the simulated viscosities more imprecise than the diffusion coefficients. Molecular dynamics simulations have been done previously for the GB model of liquid crystals in the nematic and isotropic phases [17]. The temperature dependence of shear viscosities is in good agreement with experimental data [35,36]. The viscosity obtained in that calculation is two to three times higher than our result for nearly pure GB fluid of higher aspect ratio. This may arise due to the lower aspect ratio of the ellipsoid used in this calculation compared to the previous study.

Pressure has been computed by using Equation 2.19 from the simulated pressure tensor and plotted in Figure 2.5 as a function of x_1. The simulated pressure shows a nonideal composition dependence, which can be explained in terms of packing. Pressure derives contributions from thermal energy and virial term, which is the product of interparticle distance (r_{ij}) and force (F_{ij})

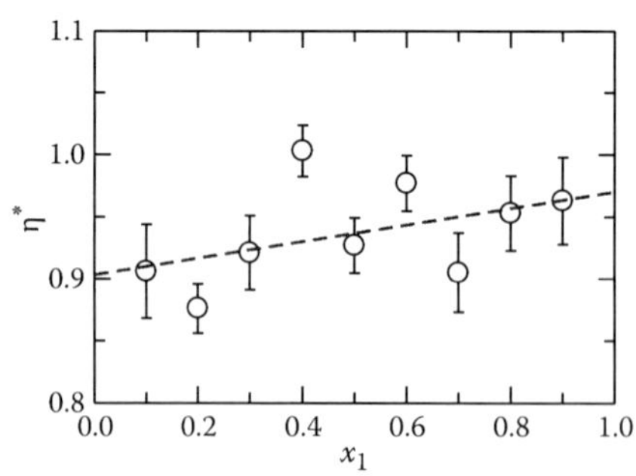

FIGURE 2.4 Plot for shear viscosity coefficient as a function of the mole fraction of the first specie computed via block average. Lines going through the data are for visual guide.

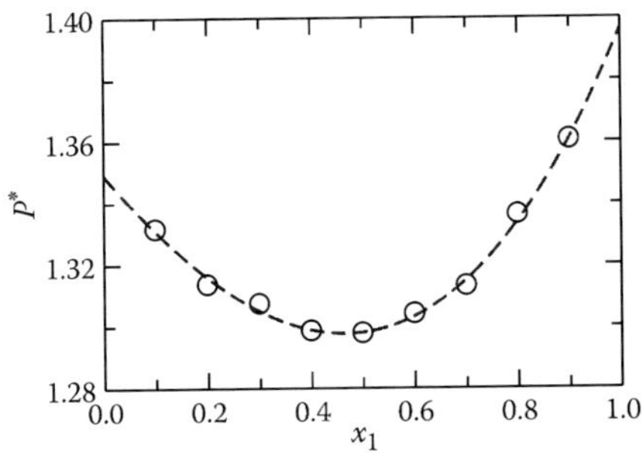

FIGURE 2.5 Composition dependence of the simulated pressure. Lines going through the data are for visual guide.

$$P = \frac{1}{3V}\sum_{\alpha}\left[\sum_{i} p_{i\alpha}p_{i\alpha}/m_i + \sum_{i}\sum_{j>i} r_{ij\alpha}F_{ij\alpha}\right]. \tag{2.20}$$

With thermal energy being constant throughout the composition range, the virial term is only responsible for variation. This means, at the equal proportion of the two species in the mixture, the total interaction is the least, and this may arise due to loose packing. As aspect ratios are different, the packing will not be as tight as in same aspect ratios case, and consequently, the presence of void space produces lower pressure. Note this aspect has not been clearly reflected in the viscosity coefficient (Figure 2.4) because of inaccuracy involved with the simulated values. Figure 2.5 also suggests that, even though the simulated pressure at $x_1 = 0.9$ is somewhat smaller (~1.36) than that (~2.0) for pure GB fluid ($x_1 = 1.0$) at comparable conditions [12], the slope of the present data indicates a very similar value for pressure at $x_1 = 1.0$.

Figure 2.6 depicts the reorientational time correlation function (RTCF) of ranks, $l = 1$ and 2 for a representative composition, $x_1 = 1.0$ (upper panel) and the product of translational diffusion coefficient and rotational correlation time ($D_T \times \tau_L$) as a function of x_1 (lower panel). RTCF has been calculated by Equation 2.17. The upper panel shows that the RTCF of first rank ($l = 1$) decays at a rate slower than that of second rank ($l = 2$). This is expected. For other compositions, this trend remains the same. Rotational correlation time constant has been obtained via time integration of RTCF as follows:

$$\tau_L = \int_0^{\infty} dt\, C_l^{(S)}(t). \tag{2.21}$$

In Table 2.1, we have shown x_1 dependence of τ_1 and τ_2, where it is observed that both the time constants are almost invariant with composition of the mixture. In the same table, ratio τ_1/τ_2 has also been shown. Interestingly, the ratio is ~1.5, which is

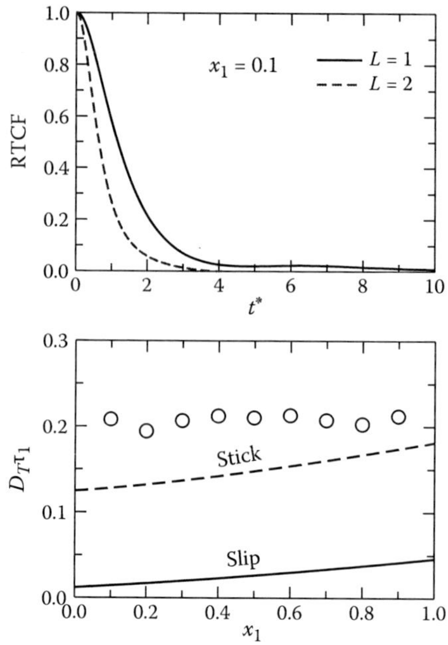

FIGURE 2.6 (Upper panel) Plot for the reorientational correlation function against time for a representative composition ($x_1 = 0.1$). (Lower panel) Product of the translational diffusion coefficient D_T and the average orientational correlation time τ_1 of the first-rank correlation function as a function of composition. Note that the solid line and dashed line indicate the hydrodynamic predictions with the stick and slip boundary conditions, respectively.

TABLE 2.1
Composition Dependence of the Orientational Correlation Time of Rank = 1 and 2, and the Ratio between the Two

x_1	τ_1^*	τ_2^*	τ_1/τ_2
0.1	1.51	0.99	1.53
0.2	1.40	0.96	1.46
0.3	1.45	0.96	1.51
0.4	1.44	0.94	1.53
0.5	1.46	0.94	1.55
0.6	1.44	0.93	1.55
0.7	1.48	0.96	1.50
0.8	1.44	0.96	1.54
0.9	1.44	0.94	1.53

Note: x_1 denotes the mole fraction of the first component ($\kappa = 2$) in the binary mixture.

half of the predicted value by Debye's law for the rank dependence of reorientational motion in normal liquids. This deviation is also seen previously for pure GB fluid near the I-N transition [11].

In the lower panel of Figure 2.6, we have plotted $D_T \times \tau_L$ as a function of x_1. Interestingly, this remains constant throughout the mole fraction range. The hydrodynamic values (combined SE and SED relation) predicted for this case are also plotted in the same figure. Tangs and Evans [37] have reported the Stokes–Einstein products $\mu = D_T\eta$ for neat hard ellipsoids of aspect ratio ranging from 1 to 10 and for both slip and stick boundary conditions. The rotational diffusion coefficient $D_R = k_BT/\xi_R$ can be computed from rotational friction ξ_R given by [12]

$$\xi_R = \pi\eta a^3\xi_z, \tag{2.22}$$

where ξ_z may be computed by using the formalism proposed in [37]. Such a calculation has been done for the neat GB fluid as a function of density [12]. Using the relation $\tau_L^{(S)} = [L(L+1)D_R]^{-1}$, we can now express the product $D_T \times \tau_L$ in terms of μ and ξ_z as follows:

$$D_T\tau_L = (\mu\xi_z)\frac{\pi}{16}\frac{\kappa^{8/3}}{L(L+1)}\sigma_0^2. \tag{2.23}$$

This expression holds for pure GB fluid for stick and slip conditions with the corresponding values for μ and ξ_z. We have modified this expression for our systems containing GB particles having different aspect ratios. We have plotted the ξ_z values as a function of effective aspect ratio $\kappa = x_1\kappa_1 + x_2\kappa_2$ from [38] and fitted with a cubic equation to get the following equation:

$$\begin{aligned} \xi_z^{Stick} &= 0.5132 + 4.0441(\kappa^{-1}) - 9.874(\kappa^{-1})^2 + 5.3315(\kappa^{-1})^3 \\ \xi_z^{Slip} &= \left[0.9336 - 1.4186(\kappa^{-1}) - 0.2804(\kappa^{-1})^2 + 0.7814(\kappa^{-1})^3\right] \times \left[\xi_z^{Stick}\right]. \end{aligned} \tag{2.24}$$

Unlike ξ_z, when we plotted the parameter μ as a function of κ from [36], we found linear dependence and obey the following relations:

$$\begin{aligned} \mu^{Stick} &= 0.1083 - 0.042(\kappa) \\ \mu^{Slip} &= 0.1427 + 0.0128(\kappa). \end{aligned} \tag{2.25}$$

Calculated products obeying stick and slip limits are denoted by the lines in the lower panel of Figure 2.6. More interestingly, the simulated product $D_T \times \tau_1$ is nearly constant and lies above the calculated product obeying the stick hydrodynamic condition.

2.4 CONCLUSION

In this chapter, the binary mixture of GB particles of different aspect ratios has been studied by molecular dynamics simulation. The composition dependence of different static and dynamic properties has been studied. The radial distribution function has been found to show some interesting features. Simulated pressure and overall diffusion coefficient exhibit nonideal composition dependence. However, simulated viscosity does not show any clear nonideality. The mole fraction dependence of self-diffusion coefficients qualitatively signals some kind of structural transition in the 50:50 mixture. The rotational correlation study shows the non-Debye behavior in its rank dependence. The product of translational diffusion coefficient and rotational correlation time (first rank) has been found to remain constant across the mixture composition and lie above the stick prediction.

ACKNOWLEDGMENTS

We thank Prof. B. Bagchi (Indian Institute of Science, Bangalore) for the encouragement. S. Daschakraborty acknowledges the CSIR, India, for the research fellowship. R. Biswas thanks Prof. H. Shirota and the Centre for Frontier Science, Chiba University, Japan, for the hospitality during a part of this work.

REFERENCES

1. Rebertus, D. W., and K. M. Sando. *J. Chem. Phys.* **1977**, *67*, 2585.
2. Allen, M. P., and D. Frenkel. *Phys. Rev. Lett.* **1987**, *58*, 1748.
3. Talbot, J., M. P. Allen, G. T. Evans, D. Frenkel, and D. Kivelson. *Phys. Rev. A* **1988**, *39*, 4330.
4. Frenkel, D., and J. F. Maguirre. *Mol. Phys.* **1983**, *49*, 503.
5. Kushik, J., and J. Berne. *J. Chem. Phys.* **1976**, *64*, 1362.
6. Frenkel, D., B. M. Mulder, and J. P. McTague. *Phys. Rev. Lett.* **1984**, *52*, 287.
7. Allen, M. P. *Phys. Rev. Lett.* **1990**, *65*, 2881.
8. Magda, J. J., H. T. Davis, and M. Tirrell. *J. Chem. Phys.* **1986**, *85*, 6674.
9. Berne, B. J., and P. Pechukas. *J. Chem. Phys.* **1972**, *56*, 4213.
10. Miguel, E. D., L. F. Rull, and K. E. Gubbins. *Phys. Rev. A* **1992**, *45*, 3813.
11. Perera, A., S. Ravichandran, M. Morean, and B. Bagchi. *J. Chem. Phys.* **1997**, *106*, 1280.
12. Ravichandran, S., A. Perera, M. Morean, and B. Bagchi. *J. Chem. Phys.* **1997**, *107*, 8469.
13. Berne, B. J., and P. Pechukas. *J. Chem. Phys.* **1981**, *74*, 3316.
14. Miguel, E. D., L. F. Rull, M. K. Chalam, K. E. Gubbins. *Mol. Phys.* **1991**, *74*, 405.
15. Luckhurst, G. R., R. A. Stephens, R. W. Phippen. *Liq. Cryst.* **1990**, *8*, 451.
16. Luckhurst, G. R., and P. S. Simmonds. *J. Mol. Phys.* **1993**, *80*, 233.
17. Smondyrev, A. M., G. B. Loriot, R. A. Pelcovits. *Phys. Rev. Lett.* **1995**, *75*, 2340.
18. Gwóźdź, E., A. Bródka, K. Pasterny. *Chem. Phys. Lett.* **1997**, *267*, 557.
19. Ravichandran, S., and B. Bagchi. *J. Chem. Phys.* **1999**, *111*, 7505.
20. Vasanthi, R., S. Ravichandran, and B. Bagchi. *J. Chem. Phys.* **2001**, *114*, 7989.
21. Vasanthi, R., S. Ravichandran, and B. Bagchi. *J. Chem. Phys.* **2001**, *115*, 10022.
22. Mukherjee, A., and B. Bagchi. *J. Phys. Chem. B* **2001**, *105*, 9581.
23. Srinivas, G., A. Mukherjee, and B. Bagchi. *J. Chem. Phys.* **2001**, *114*, 6220.

24. Jedlovszky, P., A. Idrissi, and G. Jancsó. *J. Chem. Phys.* **2009**, *130*, 124516.
25. McQuarrie, D. A. *Statistical Mechanics* **2003** (Viva Books, New Delhi).
26. Del Pópolo, M. G., Voth, G. A. *J. Phys. Chem. B* **2004**, *108*, 1744.
27. Kamala, C. R., K. G. Ayappa, and S. Yashonath. *Phys. Rev. E* **2002**, *65*, 061202.
28. Zhou, Z., B. D. Todd, K. P. Travis, and R. J. Sadus. *J. Chem. Phys.* **2005**, *123*, 054505.
29. Zhang, L., Y. Liu, and Q. Wang. *J. Chem. Phys.* **2005**, *123*, 144701.
30. Hansen, J. P., and I. R. McDonald. *Theory of Simple Liquids* **1986** (Academic Press, London).
31. Allen, M. P., and D. Tildesley. *J. Computer Simulations of Liquids* **1987** (Oxford University Press, New York).
32. Luckhurst, G. R., and P. S. Simmonds. *J. Mol. Phys.* **1993**, *80*, 233.
33. Chakraborty, D., and D. Wales. *J. Phys. Rev. E* **2008**, *77*, 051709.
34. Perera, A., K. Cassou, F. Ple, and S. Dubois. *Mol. Phys.* **2002**, *100*, 3409.
35. Miesowicz, M. *Nature (London)* **1946**, *158*, 27.
36. Chmielewski, A. G. *Mol. Cryst. Liq. Cryst.* **1986**, *132*, 339.
37. Tang, S., and T. Evans. *Mol. Phys.* **1993**, *80*, 1443.
38. Hu, C.-M., and R. Zwanzig. *J. Chem. Phys.* **1974**, *60*, 4354.

3 Time-Dependent Density Functional Theoretical Methods for Nonperturbative Treatment of Multiphoton Processes of Many-Electron Molecular Systems in Intense Laser Fields

John T. Heslar, Dmitry A. Telnov, and Shih-I Chu

CONTENTS

3.1 INTRODUCTION

In recent years, the density functional theory (DFT) has become a widely used formalism for electron structure calculations of atoms, molecules, and solids [1–5]. The DFT is based on the earlier fundamental work of Hohenberg and Kohn [6] and Kohn and Sham [7]. In the Kohn–Sham DFT formalism [7], the electron density is decomposed into a set of orbitals, leading to a set of one-electron Schrödinger-like equations to be solved self-consistently. The Kohn–Sham equations are structurally similar to the Hartree–Fock equations but include, in principle, exactly *many-body* effects through a *local* exchange-correlation (xc) potential. Thus, DFT is computationally much less expensive than traditional *ab initio* many-electron wavefunction approaches, and this accounts for its great success for large systems. However, the DFT is well developed mainly for the *ground-state* properties only. The treatment of *excited states* and time-dependent processes within the DFT is relatively less developed.

The essential element of DFT is the input of the xc-energy functional, whose exact form is unknown. The simplest approximation for the xc-energy functional is through *local spin-density approximation* (LSDA) [1,8] of homogeneous electronic gas. A severe deficiency of the LSDA is that the xc potential decays exponentially and does not have correct *long-range* behavior. As a result, the LSDA electrons are too weakly bound and, for negative ions, even unbound. More accurate forms of the xc-energy functionals are available from the *generalized gradient approximation* (GGA) [1,9,10], and hybrid-energy functionals [11], which take into account the gradient of electron density. However, the xc potentials derived from these GGA energy functionals suffer similar problems as in LSDA and do not have the proper long-range Coulombic tail $-1/r$ either. Thus, whereas the total energies of the ground states of atoms and molecules predicted by these GGA density functionals [9,12] are reasonably accurate, the excited-state energies and the ionization potentials obtained from the highest occupied orbital energies of atoms and molecules are far from satisfactory, typically 40% through 50% too low [13]. The problem of the incorrect long-range behavior of the LSDA and GGA energy functionals can be attributed to the existence of *self-interaction energy* [13].

For proper treatment of atomic and molecular dynamics such as collisions or multiphoton ionization (MPI) processes, etc., it is necessary that both the ionization potential and the excited-state properties be described more accurately. In addition, the treatment of time-dependent processes will require the use of the time-dependent density functional theory (TDDFT). The rigorous formulation of TDDFT is due to the Runge–Gross theorem [14]. For any interacting many-particle quantum system subject to a given time-dependent potential, all physical observables are uniquely determined by knowledge of the time-dependent density and the state of the system at any instant in time. In particular, if the time-dependent potential is turned on at some time t_0 and the system has been in its ground state until t_0, all observables are unique functionals of only the density. In this case, the initial state of the system at time t_0 will be a unique functional of the ground state density itself, i.e., of the density at t_0. This unique relationship allows one to derive a computational scheme in which the effect of particle–particle interaction is represented by a density-dependent single-particle potential, so that the time evolution of an interacting system can be investigated by solving a time-dependent auxiliary single-particle problem. In the last decade, there has been considerable effort and success in the use of linear response theory to the study of excitation energies [15,16]; frequency-dependent multipole polarizabilities [17,18]; optical spectra of molecules, clusters, and nanocrystals [19,20]; and autoionizing resonances [13], etc.

In this chapter, we discuss some new developments in TDDFT beyond the linear response regime for accurate and efficient nonperturbative treatment of multiphoton dynamics and very-high-order nonlinear optical processes of atomic and molecular systems in intense and superintense laser fields. In Section 2, we briefly describe the time-dependent optimized effective potential (OEP) method and its simplified version, i.e., the time-dependent Krieger–Li–Iafrate (KLI) approximation, along with self-interaction correction (SIC). In Section 3, we present the TDDFT approaches and the time-dependent generalized pseudospectral (TDGPS) methods for the accurate treatment of multiphoton processes in diatomic and triatomic molecules. In Section 4, we describe the Floquet formulation of TDDFT. This is followed by a conclusion in Section 5. Atomic units will be used throughout this chapter.

3.2 TDDFT WITH OEP/KLI-SIC FOR TREATMENT OF MANY-ELECTRON SYSTEMS IN LASER FIELDS

In this section, we discuss TDDFT with OEP and SIC for nonperturbative treatment of many-electron quantum systems in intense laser fields [21]. The steady-state OEP method [22,23] has become a practical tool in DFT after the work of Kriger, Li, and Iafrate (KLI), who suggested a simplified yet accurate procedure for determination of OEP by a set of linear equations [24,25]. This method was extended to the time domain [26], but the original formulation was computationally not efficient since it involved the construction of Hartree–Fock–like nonlocal potential at each time step. The advantage of the time-dependent OEP (TD-OEP)/KLI-SIC approach [21] is that it allows the construction of *self-interaction-free* time-dependent *local* OEP. This greatly facilitates the study of time-dependent processes of many-electron quantum systems in strong fields.

We start from the quantum mechanical *action* of a many-electron system interacting with an external field [21,26]

$$\begin{aligned}A[\{\psi_{n\sigma}\}] = &\sum_{\sigma}\sum_{n=1}^{N_\sigma}\int_{-\infty}^{t_1} dt \int \psi_{n\sigma}^{*}(\boldsymbol{r},t)\left[i\frac{\partial}{\partial t}+\frac{1}{2}\nabla^2\right]\psi_{n\sigma}(\boldsymbol{r},t)d\boldsymbol{r}\\ &-\sum_{\sigma}\int_{-\infty}^{t_1} dt\int \rho_\sigma(\boldsymbol{r},t)[v_{\mathrm{n}}(\boldsymbol{r})+v_{\mathrm{ext}}(\boldsymbol{r},t)]d\boldsymbol{r}\\ &-\frac{1}{2}\int_{-\infty}^{t_1} dt\int\int \frac{\rho(\boldsymbol{r},t)\rho(\boldsymbol{r}',t)}{|\boldsymbol{r}-\boldsymbol{r}'|}d\boldsymbol{r}d\boldsymbol{r}' - A_{\mathrm{xc}}[\{\psi_{n\sigma}\}],\end{aligned} \tag{3.1}$$

where $\psi_{n\sigma}(\boldsymbol{r},t)$ are the time-dependent spin orbitals, $N = \sum_{\sigma} N_{\sigma}$ is the total number of electrons, $v_{n}(\boldsymbol{r})$ is the electron–nucleus Coulomb interaction, $v_{ext}(\boldsymbol{r},t)$ describes the coupling of the electron to the external laser fields, and $A_{xc}[\{\psi_{n\sigma}\}]$ is the xc-action functional. The electron spin densities $\rho_{\sigma}(\boldsymbol{r},t)$ are defined as

$$\rho_{\sigma}(\boldsymbol{r},t) = \sum_{n=1}^{N_{\sigma}} \left|\psi_{n\sigma}(\boldsymbol{r},t)\right|^{2} \tag{3.2}$$

and the total electron density $\rho(\boldsymbol{r},t)$ is obtained by summation of the spin densities $\rho(\boldsymbol{r},t) = \sum_{\sigma} \rho_{\sigma}(\boldsymbol{r},t)$.

The spin orbitals satisfy the one-electron Schrödinger-like equation

$$i\frac{\partial}{\partial t}\psi_{n\sigma}(\boldsymbol{r},t) = \left(-\frac{1}{2}\nabla^{2} + V_{\sigma}(\boldsymbol{r},t)\right)\psi_{n\sigma}(\boldsymbol{r},t), \tag{3.3}$$

where $V_{\sigma}(\boldsymbol{r},t)$ will be the TD-OEP if we choose the set of spin orbitals $\{\psi_{n\sigma}\}$, which render the total action functional $A[\{\psi_{n\sigma}\}]$ *stationary* $\dfrac{\delta A[\{\psi_{n\sigma}\}]}{\delta V_{\sigma}(\boldsymbol{r},t)} = 0.$

Generally, $V_{\sigma}(\boldsymbol{r},t)$ contains the memory effect. It depends on not only the density at time moment t but also the densities at preceding times. However, if we use the following explicit SIC expression for the xc-action functional [21]

$$\begin{aligned} A_{xc}[\{\psi_{n\sigma}\}] = & \int_{-\infty}^{t_1} dt E_{xc}\left[\rho_{\uparrow}(\boldsymbol{r},t),\rho_{\downarrow}(\boldsymbol{r},t)\right] \\ & -\sum_{\sigma}\sum_{n=1}^{N_{\sigma}}\int_{-\infty}^{t_1} dt\left\{J[\rho_{n\sigma}] + E_{xc}[\rho_{n\sigma},0]\right\}, \end{aligned} \tag{3.4}$$

then the memory term vanishes identically. Similar results are obtained as long as one uses an explicit E_{xc} form (such as that in LSDA or GGA) of energy functional and the adiabatic approximation. The use of the SIC form in Equation 3.4 removes spurious self-interaction terms in conventional TDDFT and results in a proper *long-range* asymptotic potential. Another major advantage of this procedure is that only *local* potential is required to construct the *orbital-independent* OEP. This facilitates considerably the numerical computation. The TDDFT/KLI-SIC approach and other TDDFT approaches with proper long-range potential have been applied successfully to a number of multiphoton processes of atomic and molecular systems in intense laser fields in the last decade, including MPI [27–34], high harmonic generation [27,32,33,35–37], and VUV/XUV frequency comb laser generation [38], etc. For a review, see [39].

3.3 TDDFT FOR MULTIPHOTON PROCESSES IN DIATOMIC AND TRIATOMIC MOLECULES

3.3.1 TDGPS METHOD FOR ACCURATE NUMERICAL SOLUTION OF TDDFT EQUATIONS IN TWO-CENTER SYSTEMS

In the following, we discuss the extension of the TDGPS procedure [40] to the accurate numerical solution of the TDDFT equations in two-center molecular systems. In the spin-polarized theory, the spin orbitals $\psi_{n\sigma}(\boldsymbol{r},t)$ corresponding to different spin projections σ satisfy equations with different effective potentials $v_{\text{eff},\sigma}(\boldsymbol{r},t)$, i.e.,

$$i\frac{\partial}{\partial t}\psi_{n\sigma}(\boldsymbol{r},t)=\left[-\frac{1}{2}\nabla^2+v_{\text{eff},\sigma}(\boldsymbol{r},t)\right]\psi_{n\sigma}(\boldsymbol{r},t),\quad n=1,2,\ldots,N_{\sigma}. \tag{3.5}$$

The time-dependent effective potential $v_{\text{eff},\sigma}(\boldsymbol{r},t)$ is a functional of both electron spin densities $\rho_{\uparrow}(\boldsymbol{r},t)$ and $\rho_{\downarrow}(\boldsymbol{r},t)$. The potential $v_{\text{eff},\sigma}(\boldsymbol{r},t)$ can be written in the general form

$$v_{\text{eff},\sigma}(\boldsymbol{r},t) = v_{\text{n}}(\boldsymbol{r}) + v_{\text{H}}(\boldsymbol{r},t) + v_{\text{xc},\sigma}(\boldsymbol{r},t) + v_{\text{ext}}(\boldsymbol{r},t), \tag{3.6}$$

where $v_{\text{n}}(\boldsymbol{r})$ is the electron interaction with the nuclei

$$V_{\text{n}}(\boldsymbol{r})=-\frac{Z_1}{|R_1-r|}-\frac{Z_2}{|R_2-r|}, \tag{3.7}$$

Z_1 and Z_2 are the charges of the nuclei, $\boldsymbol{R}_1$ and $\boldsymbol{R}_2$ are the positions of the nuclei, and $v_{\text{H}}(\boldsymbol{r},t)$ is the Hartree potential due to electron–electron Coulomb interaction

$$v_{\text{H}}(\boldsymbol{r},t)=\int d^3r'\,\frac{\rho(\boldsymbol{r}',t)}{|\boldsymbol{r}-\boldsymbol{r}'|}. \tag{3.8}$$

In the time-dependent calculations, we adopt the commonly used adiabatic approximation, where the xc potential is calculated with the time-dependent density. The adiabatic approximation had many successful applications to atomic and molecular processes in intense external fields [39,41]. For the studies of the diatomic molecules [29,32,42], we utilize the LBα (van Leeuwen–Baerends) xc potential [43]:

$$v_{\text{xc},\sigma}^{\text{LB}\alpha}(\boldsymbol{r},t)=\alpha v_{\text{x},\sigma}^{\text{LSDA}}(\boldsymbol{r},t)+v_{\text{c},\sigma}^{\text{LSDA}}(\boldsymbol{r},t)-\frac{\beta x_{\sigma}^2(\boldsymbol{r},t)\rho_{\sigma}^{1/3}(\boldsymbol{r},t)}{1+3\beta x_{\sigma}(\boldsymbol{r},t)\ln\left\{x_{\sigma}(\boldsymbol{r},t)+[x_{\sigma}^2(\boldsymbol{r},t)+1]^{1/2}\right\}}. \tag{3.9}$$

The LBα potential contains two parameters α and β, which have been adjusted in time-independent DFT calculations of several molecular systems and have the values $\alpha = 1.19$ and $\beta = 0.01$ [43]. The first two terms in Equation 3.9, i.e., $v_{\text{x},\sigma}^{\text{LSDA}}$ and

$v_{c,\sigma}^{LSDA}$, are the exchange and correlation potentials within the LSDA. The last term in Equation 3.9 is the gradient correction with $x_\sigma(\boldsymbol{r}) = |\nabla\rho_\sigma(\boldsymbol{r})|/\rho_\sigma^{4/3}(\boldsymbol{r})$, which ensures the proper long-range asymptotic behavior $v_{xc,\sigma}^{LB\alpha} \rightarrow -1/r$ as $r \rightarrow \infty$. The potential (Equation 3.9) has proved to be reliable in molecular TDDFT studies [32,34]. The correct long-range asymptotic behavior of the LBα potential is crucial in photoionization problems since it allows reproducing accurate molecular orbital (MO) energies and proper treatment of the molecular continuum. The potential $v_{ext}(\boldsymbol{r},t)$ in Equation 3.6 describes the interaction with the laser field.

The Hamiltonian and the coordinates are discretized by means of the generalized pseudospectral (GPS) method in prolate spheroidal coordinates [44–47], allowing optimal and nonuniform spatial grid distribution and accurate solution of the wave functions. The time-dependent Kohn–Sham Equation 3.5 can be solved accurately and efficiently by means of the split-operator method in the *energy* representation with spectral expansion of the propagator matrices [44–46,48]. We employ the following split operator, second-order short-time propagation formula [40]:

$$\psi_{n\sigma}(\boldsymbol{r},t+\Delta t) = \exp\left(-\frac{i}{2}\Delta t\hat{H}_0\right)\exp\left(-i\Delta tV\left(\boldsymbol{r},t+\frac{1}{2}\Delta t\right)\right) \times \exp\left(-\frac{i}{2}\Delta t\hat{H}_0\right)\psi_{n\sigma}(\boldsymbol{r},t) + O((\Delta t)^3). \quad (3.10)$$

Note that Equation 3.10 is different from the conventional split-operator techniques [49,50], where $\hat{H}_0$ is usually chosen to be the kinetic energy operator and $\hat{V}$ is the remaining Hamiltonian, depending on the spatial coordinates only. The use of the *energy* representation in Equation 3.10 allows explicit *elimination* of the undesirable fast-oscillating *high-energy* components and speeds up considerably the time propagation [40,44,48]. For the given Δt, the propagator matrix $\exp\left(-\frac{i}{2}\Delta t\hat{H}_0\right)$ is time independent and constructed only once from the spectral expansion of the unperturbed Hamiltonian $\hat{H}_0$ before the propagation process starts. The matrix $\exp\left(-i\Delta tV\left((\boldsymbol{r},t+\frac{1}{2}\Delta t\right)\right)$ is time dependent and must be calculated at each time step. However, for interaction with the laser field in the length gauge, this matrix is diagonal, and its calculation is straightforward.

3.3.2 MPI and HHG of Diatomic Molecules

3.3.2.1 MPI

For a case study, we present the study of the orientation-dependent MPI probabilities for N_2 molecules at the peak intensity 2×10^{14} W/cm^2 and wavelength 800 nm (Figure 3.1). The orientation dependence of our calculated total MPI probability is in a good accord with recent experimental observations [51,52] for this molecule and

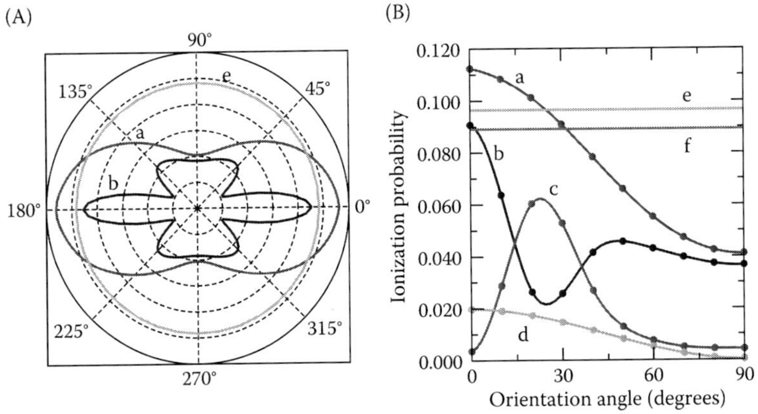

FIGURE 3.1 MPI probabilities of the N_2 molecule and Ar atom for the peak intensity 2×10^{14} W/cm^2. Panels A (polar coordinates) and B (Cartesian coordinates). (a) Total probability for N_2. (b) $3\sigma_g$ (HOMO) probability for N_2. (c) $1\pi_u$ (HOMO-1) probability for N_2. (d) $2\sigma_u$ probability for N_2. (e) Total probability for Ar. (f) $3p_0$ probability for Ar.

reflects the symmetry of its HOMO: the maximum MPI corresponds to the parallel orientation. However, multielectron effects are quite important for N_2, particularly at intermediate orientation angles. In the angle range around 30°, the MPI probability of HOMO-1 ($1\pi_u$) is larger than that of HOMO ($3\sigma_g$). Despite the orbital probabilities have local minima and maxima, the total probability shows monotonous dependence on the orientation angle. For comparison, we show also the ionization probability of the Ar atom, whose ionization potential is similar to that of N_2 (HOMO). As one can see from Figure 3.1, the absolute values of the ionization probabilities of N_2 and Ar are close to each other. However, the inner shell contributions are less important for Ar: the total probability is dominated by the highest occupied ($3p$) shell contribution.

3.3.2.2 HHG

For nonmonochromatic fields, the spectral density of the radiation energy emitted for all the time is given by the following expression [53]:

$$S(\omega) = \frac{4\omega^4}{6\pi c^3} \left| \tilde{\boldsymbol{D}}(\omega) \right|^2 . \tag{3.11}$$

Here, ω is the frequency of radiation, c is the velocity of light, and $\tilde{\boldsymbol{D}}(\omega)$ is a Fourier transform of the time-dependent induced dipole moment $D(t)$, i.e.,

$$\tilde{\boldsymbol{D}}(\omega) = \int_{-\infty}^{\infty} dt \boldsymbol{D}(t) \exp(i\omega t), \quad \text{and} \quad \boldsymbol{D}(t) = \int d^3 r \, \boldsymbol{r} \rho(\boldsymbol{r}, t). \tag{3.12}$$

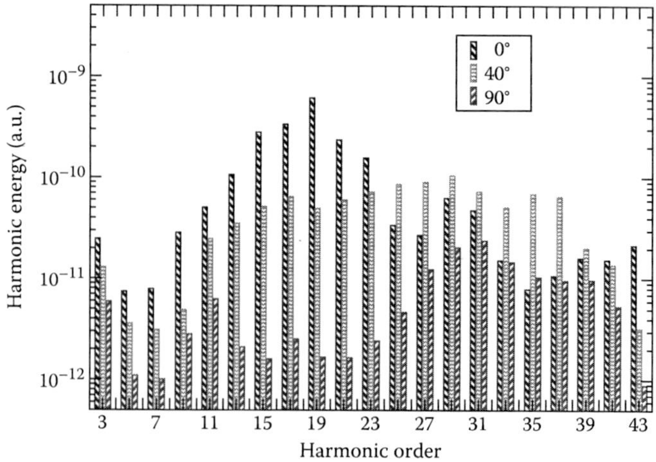

FIGURE 3.2 Energy emitted in harmonic radiation by the N_2 molecule for the peak intensity 2×10^{14} W/cm^2: left bar, orientation angle $\gamma = 0°$; middle bar, orientation angle $\gamma = 90°$; and right bar, orientation angle $\gamma = 90°$.

In Figure 3.2, we present a case study of the high-order harmonic generation (HHG) of the N_2 molecule at the peak intensity 2×10^{14} W/cm^2 and wavelength 800 nm [42]. The cutoff position in the HHG spectrum for this intensity is expected at the harmonic order 35, in fair agreement with the computed data. To show the orientation dependence of the HHG spectra, we choose three values of the orientation angle γ: 0°, 40°, and 90°, which represent the limiting cases of the parallel and perpendicular orientation, as well as the intermediate angle case. The orientation dependence of HHG also resembles that of MPI: HHG is more intense for the orientations where MPI reaches its maximum. For N_2, the HHG signal at 0° is dominant in the low-order part of the spectrum, whereas in the central part, a stronger signal is observed at 40°. One can also see that the emission of the harmonic radiation at the perpendicular orientation ($\gamma = 90°$) is suppressed for N_2 in the low-order and central parts of the HHG spectra. The maximum in the harmonic energy distribution at 90° is shifted to higher orders. This result is in a good accord with the recent experimental measurements on N_2 [54].

3.3.2.3 MPI and HHG of Heteronuclear and Homonuclear Diatomic Molecular Systems: Exploration of Multiple Orbital Contributions

In this section, we discuss all-electron TDDFT MPI calculations of homonuclear (N_2 and F_2) and heteronuclear (BF, HF, and CO) diatomic molecules [32,55]. In Figure 3.3, we present the HHG power spectra (Equation 3.11) of the N_2 and CO molecules. An important difference between the N_2 and CO spectra is that the latter contain even and odd harmonics. Generation of even harmonics is forbidden in systems with inversion symmetry, such as atoms and homonuclear diatomic molecules. This selection rule does not apply to the heteronuclear molecules with no inversion center (CO, BF, and HF). From Figure 3.3, one can see that, in general, HHG is more efficient

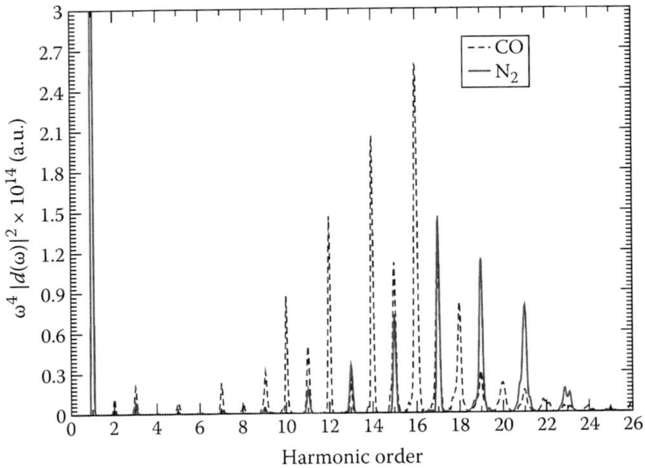

FIGURE 3.3 Comparison of the HHG power spectra of CO and N_2 in 800 nm 1×10^{14} W/cm^2 sin^2 pulse laser field.

in CO than in N_2. To investigate the detailed spectral and temporal structure of HHG for homonuclear and heteronuclear systems, we perform the time–frequency analysis by means of the wavelet transform of the total induced dipole moment $d(t)$ [32,44,56]:

$$d_{\omega}(t) = \int d(t)\sqrt{\frac{\omega}{\tau}}e^{i\omega(t-t_0)}e^{-(\omega(t-t_0))^2/2\tau^2}dt. \tag{3.13}$$

For the case of the N_2 molecule, the time profiles of the 19th to 25th harmonic orders are shown in Figure 3.4a. There are two emissions occurring at each optical cycle, and the most prominent bursts take place at the center of the laser field envelope. More importantly, for the CO molecule, a distinct feature possibly characteristic of all heteronuclear diatomic systems is observed in Figure 3.4b for harmonic orders 22 through 26. The number of dominant emissions per optical cycle is now limited to only one. This finding is in contrast with results normally obtained in the HHG for atoms and homonuclear molecules, in which two bursts per optical cycle are observed. We further explore the nonlinear response of individual MOs to the laser field and their dynamic role in formation of the HHG spectra of multielectron heteronuclear and homonuclear diatomic molecules. We also analyze the effect of asymmetry of the heteronuclear molecules on their HHG spectra. It is found that the interference of contributions from the MO is mostly constructive for the heteronuclear molecules and destructive for the homonuclear molecules. A more detailed investigation and comparison of the very high-order nonlinear optical response of the homonuclear (N_2 and F_2) and heteronuclear (CO, BF, and HF) diatomic molecules in intense ultrashort laser fields can be found in Ref. [55].

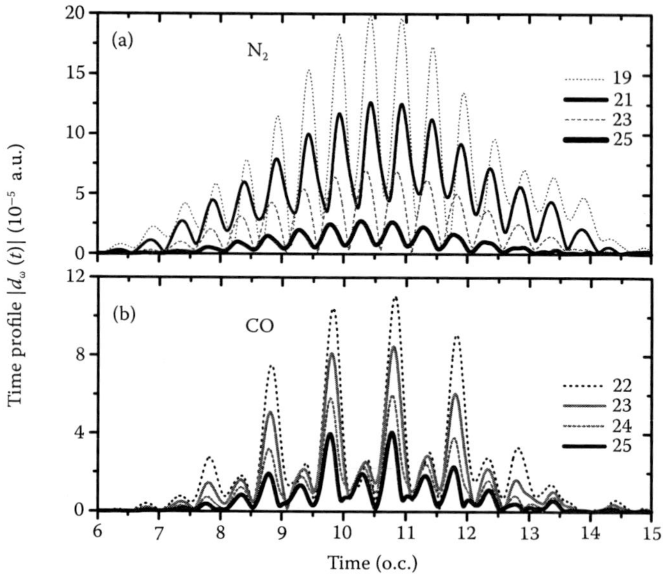

FIGURE 3.4 Wavelet time profiles for (a) N_2 and (b) CO. The laser intensity used is 5×10^{13} W/cm^2, and the wavelength used is 800 nm, with 20 optical cycles in pulse duration.

3.3.3 Multielectron Effects on Orientation Dependence of MPI of Small Polyatomic Molecules

In this section, we present all-electron TDDFT calculations of the orientation-dependent MPI of the three-center CO_2 molecule [31]. The electronic structure of CO_2 is solved with the help of the Voronoi cell finite difference (VFD) method [31]. In contrast to the ordinary finite difference method with regular uniform grids, the VFD method can accommodate any type of grid distributions, so-called unstructured grids, with the help of geometrical flexibility of the Voronoi diagram. To attack multicenter Coulombic singularity in all-electron calculations of polyatomic molecules, highly adaptive molecular grids are used [31] in this study.

Table 3.1 compares experimental vertical ionization potentials [57] of CO_2 and absolute values of orbital binding energies computed with the LBα potential. Molecular grids are constructed by a combination of spherical atomic grids covering large distances (r_{max} ~ 20 Å). The C–O bond length is fixed at 1.162 Å [58]. As one can see from Table 1, the calculated orbital binding energies are in fairly good agreement with the experimental data, particularly those for HOMO ($1\pi_g$) and HOMO-1 ($1\pi_u$).

Figure 3.5 shows the orientation dependence of the total MPI probability. The laser parameters used are 20-optical-cycle sin^2-envelope laser pulses with two different sets of the wavelength and the peak intensity: (a) 820 nm and 1.1×10^{14} W/cm^2 and (b) 800 nm and 5×10^{13} W/cm^2. For comparison, Figure 3.5 includes experimental measurements [51,52] and MO–ADK model results [51,59]. All data sets are normalized to their maximum value. In Figure 3.5a, two dashed lines of experiment

TABLE 3.1
Absolute Values of Spin-Orbital Energies of CO_2

Spin Orbital	A	B
$1\pi_g$ (HOMO)	13.9	13.8 [57]
$1\pi_u$	17.5	17.6 [57]
$3\sigma_u$	17.2	18.1 [57]
$4\sigma_g$	18.5	19.4 [57]

Note: (A) Present DFT calculations (in electron volts).
(B) Experimental vertical ionization energies (in electron volts).

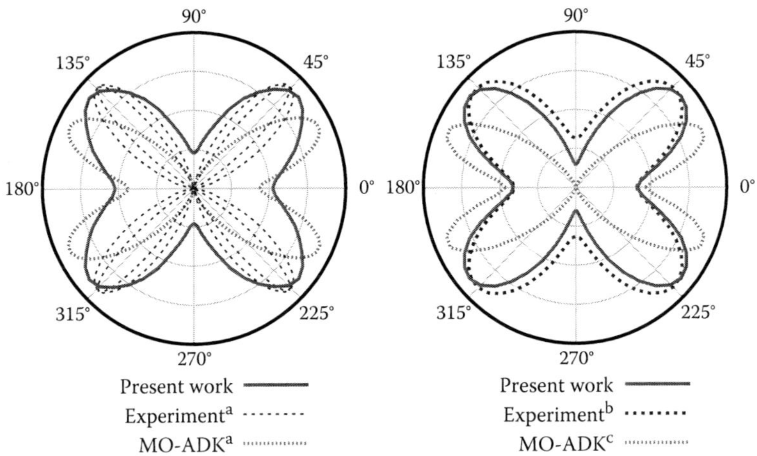

(a) 820 nm, 1.1 × 10^{14} W/cm^2 (b) 800 nm, 5 × 10^{13} W/cm^2

FIGURE 3.5 Orientation dependence of the total ionization probability of CO_2 ([a][51], [b][52], [c][59]).

are due to uncertainty of the measured alignment distribution [51]. The total ionization probability computed by TDDFT manifests the center-fat propeller shape with the peak at 40°. The position of the peak agrees well with both experiments [51,52], which give it at 45°. As for the broadness of the central pattern, our results agree well with the data of [52] but are different from that of [51], with the latter showing a narrower pattern. This discrepancy may be related to the experimental uncertainty in the molecular alignment processes. We now examine contributions of the individual orbitals to the total ionization probability. Thus, the center-fat propeller shape of the total ionization probability in Figure 3.5 is mostly reflected by contributions of the two HOMOs ($1\pi_{g,x}$ and $1\pi_{g,y}$). On the other side, the MO–ADK model predicts the butterfly shape with the peak at 25° [51,59], which is in large disagreement with experimental data. In contrast with the MO–ADK model, the self-interaction-free TDDFT approach [31] incorporates multielectron correlation and multiple orbital effects, and the results are in excellent agreement with experimental observation. The TDDFT [31] results show the significance of the electron correlations and suggest that all the valence orbitals should be taken into account, even when HOMO dominates the ionization process.

3.4 GENERALIZED FLOQUET FORMULATION OF TDDFT

In this section, we discuss briefly the generalized Floquet formulation of TDDFT [28,60–64]. It can be applied to the nonperturbative study of multiphoton processes of many-electron atoms and molecules in intense periodic or quasi-periodic (multicolor) time-dependent fields, allowing the transformation of time-dependent Kohn–Sham equations to an equivalent *time-independent* generalized Floquet matrix eigenvalue problems.

For time-periodic Hamiltonians $\hat{H}(t) = \hat{H}(t+T)$, the Floquet theorem allows a solution of the time-dependent Schrödinger equation (TDSE) in the following form:

$$\Psi(\boldsymbol{R},t) = \exp(-i\varepsilon t)\Phi(\boldsymbol{R},t), \tag{3.14}$$

where ε is the *quasi-energy*, $\Phi(\boldsymbol{R},t)$ is a periodic function of time, and the TDSE can be recast into the form of the quasi-energy eigenvalue equation:

$$\hat{\mathcal{H}}\Phi(\boldsymbol{R},t) = \varepsilon\Phi(\boldsymbol{R},t), \quad \text{with} \quad \hat{\mathcal{H}} = \hat{H}(t) - i\frac{\partial}{\partial t}, \tag{3.15}$$

in the *extended* Hilbert space S containing all square-integrable time-periodic functions $\Phi(\boldsymbol{R},t)$. The inner product in this space is defined as $\langle\langle\Phi\,|\,\Xi\rangle\rangle = \frac{1}{T}\int_0^T dt\langle\Phi\,|\,\Xi\rangle$. Since Equation 3.15 resembles the *steady-state* problem, it can be rigorously justified that $\hat{\mathcal{H}}(t)$, $\hat{H}(t)$, $\Phi(t)$ and the quasi-energy ε are all unique functionals of the electron density (spin densities in the spin-polarized theory) [65], with the latter being periodic in time. Thus, the quasi-energy functional Equation 3.16 can be expressed as a functional of the density.

In the Floquet formulation of TDDFT, the main role is played by the *quasi-energy functional* (compare with the *action functional* in the general time-dependent formulation [14])

$$F[\Phi] = \left\langle\left\langle\Phi\middle|\hat{\mathcal{H}}\middle|\Phi\right\rangle\right\rangle. \tag{3.16}$$

Variation of the functional (Equation 3.16) under the normalization condition $\langle\langle\Phi|\Phi\rangle\rangle = 1$, leads to Equation 3.15 for the time-periodic multielectron wave function $\Phi(\boldsymbol{r},t)$. The solution brings the stationary value (equal to the quasi-energy ε) to the functional (Equation 3.16). For TDSE, one normally has an initial state problem:

once the initial state is specified at $t = t_0$, then the solution of the TDSE is unique, and the wave function can be constructed at any time t. For the Floquet states, however, one does not have an initial state problem. Instead, one has an *eigenvalue* problem in the extended Hilbert space S.

Consider now the corresponding Kohn–Sham system of noninteracting particles with the same electron spin densities. The quasi-energy Kohn–Sham equations for the time-periodic spin orbitals ϕ_k^σ can be obtained from the stationary principle for the quasi-energy functional (Equation 3.16)

$$\left[-\frac{1}{2}\nabla^2 + u(\boldsymbol{r}) + v_H(\boldsymbol{r},t) + v_{\text{xc}}(\boldsymbol{r},t) + v_{\text{ext}}(\boldsymbol{r},t) - i\frac{\partial}{\partial t}\right]\phi_k^\sigma = \epsilon_k^\sigma \phi_k^\sigma. \tag{3.17}$$

Here, $v_H(\boldsymbol{r},t)$ is the Hartree potential, $v_{\text{xc}}(\boldsymbol{r},t)$ is the exchange-correlation potential, $v_{\text{ext}}(\boldsymbol{r},t)$ is the external field potential, and ϵ_k^σ is the orbital quasi-energy.

In the presence of intense external electromagnetic fields, atoms (molecules) can be ionized (dissociated) by the absorption of multiple photons, and all the bound states become *shifted* and *broadened* resonance states possessing *complex quasi-energies* $\varepsilon = \varepsilon_r - i\Gamma/2$. To determine these complex quasi-energy states, the *non-Hermitian Floquet Hamiltonian formalisms* previously developed [66,67], which employ the use of complex scaling transformation methods [68], can be extended to TDDFT [60–63]. The total quasi-energy ε can be expressed through the orbital quasi-energies, Hartree ($J(t)$), and exchange-correlation ($E_{xc}(t)$) energies, as well as the expectation values of the exchange-correlation potentials [60–63]:

$$\varepsilon = \sum_{k,\sigma} \epsilon_k^\sigma + \frac{1}{T}\int_0^T dt \left[E_{\text{xc}}(t) - J(t) - \sum_\sigma \int d^3r\, v_{\text{xc}}^\sigma(\boldsymbol{r},t)\rho^\sigma(\boldsymbol{r},t)\right]. \tag{3.18}$$

The analytical continuation in the complex plane of the radial coordinate r corresponding to the complex scaling transformation preserves that the spin densities remain real quantities for the real values of r [61]. That is why all the contributions to the right-hand side of Equation 3.18 are real, except the eigenvalues ϵ_k^σ. Thus, the *total* ionization rate can be expressed as a sum of *spin-orbital* ionization rates [61]:

$$\Gamma = \sum_{k,\sigma} \Gamma_k^\sigma \equiv -2\,\text{Im} \sum_{k,\sigma} \epsilon_k^\sigma. \tag{3.19}$$

3.4.1 Multiphoton Above-Threshold Detachment of Li^-

In this section, we apply the Floquet–TDDFT approach with exterior complex scaling to the calculation of multiphoton detachment of Li^- in monochromatic linearly polarized laser field [28,64]. When applying a complex scaling transformation, a delicate task is to perform the analytical continuation of the exchange-correlation potential to the complex plane. Usually, this potential exhibits very complicated functional form that may or may not allow obtaining accurate results with the uniform complex scaling technique. The exterior complex [69] scaling procedure allows the overcome of these difficulties.

We make use of the (spin polarized) Becke exchange [72] and Lee–Yang–Parr correlation [9] functionals (BLYP exchange-correlation). For the SIC, we extend the KLI procedure [25,73] with the implementation of an explicit self-interaction correction term [13]. The combination of BLYP exchange-correlation and KLI SIC (BLYP–KLI/SIC) has proved its accuracy in extensive atomic structure calculations [13,74]. With no external field, the electron affinity of Li as calculated by the BLYP-KLI/SIC method is 0.02294 a.u., which is in good agreement with the experimental value of 0.02271 a.u. [75]

In Figure 3.6, we show the results for the one-photon detachment cross section obtained from weak-field calculations. For comparison, also shown are the results of multichannel *R*-matrix calculation [70] and experimental data [71]. As one can see, our results are in fair agreement with the more sophisticated *R*-matrix multichannel calculations [70] and with the experiment [71]. We have also calculated electron angular distributions after the two-photon detachment of Li^- by the linearly polarized infrared laser field, for the laser frequencies range between the one- and two-photon detachment thresholds [28,64]. The results are shown in Figure 3.7 for the laser field intensity 1×10^{10} W/cm^2 and several frequencies. As our analysis reveals [28,64], the angular distributions show the dramatic interference of *s*- and *d*-waves in the detachment amplitude. For higher frequencies (0.016 and 0.020 a.u. in our calculations), the *d*-wave dominates the amplitude, and the angular distributions show the strongly anisotropic pattern as in Figure 3.7c or d, with the maximum pointing at the field direction. For smaller frequencies (0.012 and 0.014 a.u.), the quasi-energy level of the ground state is brought closer to the threshold, and the relative weight of the *s*-wave increases, in accordance with the Wigner threshold law [76]. For the frequency 0.012 a.u., the electron angular distribution is nearly isotropic (Figure 3.7a).

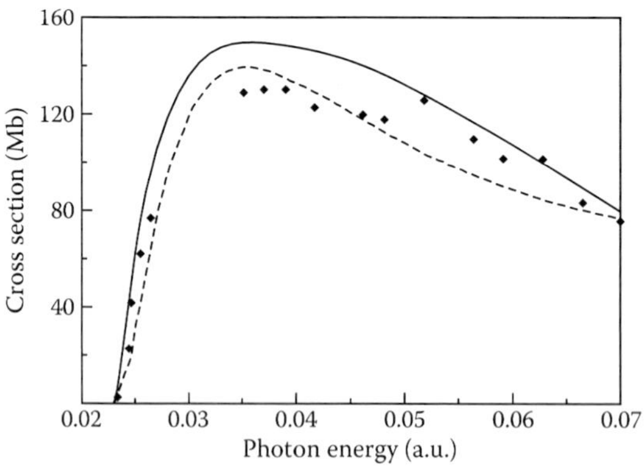

FIGURE 3.6 Cross section of one-photon detachment of Li^-. (Full curve) Floquet–TDDFT calculation [28,64]. (Dashed curve) Multichannel *R*-matrix calculation [70]. (Diamond) Experiment [71].

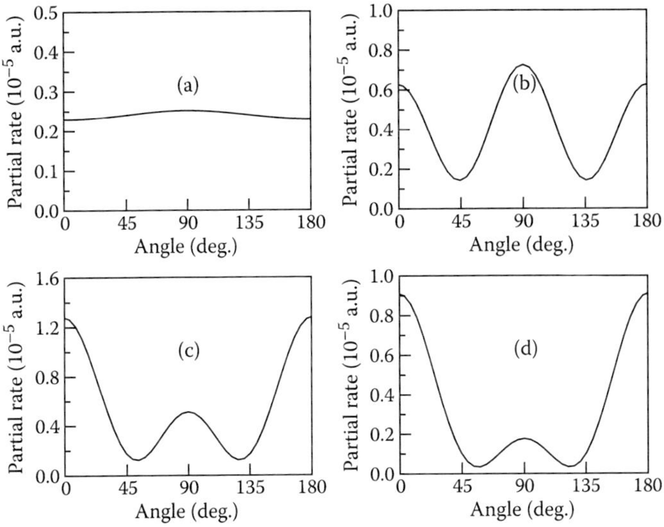

FIGURE 3.7 Angular distributions for two-photon detachment of Li^-. The laser field intensity is 1×10^{10} W/cm^2. The laser field frequency is (a) 0.012 a.u., (b) 0.014 a.u., (c) 0.016 a.u., and (d) 0.020 a.u.

3.5 CONCLUSIONS

In this chapter, we have presented several TDDFT approaches with proper long-range potentials for accurate and efficient treatment of the time-dependent multiphoton dynamics of many electron atomic and molecular systems. They allow the construction of orbital-independent single-particle local xc potential, which possesses the correct $(-1/r)$ long-range asymptotic behavior. With the asymptotically correct potential, the energy of the highest occupied spin orbital provides good approximation to the ionization potential. The GPS technique allows the construction of nonuniform and optimal spatial grids, denser mesh nearby each nucleus, and sparser mesh at longer range, leading to efficient and accurate solution of both electronic structure and time-dependent quantum dynamics with the use of only a modest number of spatial grid points. The TDDFT formalism along with the use of the time-dependent GPS numerical technique provides a powerful new nonperturbative time-dependent approach for exploration of the electron correlation and multiple orbitals effects on strong field multiphoton processes.

Like the steady-state case, the exact form of time-dependent xc-energy functional is unknown. Most of the strong-field calculations so far (including those discussed in this chapter) have used the *adiabatic* approximation, neglecting the memory-effect terms in the xc potential. As shown by the recent study [77], the adiabatic approximation is well justified in the case of medium–strong low-frequency laser fields. However, its validity in very strong fields still remains to be investigated. More rigorous nonadiabatic treatment of the time-dependent xc-energy functional can be facilitated if some information regarding the electron density for N-electron systems can be determined by means of the *ab initio* wavefunction approach. However, this task is not feasible at the current time for $N > 2$. Since the exact time-dependent xc-energy functional form is supposed to be universal and independent of N, the information of the strong-field behavior of the simplest but nontrivial two-electron systems will be very valuable for the future construction of time-dependent xc-energy functional.

Applicability of the modern TDDFT approaches for the treatment of multiple electron ionization processes is another problem related to the quality of time-dependent xc-energy functionals. Most of approximate xc functionals lack the important property of the exact functional, the discontinuity of its derivative with respect to the number of particles N, when N passes through integer values [78]. Several attempts to apply TDDFT with such approximate functionals for calculations of nonsequential double ionization were unsuccessful [79,80]. Recently it was shown [81] that the derivative discontinuity is crucial for correct description of double ionization.

At this time, the TDDFT is the primary approach available for the treatment of time-dependent processes of many-electron quantum systems in strong fields. Further extension of the self-interaction-free TDDFT approaches to larger molecular systems will be valuable and can lead to significant advancement in the understanding of strong-field chemical physics and atomic and molecular physics in the future.

ACKNOWLEDGMENTS

This work was partially supported by the U.S. Department of Energy. We also would like acknowledge the partial support of the National Science Council of Taiwan (Grant No. 100C6156-2) and National Taiwan University (Grant No. 101R8700-2, 10R10040-4, 101R104021).

REFERENCES

1. Parr, R. G., and W. Yang. *Density-Functional Theory of Atoms and Molecules*. Oxford University Press: Oxford, 1989.
2. *Density Functional Theory*. Gross, E. K. U., and R. M. Dreizler, Eds. Vol. 337. Plenum: New York, 1995.
3. March, N. H. *Electron Density Theory of Atoms and Molecules*. Academic: San Diego, 1992.
4. Parr, R. G., and Yang. *Annu. Rev. Phys. Chem.* **1995**, *46*, 701.
5. Marques, M. A. L., N. T. Maitra, F. M. S. Nogueira, E. K. U. Gross, and A. Rubio. *Fundamentals of Time-Dependent Density Functional Theory: Lecture Notes in Physics*. Vol. 837. Springer: Berlin, 2012.
6. Hohenberg, P., and W. Kohn. *Phys. Rev.* **1964**, *136*, B864.
7. Kohn, W., and L. Sham. *J. Phys. Rev.* **1965**, *140*, A1113.
8. Vosko, S. J., L. Wilk, and M. T. Nusair. *Can. J. Phys.* **1980**, *58*, 1200.
9. Lee, C., W. Yang, and R. G. Parr. *Phys. Rev. B* **1988**, *37*, 785.
10. Perdew, J. P., K. Burke, and M. Ernzerhof. *Phys. Rev. Lett.* **1996**, *77*, 3865.
11. Becke, A. D. *J. Chem. Phys.* **1993**, *98*, 1372.
12. Zhao, Q., and R. G. Parr. *Phys. Rev. A* **1992**, *46*, R5320.
13. Tong, X. M., and S. I. Chu. *Phys. Rev. A* **1997**, *55*, 3406.
14. Runge, E., and E. K. U. Gross. *Phys. Rev. Lett.* **1984**, *52*, 997.

15. Casida, M. E. In *Recent Developments and Applications of Modern Density Functional Theory*; Seminario, J. M., Ed.; Elsevier: Amsterdam, 1996; p. 391.
16. Onida, G., L. Reining, and A. Rubio. *Rev. Mod. Phys.* **2002**, *74*, 601.
17. Osinga, V. P., S. J. A. van Gisbergen, J. G. Snijders, and E. J. Baerends. *J. Chem. Phys.* **1997**, *106*, 5091.
18. Hohm, U., D. Goebel, and S. Grimme. *Chem. Phys. Lett.* **1997**, *272*, 1059.
19. Chelikowsky, J. R., L. Kronik, and I. Vasiliev. *J. Phys. Condens. Matter* **2003**, *123*, 062207.
20. Burke, K., J. Werschnik, and E. K. U. Gross. *J. Chem. Phys.* **2005**, *123*, 062206.
21. Tong, X. M., and S. I. Chu. *Phys. Rev. A* **1998**, *57*, 452.
22. Sharp, R. T., and G. K. Horton. *Phys. Rev.* **1953**, *90*, 317.
23. Talman, J. D., and W. F. Shadwick. *Phys. Rev. A* **1976**, *14*, 36.
24. Krieger, J. B., Y. Li, and G. F. Iafrate. *Phys. Lett. A* **1990**, *146*, 256.
25. Krieger, J. B., Y. Li, and G. F. Iafrate. *Phys. Rev. A* **1992**, *45*, 101.
26. Ullrich, C. A., U. J. Gossmann, and E. K. U. Gross. *Phys. Rev. Lett.* **1995**, *74*, 872.
27. Telnov, D. A., J. Heslar, and S. I. Chu. *Chem. Phys.* **2011**, *391*, 88.
28. Telnov, D. A., and S. I. Chu. *Phys. Rev. A* **2002**, *66*, 043417.
29. Telnov, D. A., and S. I. Chu. *Phys. Rev. A* **2009**, *79*, 041401(R).
30. Son, S. K., and S. I. Chu. *Chem. Phys.* **2009**, *366*, 91.
31. Son, S. K., and S. I. Chu. *Phys. Rev. A* **2009**, *80*, 011403(R).
32. Heslar, J., J. J. Carrera, D. A. Telnov, and S. I. Chu. *Int. J. Quant. Chem.* **2007**, *107*, 3159.
33. Telnov, D. A., and S. I. Chu. *Comput. Phys. Comm.* **2011**, *182*, 18.
34. Chu, X., and S. I. Chu. *Phys. Rev. A* **2004**, *70*, 061402(R).
35. Son, S. K., D. A. Telnov, and S. I. Chu. *Phys. Rev. A* **2010**, *82*, 043829.
36. Carrera, J. J., and S. I. Chu. *Phys. Rev. A* **2007**, *75*, 033807.
37. Carrera, J. J., S. I. Chu, and X. M. Tong. *Phys. Rev. A* **2005**, *71*, 063813.
38. Carrera, J. J., and S. I. Chu. *Phys. Rev. A* **2009**, *79*, 063410.
39. Chu, S. I. *J. Chem. Phys.* **2005**, *15*, R1517.
40. Tong, X. M., and S. I. Chu. *Chem. Phys.* **1997**, *217*, 119.
41. *Time-Dependent Density Functional Theory*. Marques, M. A. L., C. A. Ullrich, F. Nogueira, A. Rubio, K. Burke, and E. K. U. Gross, Eds. Springer: Berlin, 2006.
42. Telnov, D. A., and S. I. Chu. *Phys. Rev. A* **2009**, *80*, 043412.
43. Schipper, P. R. T., O. V. Gritsenko, S. J. A. van Gisbergen, and E. J. J. Baerends. *J. Chem. Phys.* **2000**, *112*, 1344.
44. Chu, X., and S. I. Chu. *Phys. Rev. A* **2001**, *63*, 023411.
45. Telnov, D. A., and S. I. Chu. *Phys. Rev. A* **2007**, *76*, 043412.
46. Telnov, D. A., and S. I. Chu. *Phys. Rev. A* **2005**, *71*, 013408.
47. Chu, X., and S. I. Chu. *Phys. Rev. A* **2001**, *64*, 063404.
48. Chu, X., and S. I. Chu. *Phys. Rev. A* **2001**, *63*, 013414.
49. Hermann, M. R., and J. A. Fleck. *Phys. Rev. A* **1988**, *38*, 6000.
50. Feit, M. D., J. A. Fleck, Jr., and A. Steiger. *J. Comput. Phys.* **1982**, *47*, 412.
51. Pavičić, D., K. F. Lee, D. M. Rayner, P. B. Corkum, and D. M. Villeneuve. *Phys. Rev. Lett.* **2007**, *98*, 243001.
52. Thomann, I., R. Lock, V. Sharma, E. Gagnon, S. T. Pratt, H. C. Kapteyn, M. M. Murnane, and W. Li. *J. Phys. Chem. A* **2007**, *112*, 9382.
53. Landau, L. D., and E. M. Lifshitz. *The Classical Theory of Fields*. Pergamon Press: Oxford, 1975.
54. McFarland, B. K., J. P. Farrell, P. H. Bucksbaum, and M. Gühr. *Science* **2008**, *322*, 1232.
55. Heslar, J., D. A. Telnov, and S. I. Chu. *Phys. Rev. A* **2011**, *043414*.
56. Tong, X. M., and S. I. Chu. *Phys. Rev. A* **2000**, *61*, 021802(R).

57. Turner, D. W., C. Baker, A. D. Baker, and C. R. Brundle. *Molecular photoelectron spectroscopy*; Wiley: London, 1970.
58. Herzberg, G. *Molecular spectra and molecular structure. III. Electronic spectra and electronic structure of polyatomic molecules*; Van Nostrand: New York, 1966.
59. Le, A. T., X. M. Tong, and C. D. Lin. *J. Mod. Opt.* **2007**, *54*, 967.
60. Telnov, D. A., and S. I. Chu. *Chem. Phys. Lett.* **1997**, *264*, 466.
61. Telnov, D. A., and S. I. Chu. *Phys. Rev. A* **1998**, *58*, 4749.
62. Telnov, D. A., and S. I. Chu. *Int. J. Quant. Chem.* **1998**, *69*, 305.
63. Telnov, D. A., and S. I. Chu. *Phys. Rev. A* **2001**, *63*, 012514.
64. Telnov, D. A., and S. I. Chu. *Phys. Rev. A* **2003**, *67*, 059903.
65. Deb, B. M., and S. K. Ghosh. *J. Chem. Phys.* **1982**, *77*, 342.
66. Chu, S. I., and W. P. Reinhardt. *Phys. Rev. Lett.* **1977**, *39*, 1195.
67. Chu, S. I., and D. A. Telnov. *Phys. Rep.* **2004**, *390*, 1.
68. Aguilar, A., and J. M. Combes. *Comm. Math. Phys.* **1971**, *22*, 265.
69. Simon, B. *Phys. Lett. A* **1979**, *71*, 211.
70. Ramsbottom, C. A., K. L. Bell, and K. A. Berrington. *J. Phys. B* **1994**, *27*, 2905.
71. Kaiser, H. J., E. Heinicke, R. Rackwitz, and D. Feldmann. *Z. Phys.* **1974**, *270*, 259.
72. Becke, A. D. *J. Chem. Phys.* **1992**, *96*, 2155.
73. Li, Y., J. B. Krieger, and G. F. Iafrate. *Phys. Rev. A* **1993**, *47*, 165.
74. Tong, X. M., and S. I. Chu. *Phys. Rev. A* **1998**, *57*, 855.
75. Haeffler, G., D. Hanstorp, I. Kiyan, A. E. Klinkmüller, U. Ljungbad, and D. Pegg. *J. Phys. Rev. A* **1996**, *53*, 4127.
76. Wigner, E. P. *Phys. Rev.* **1948**, *73*, 1002.
77. Thiele, M., E. K. U. Gross, and S. Kümmel. *Phys. Rev. Lett.* **2008**, *100*, 153004.
78. Perdew, J. P., R. G. Parr, M. Levy, and J. L. Balduz, Jr. *Phys. Rev. Lett.* **1982**, *49*, 1691.
79. Petersilka, M., and E. K. U. Gross. *Laser Phys.* **1999**, *9*, 105.
80. Lappas, D. G., and R. van Leeuwen. *J. Phys. B* **1998**, *31*, L249.
81. Lein, M., and S. Kümmel. *Phys. Rev. Lett.* **2005**, *94*, 143003.

4 Symmetries and Conservation Laws in the Lagrangian Picture of Quantum Hydrodynamics

Peter Holland

CONTENTS

4.1 INTRODUCTION

The similarity in form between the two real equations implied by the single-body spin-0 Schrödinger equation in the position representation (wave mechanics) and the equations of fluid mechanics with potential flow in its Eulerian formulation was first pointed out by Madelung in 1926 [1]. In this analogy, the probability density is proportional to the fluid density, and the phase of the wave function is a velocity potential. A novel feature of the quantum fluid is the appearance of quantum stresses, which are usually represented through the quantum potential. To achieve mathematical equivalence of the models, the hydrodynamic variables have to satisfy conditions inherited from the wave function. These, in turn, provide physical insight into the original conditions. For example, the single-valuedness requirement on the wave function corresponds to the appearance of quantized vortices in the fluid. The hydrodynamic model has inspired several computational advances driven especially by theoretical chemists. (For a comprehensive review up to the early 1980s, see [2], and for more recent developments, see [3].)

Madelung's approach was based on the Eulerian picture, and no reference was made to the Lagrangian picture of hydrodynamics, which conceives the motion of a fluid in terms of a continuum of trajectories and potentially opens the route to a new class of computational schemes. Contemporaneous with Madelung's work, de Broglie was developing a theory of trajectories in quantum mechanics, but at this historic point, an unfortunate digression occurred that, in the context of quantum hydrodynamics, took nearly 80 years to fully rectify. In the 1920s, the notion of trajectory in quantum theory became irretrievably embroiled in issues of interpretation, and its potential value as a tool in physics, regardless of its interpretational provenance, was not examined. The trajectory was so unacceptable that it was essentially banished from quantal discourse until Bohm reintroduced it in 1952 [4]. However, this restoration was still construed in a context circumvented by interpretational disputes. The earliest reference to the Lagrangian picture in quantum hydrodynamics appears to be due to Takabayasi [5], who, in the course of an early paper exploring the ramifications of Bohm's 1952 papers, mentioned in a footnote that obtaining the trajectories from the wave function corresponds to the transition from the Eulerian to the Lagrangian picture and that in order to obtain the trajectories "directly," one should start from the equations of motion in the Lagrangian form (see Equation 4.15), subject to the subsidiary condition that the velocity field is irrotational. Takabayasi did not elaborate on what he meant by obtaining the trajectories "directly," and to our knowledge, neither he nor any other writer remarked upon the footnote subsequently. It took a further 20 years for the idea of employing the trajectories in methods to solve the wave equation to emerge. Subsequently, and especially since around 2000, many workers have contributed to the development of approximation techniques that actively employ the trajectories alongside the Eulerian equations (for references, see [3,6]). It appears that the first clear demonstration that a congruence of trajectories—computed independently of the wave function (only the initial wave function is needed)—may exhibit sufficient structure to provide an exact method to deduce the time dependence of the wave function, thereby laying the foundation for an alternative picture of quantum mechanics, was given in 2005 [6]. The exact constructive method has since been extended to flows in an arbitrary-dimensional Riemannian manifold [7,8], which has proved fruitful as it includes, as special cases, the many-body system, the inclusion of an external vector potential, relativistic spin theory including spin 1/2 [9] and spin-1 electromagnetism [7], and quantum fields [8]. The method has also been extended to a multiphase-flow model of quantum evolution [10] and to second-order (in time) field theories [11].

The Lagrangian approach introduces a new conception of state into quantum mechanics complementary to the wave function, i.e., the time-dependent *position of a particle* $q_i(a, t)$, and a new degree of freedom, i.e., the *particle label* a_i. More precisely, the state is represented by the collective motion of the continuum of fluid particles obtained by continuously varying the label. Thus, unlike the Madelung (Eulerian) formulation, the Lagrangian picture adds variables to the quantum formalism, but none is singled out as special. As has been pointed out [6], this implies that the Lagrangian picture exhibits a new quantum symmetry or gauge freedom *viz.* a continuous particle-relabeling covariance group of the dynamics with respect to which the Eulerian variables (position, density, and velocity) are invariant. This is an analog of the classical symmetry that is connected with vorticity conservation [12–16]. The origin of the relabeling symmetry is that the deformation coefficients (derivatives of the current position with respect to the label) appear in the field equations only through the Jacobian. This is a characteristic feature of fluid mechanics that is not displayed in other continuum theories (such as elasticity [15]) and reinforces the hydrodynamic analogy.

In this paper, we examine the relabeling symmetry as a component of a general investigation of symmetries and conservation laws in the Lagrangian picture of quantum hydrodynamics, emphasizing their relation with symmetries and conservation laws in the Eulerian picture (which, being just wave mechanics, are well known). The emphasis is on general principles; whether the results will aid computational work remains to be investigated. The fact that the quantum Lagrangian picture has a variational basis gives us an opportunity to explore these connections through Noether's (first) theorem, which we formulate in the Lagrangian language. Analogous work has been done in classical Galilean-covariant hydrodynamics, but no sufficiently general treatment is available from which to draw ready-made formulas suitable for application to the quantum regime. This is partly because the latter displays in the quantum internal energy higher-order derivatives of the fields (the position coordinates) than are customarily considered. Alongside the infinite relabeling group, the 12-parameter kinematical covariance group of the Schrödinger equation is derived within the Noetherian approach. Two methods to connect with the Eulerian treatment are described. A point we wish to stress is that the role of label transformations extends beyond the class corresponding to Eulerian invariance. As an example, it is shown that the linear superposition of waves, a fundamental symmetry of the quantum Eulerian picture, can be generated by a deformation-dependent label transformation in the Lagrangian picture.

4.2 DEDUCTION OF SCHRÖDINGER'S EQUATION FROM THE LAGRANGIAN-PICTURE TRAJECTORY THEORY

4.2.1 Lagrangian Picture

We here summarize the constructive method of obtaining the wave equation from a deterministic particle model given in [6].

In the Lagrangian-picture of a fluid, the history of the system is encoded in the state variables $q_i(a, t)$ (indices $i, j, k, \ldots = 1,2,3$), i.e., the positions of all the distinct fluid elements at time t, each particle being distinguished by a continuously variable vector label a_i. The particle label may be chosen to be the initial position or more abstractly as a point in some continuum, e.g., it may be a color in a continuous spectrum. In order to have the freedom to choose physical position as a label, we shall always assume that the label is a vector in a three-dimensional (Euclidian) space. Since only one particle can occupy a point at each time, this label exhausts its identification. For labeling purposes, we do not need to give, for example, the initial velocity. (This information is, however, necessary for a fully posed dynamics.) The motion is *continuous* in that the mapping from a-space to q-space is single valued and differentiable with respect to a_i and t to whatever order is necessary, and the inverse mapping $a_i(q, t)$ exists and has the same properties. These assumptions are in accord with the properties of the single-valued velocity field implied by quantum mechanics (i.e., the ratio of current to density). The entire set of motions for all a_i is termed a *flow*. The vectors q_i and a_i are referred to the same set of Cartesian space axes, but they may also be regarded as related to one another by a time-dependent coordinate transformation. To each particle, there is associated an elementary volume whose mass is conserved by the flow. The whole is structured by an internal potential derived from the density that represents a certain kind of particle interaction, and each particle responds to the potential via a force whose action is described by a form of Newton's second law. For all these reasons, we may regard the model as providing a "particle" picture.

Let $\rho_0(a)$ be the initial quantal probability density. In the hydrodynamic model, $\rho_0(a)$ is identified with the initial number density (which is normalized: $\int \rho_0(a)d^3a = 1$). Then, introducing a mass parameter m, the mass of an elementary volume d^3a attached to the point a_i is given by $m\rho_0(a)d^3a$. The significance of the parameter m, which is conventionally described as the "mass of the quantum system," is that it is the total mass of the fluid since $\int m\rho_0(a)d^3a = m$. In this picture, the conservation of the mass of a fluid element in the course of its motion is expressed through the relation

$$m\rho(a, t)d^3q(a, t) = m\rho_0(a)d^3a \tag{4.1}$$

or

$$\rho(a, t) = J^{-1}(a, t)\rho_0(a), \tag{4.2}$$

where J is the Jacobian of the transformation between the two sets of coordinates:

$$J = \det\left(\partial q_i/\partial a_j\right) = \frac{1}{3!}\varepsilon_{ijk}\varepsilon_{lmn}\frac{\partial q_i}{\partial a_l}\frac{\partial q_j}{\partial a_m}\frac{\partial q_k}{\partial a_n}, \quad 0 < J < \infty. \tag{4.3}$$

Here ε_{ijk} is the completely antisymmetric tensor with $\varepsilon_{123} = 1$, and summation over repeated indices is always assumed.

Let V be the potential of an external (classical) conservative body force and U be the internal potential energy of the fluid due to interparticle interactions. We assume that the Lagrangian has the same form as in the classical theory of ideal fluids, except for the functional dependence of U: this depends on $\rho(q)$ and its first derivatives and, hence, from Equation 4.2, on the second-order derivatives of q_i with respect to a_i, and is independent of other variables such as entropy. The Lagrangian is then

$$L = \int \ell\left(q, \partial q/\partial t, \partial q/\partial a, \partial^2 q/\partial a^2, t\right) d^3a$$

$$= \int \left[\frac{1}{2} m\rho_0(a) \frac{\partial q_i}{\partial t}\frac{\partial q_i}{\partial t} - \rho_0(a)U(\rho) - \rho_0(a)V(q(a),t)\right] d^3a. \tag{4.4}$$

Here, $\rho_0(a)$ and V are prescribed functions, and we substitute for ρ from Equation 4.2. We assume that ρ_0 and its derivatives vanish at infinity, which ensures that the surface terms in the variational principle vanish.

It is the action of the conservative force derived from U on the trajectories that represents the quantum effects in this theory. As we shall see, these effects are characterized by the following choice for U:

$$U = \frac{\hbar^2}{8m}\frac{1}{\rho^2}\frac{\partial\rho}{\partial q_i}\frac{\partial\rho}{\partial q_i} = \frac{\hbar^2}{8m}\frac{1}{\rho_0^2} J_{ij}J_{ik}\frac{\partial}{\partial a_j}\left(\frac{\rho_0}{J}\right)\frac{\partial}{\partial a_k}\left(\frac{\rho_0}{J}\right), \tag{4.5}$$

where we have substituted from Equation 4.2 and used

$$\frac{\partial}{\partial q_i} = J^{-1}J_{ij}\frac{\partial}{\partial a_j}, \tag{4.6}$$

where

$$J_{il} = \frac{\partial J}{\partial\left(\partial q_i/\partial a_l\right)} = \frac{1}{2}\varepsilon_{ijk}\varepsilon_{lmn}\frac{\partial q_j}{\partial a_m}\frac{\partial q_k}{\partial a_n} \tag{4.7}$$

is the cofactor of $\partial q_i/\partial a_i$. The latter satisfies

$$\frac{\partial q_k}{\partial a_j} J_{ki} = J\delta_{ij}. \tag{4.8}$$

Clearly, U has a local dependence on ρ and its derivatives, and the coordinates q_i enter only through the deformation gradients $\partial q_i/\partial a_j$ and their derivatives with respect to a_i.

The Euler–Lagrange equations for the coordinates

$$\frac{\partial}{\partial t}\frac{\partial L}{\partial\left(\partial q_i(a)/\partial t\right)} - \frac{\delta L}{\delta q_i(a)} = 0, \tag{4.9}$$

where

$$\frac{\delta L}{\delta q_i}=\frac{\partial l}{\partial q_i}-\frac{\partial}{\partial a_j}\frac{\partial l}{\partial\left(\partial q_i/\partial a_j\right)}+\frac{\partial^2}{\partial a_j\partial a_k}\frac{\partial l}{\partial\left(\partial^2 q_i/\partial a_j\partial a_k\right)} \tag{4.10}$$

give the equation of motion of the ath fluid particle due to interparticle forces and the external force:

$$m\rho_0(a)\frac{\partial^2 q_i(a)}{\partial t^2}=-\rho_0(a)\frac{\partial V}{\partial q_i}-J_{kj}\frac{\partial\sigma_{ik}}{\partial a_j}, \tag{4.11}$$

where

$$\sigma_{ij}=\frac{\hbar^2}{4m}\left(\frac{1}{\rho}\frac{\partial\rho}{\partial q_i}\frac{\partial\rho}{\partial q_j}-\frac{\partial^2\rho}{\partial q_i\partial q_j}\right) \tag{4.12}$$

is a symmetric stress tensor, which has been written in simplified form in terms of the dependent variables using Equations 4.2 and 4.6. This (second order in t and fourth order in a_i) local nonlinear partial differential equation is the principal analytical result of the quantum Lagrangian method. For we shall see that, from its solutions $q_i(a, t)$, subject to specification of $\partial q_{i0}/\partial t$, we may derive solutions to Schrödinger's equation. *Motion in quantum mechanics may be regarded as the unraveling of a time-dependent coordinate transformation* $q_i(a, t)$.

To obtain a flow that is representative of quantum mechanics, we need to restrict the initial conditions of Equation 4.11 to those that correspond to what we shall term "quasi-potential" flow. This means that the initial velocity field is of the form (we introduce the mass factor for later convenience)

$$\frac{\partial q_{i0}}{\partial t}=\frac{1}{m}\frac{\partial S_0(a)}{\partial a_i}, \tag{4.13}$$

but the flow is not irrotational everywhere because the potential $S_0(a)$ (the initial quantal phase) obeys the quantization condition

$$\oint_{C_0}\frac{\partial q_{i0}}{\partial t}da_i=\oint_{C_0}\frac{1}{m}\frac{\partial S_0(a)}{\partial a_i}da_i=\frac{nh}{m},\quad n\in\mathbb{Z}, \tag{4.14}$$

where C_0 is a closed curve. If it exists, vorticity occurs in nodal regions where the density vanishes, and it is assumed that C_0 passes through a region of "good" fluid, where $\rho_0 \neq 0$. To show that these assumptions imply motion characteristic of quantum mechanics, we first demonstrate that they are preserved by the dynamical equation. To this end, we use a method based on Weber's transformation applied to the law of motion (Equation 4.11) in its "Lagrangian" form:

$$m\frac{\partial^2 q_i}{\partial t^2}\frac{\partial q_i}{\partial a_k} = -\frac{\partial}{\partial a_k}(V+V_Q), \tag{4.15}$$

where

$$V_Q = \frac{\hbar^2}{4m\rho}\left(\frac{1}{2\rho}\frac{\partial\rho}{\partial q_i}\frac{\partial\rho}{\partial q_i} - \frac{\partial^2\rho}{\partial q_i\partial q_i}\right) \tag{4.16}$$

is the de Broglie–Bohm quantum potential [17]. Integrating this equation between the time limits $(0,t)$ and substituting Equation 4.13 give

$$\frac{\partial q_i}{\partial t}\frac{\partial q_i}{\partial a_k} = \frac{1}{m}\frac{\partial S}{\partial a_k}, \quad S(a,t) = S_0(a) + \chi(a,t), \tag{4.17}$$

where

$$\chi(a,t) = \int_0^t \left(\frac{1}{2}m\left(\frac{\partial q_i}{\partial t}\right)^2 - V - V_Q\right)dt \tag{4.18}$$

with initial conditions $q_{i0} = a_i$, $\chi_0 = 0$. The left-hand side of Equation 4.17 gives the velocity at time t with respect to the a-coordinates, and this is obviously a gradient. To obtain the q-components, we multiply by $J^{-1}J_{ik}$ and use Equations 4.6 and 4.8 to get

$$\frac{\partial q_i}{\partial t} = \frac{1}{m}\frac{\partial S}{\partial q_i}, \tag{4.19}$$

where $S = S(a(q, t), t)$. Thus, for all time, the velocity of each particle is the gradient of a potential with respect to the current position.

To complete the demonstration, we note that the motion is quasi-potential since the value (Equation 4.14) of the circulation is preserved following the flow:

$$\frac{\partial}{\partial t}\oint_C \frac{\partial q_i}{\partial t}dq_i = 0, \tag{4.20}$$

where C is a curve moving with the flow. This theorem has been stated previously in the quantum context [18] and will be rederived below using relabeling invariance.

We conclude that each particle retains forever the quasi-potential property if it possesses it at any moment.

4.2.2 Eulerian Picture

The fundamental link between the particle (Lagrangian) and wave-mechanical (Eulerian) pictures is defined by the following expression for the Eulerian density:

$$\rho(x,t) = \int \delta(x - q(a,t))\rho_0(a)d^3a. \tag{4.21}$$

The corresponding formula for the Eulerian velocity is contained in the expression for the current

$$\rho(x,t)v_i(x,t) = \int \frac{\partial q_i(a,t)}{\partial t}\delta(x - q(a,t))\rho_0(a)d^3a. \tag{4.22}$$

Evaluating the integrals, Equations 4.21 and 4.22 are equivalent to the following local expressions:

$$\rho(x,t) = J^{-1}\Big|_{a(x,t)} \rho_0(a(x,t)) \tag{4.23}$$

$$v_i(x,t) = \frac{\partial q_i(a,t)}{\partial t}\bigg|_{a(x,t)}. \tag{4.24}$$

Equation 4.23 restates the conservation equation (Equation 4.2), and Equation 4.24 gives the relation between the velocities in the two pictures.

These formulas enable us to translate the Lagrangian flow equations into Eulerian language. Differentiating Equation 4.21 with respect to t and using Equation 4.22, we easily deduce the continuity equation

$$\frac{\partial \rho}{\partial t} + \frac{\partial}{\partial x_i}(\rho v_i) = 0. \tag{4.25}$$

Next, differentiating Equation 4.22 and using Equations 4.15 and 4.25, we get the quantum analog of Euler's equation:

$$\frac{\partial v_i}{\partial t} + v_j \frac{\partial v_i}{\partial x_j} = -\frac{1}{m}\frac{\partial}{\partial x_i}(V + V_Q). \tag{4.26}$$

Finally, the quasi-potential condition (Equation 4.19) becomes

$$v_i = \frac{1}{m}\frac{\partial S(x,t)}{\partial x_i}. \qquad (4.27)$$

Equations 4.23 and 4.24 give the general solution of the continuity equation (Equation 4.25) and Euler equation (Equation 4.26) in terms of the trajectories and the initial density.

To establish the connection between the Eulerian equations and Schrödinger's equation, it is a simple matter to deduce from Equations 4.26 and 4.27 that

$$\frac{\partial S}{\partial t} + \frac{1}{2m}\frac{\partial S}{\partial x_i}\frac{\partial S}{\partial x_i} + V + V_Q = 0, \qquad (4.28)$$

where we have absorbed a function of t in S. Combining Equations 4.25, 4.27, and 4.28, the function $\psi(x,t) = \sqrt{\rho}\exp\left(iS/\hbar\right)$ obeys Schrödinger's equation:

$$i\hbar\frac{\partial \psi}{\partial t} = -\frac{\hbar^2}{2m}\frac{\partial^2 \psi}{\partial x_i \partial x_i} + V\psi. \qquad (4.29)$$

We have deduced this from the Lagrangian particle equation (Equation 4.11) subject to the quasi-potential requirement.

This procedure enables us to write down an explicit formula for the time-dependent wave function in terms of the trajectories, up to a global phase, given the initial wave function $\psi_0(a) = \sqrt{\rho_0}\exp\left(iS_0/\hbar\right)$. First, solve Equation 4.11 subject to the initial conditions $q_{i0}(a) = a_i, \partial q_{i0}/\partial t = m^{-1}\,\partial S_0/\partial a_i$ to get the ensemble of trajectories for all a_i, t. Next, substitute $q_i(a, t)$ in Equation 4.23 to find ρ and $\partial q_i/\partial t$ in Equation 4.24 to get $\partial S/\partial x_i$. This gives S up to an additive function of time $f(t)$. To fix this function, apart from an additive constant, use Equation 4.28. We obtain finally for the wave function

$$\psi(x,t) = \sqrt{\left(J^{-1}\rho_0\right)\Big|_{a(x,t)}}\exp\left[\frac{i}{\hbar}\left(\int m\,\partial q_i/\partial t\Big|_{a(x,t)} dx_i + f(t)\right)\right]. \qquad (4.30)$$

Note that this method of solution does not dispense with the wave function-the initial form of the latter is integral to the dynamical equation of the trajectories (through the density) and to its initial conditions (through the phase).

The Eulerian Equations 4.25 and 4.26 form a closed system of four first-order coupled partial differential equations to determine the four independent "basic" fields $\rho(x)$, $v_i(x)$. The erasure from them of the particle variables is part of the reason the Eulerian language is particularly suited to represent the wave-mechanical formalism, which likewise, of course, makes no reference to the trajectory concept. The passage from the

Lagrangian to the Eulerian picture is a reductive process in which the number of dependent variables decreases [19]. In this connection, we note that any relation between Lagrangian variables can be recast formally in the Eulerian *language* by writing $a_i(x, t)$, but it will not always be a statement in the Eulerian *picture* since the derived functions of x_i and t may not be reducible to the basic Eulerian set ρ, v and their derivatives. We find this in the canonical phase space formulation of the theory where extraneous advected variables appear [6] and in the example of Section 4.5 (cf. Equation 4.60).

4.2.3 Equivalence of First-Order and Second-Order Trajectory Laws

We have established in the preceding discussion that the second-order trajectory Equation 4.11 implies the first-order law (Equation 4.17), where the potential S is determined by Equation 4.18. The converse is trivially proved: Given Equations 4.17 and 4.18, we can deduce Equation 4.11 by differentiation. Hence, the first- and second-order formulations are formally *equivalent*. Each may have advantages where the other is deficient. For example, if we wish to compute the trajectories, knowing only ψ_0 and without invoking the Eulerian equations as aids, the second-order version must be used. On the other hand, if we desire to compute the trajectories from a known wave function, the first-order version may be preferable. It would artificially restrict the insights and opportunities afforded by quantum hydrodynamics to treat one law as more fundamental than the other, either conceptually or computationally.

4.3 EULERIAN IDENTITY TRANSFORMATION

We commence our analysis of symmetries in quantum hydrodynamics by establishing a simple but fundamental result. A continuous transformation of the independent and dependent variables in one picture, particularly a symmetry (a transformation that leaves the dynamical equations covariant), will have a continuous image in the other picture. However, this correspondence will not be 1–1. Since the Eulerian picture is a reduction of the Lagrangian one, a unique transformation of the Eulerian variables generally corresponds to a *class* of transformations of the Lagrangian variables. More specifically, *a Lagrangian-picture symmetry corresponding to an Eulerian-picture symmetry will be unique only up to a relabeling transformation, independent of the parameters defining the Eulerian transformation.*

To demonstrate this property, it is sufficient to examine the identity transformation in the Eulerian picture, which is, for any fluid described by the density and velocity fields,

$$x_i' = x_i, \quad t' = t, \quad v_i'(x', t') = v_i(x, t), \quad \rho'(x', t') = \rho(x, t). \tag{4.31}$$

This corresponds to the following transformation of the Lagrangian variables:

$$q_i'(a', t') = q_i(a, t), \quad t' = t, \quad \frac{\partial q_i'}{\partial t'} = \frac{\partial q_i}{\partial t}, \quad \rho_0'(a')J'^{-1} = \rho_0(a)J^{-1}, \tag{4.32}$$

where $a_i' = a_i'(a,t)$. The latter is therefore not generally an identity transformation. To discover the function a_i', we write $d = \det\left(\partial(t',a_i')/\partial(t,a_j)\right) \neq 0$ and $D = \det\left(\partial a_i'/\partial a_j\right) \neq 0$, with $d(t' = t) = D$, and use the first two members of Equation 4.32 to get

$$\frac{\partial q_i'}{\partial a_j'} = \frac{1}{D}\frac{\partial d}{\partial\left(\partial a_j'/\partial a_k\right)}\bigg|_{t'=t}\frac{\partial q_i}{\partial a_k}, \quad \frac{\partial q_i'}{\partial t'} = \frac{\partial q_i}{\partial t} + \frac{1}{D}\frac{\partial d}{\partial\left(\partial t'/\partial a_j\right)}\bigg|_{t'=t}\frac{\partial q_i}{\partial a_j}. \tag{4.33}$$

Comparing with the last two members of Equation 4.32, we obtain

$$\frac{\partial a_i'}{\partial t} = 0, \quad \rho_0'(a')D = \rho_0(a). \tag{4.34}$$

Thus, the corresponding transformation in the Lagrangian picture constitutes a time-independent diffeomorphism $a_i'(a)$, or relabeling of the fluid particles, with respect to which the reference density transforms as a tensor density. Invariants of the transformation include the number of particles in an elementary volume of label space, i.e.,

$$\rho_0'(a')d^3a' = \rho_0(a)d^3a, \tag{4.35}$$

and their position and velocity. The relabeling is arbitrary if no other conditions are required on the transformation of ρ_0. Otherwise, Equation 4.34 may involve a constraint on the relabeling (see Section 4.5).

A Lagrangian-picture theory that is reducible to the basic Eulerian variables will be covariant with respect to the transformation (Equation 4.34). The Lagrangian Equation 4.4 can be written as $L = \int\left[\frac{1}{2}m\rho v^2 - \rho U\left(\rho, \partial\rho/\partial x\right) - \rho V(x)\right]d^3x$ and hence falls into this category (note that this Lagrangian cannot be used in a variational principle to derive Equations 4.25 and 4.26 as Euler–Lagrange equations). Label transformations other than the class Equation 4.34 do not generally leave the Eulerian functions invariant.

As mentioned previously, a common choice for labeling is the initial particle position: $q_{i0}(a) = a_i$. This is intuitively reasonable and implies that the element d^3a coincides with an elementary spatial volume. Since, according to Equation 4.32, $q_i'(a',t'=0) = q_i(a,t=0)$ and, in general, $a_i \neq a_i'$, the transformation Equation 4.34 expresses the freedom to choose a label other than the initial position. A natural choice is to distort the coordinates a_i so that the density is uniform with respect to them: $\rho_0'(a') = k = \text{constant}$ [11]. In one dimension, this is achieved by the labeling

$$a' = k^{-1}\int_{-\infty}^{a}\rho_0(a)da. \tag{4.36}$$

Starting from an arbitrary labeling, we can transform to a labeling that coincides with initial position by observing that, in general, $q_i(a, t = 0) = f_i(a)$, so we just set $a_i' = f_i(a)$, requiring only that the Jacobian of the transformation is positive.

Some benefits of the label symmetry include a means to simplify the Lagrangian-picture problem and, as shown below, an effective way of generating a class of conservation laws.

4.4 NOETHER'S THEOREM IN LABEL SPACE

4.4.1 Conservation in the Lagrangian Picture

A typical textbook illustration of Noether's first theorem employs a known continuous symmetry of a system of differential equations (the Euler–Lagrange equations deduced from a variational principle) that leaves the action functional invariant to derive an associated conserved charge. Actually, the theorem allows one to do more than this since the conservation laws it provides determine a class of symmetries of the differential equations (corresponding to the chosen Lagrangian), which need not then be known in advance. The class of symmetry transformations so obtained will not generally be exhaustive of all the continuous symmetries admitted by the dynamical equations since some dynamical symmetries may not be Noetherian symmetries with respect to the chosen Lagrangian, and those obtained are contingent on that choice. For these reasons, the implied set of conserved charges may likewise not be comprehensive. Nevertheless, as we shall see in our example, a sufficiently broad class of symmetries and associated conserved charges including all the principal expected ones can be generated by a straightforward application of this method.

This approach of deriving rather than assuming the transformation functions has been pursued previously in a similar context, that of the Lagrangian-picture of a classical ideal fluid [20]. We shall generalize the previous treatment to allow for the appearance of higher field derivatives in the internal potential energy, and for the external potential, and will include symmetries that were missed in the cited prior work. A version of Noether's theorem that is general enough for our purposes follows (for more details, see [21]).

Consider a Lie group of transformations of the independent and dependent variables

$$t' = t + \varepsilon\xi_0(q,a,t), \quad a_i' = a_i + \varepsilon\xi_i(q,a,t), \quad q_i'(a',t') = q_i(a,t) + \varepsilon\eta_i(q,a,t), \tag{4.37}$$

where ε is a dimensionless infinitesimal parameter. Note that the functional dependence of the transformation functions ξ_0, ξ_i, and η_i on a_i and t may be explicit or implicit via $q_i(a, t)$. To take account of this dependence, we use the following notation for the derivatives with respect to the independent variables:

$$\left.\begin{aligned} \frac{D}{\partial t} &= \frac{\partial}{\partial t} + \frac{\partial q_i}{\partial t}\frac{\partial}{\partial q_i} + \frac{\partial^2 q_i}{\partial t^2}\frac{\partial}{\partial\left(\partial q_i/\partial t\right)} \\ \frac{D}{\partial a_i} &= \frac{\partial}{\partial a_i} + \frac{\partial q_j}{\partial a_i}\frac{\partial}{\partial q_j} + \frac{\partial^2 q_j}{\partial a_i \partial a_k}\frac{\partial}{\partial\left(\partial q_j/\partial a_k\right)} + \ldots \end{aligned}\right\}. \tag{4.38}$$

The induced infinitesimal transformations of the derivatives of the dependent variables are

$$\left.\begin{aligned}
\frac{\partial q_i'}{\partial t'} &= \frac{\partial q_i}{\partial t} + \varepsilon\frac{D\eta_i}{\partial t} - \varepsilon\frac{D\xi_0}{\partial t}\frac{\partial q_i}{\partial t} - \varepsilon\frac{D\xi_k}{\partial t}\frac{\partial q_i}{\partial a_k} \\
\frac{\partial q_i'}{\partial a_j'} &= \frac{\partial q_i}{\partial a_j} + \varepsilon\frac{D\eta_i}{\partial a_j} - \varepsilon\frac{D\xi_0}{\partial a_j}\frac{\partial q_i}{\partial t} - \varepsilon\frac{D\xi_l}{\partial a_j}\frac{\partial q_i}{\partial a_l} \\
\frac{\partial^2 q_i'}{\partial a_k'\partial a_j'} &= \frac{\partial^2 q_i}{\partial a_k\partial a_j} + \varepsilon\frac{D^2\eta_i}{\partial a_k\partial a_j} - \varepsilon\frac{D^2\xi_0}{\partial a_k\partial a_j}\frac{\partial q_i}{\partial t} - \varepsilon\frac{D\xi_0}{\partial a_k}\frac{\partial^2 q_i}{\partial a_j\partial t} - \varepsilon\frac{D\xi_0}{\partial a_j}\frac{\partial^2 q_i}{\partial a_k\partial t} \\
&\quad - \varepsilon\frac{D^2\xi_l}{\partial a_k\partial a_j}\frac{\partial q_i}{\partial a_l} - \varepsilon\frac{D\xi_l}{\partial a_k}\frac{\partial^2 q_i}{\partial a_j\partial a_l} - \varepsilon\frac{D\xi_l}{\partial a_j}\frac{\partial^2 q_i}{\partial a_k\partial a_l}
\end{aligned}\right\} . \tag{4.39}$$

The invariance of the action $\int \ell d^3a\,dt$ under the transformation Equation 4.37 entails the local condition

$$\begin{aligned}
&\ell\left(q', \partial q'/\partial t', \partial q'/\partial a', \partial^2 q'/\partial a'^2, t', a'\right)\left(1 + \varepsilon\frac{D\xi_0}{\partial t} + \varepsilon\frac{D\xi_i}{\partial a_i}\right) \\
&\quad = \ell\left(q, \partial q/\partial t, \partial q/\partial a, \partial^2 q/\partial a^2, t, a\right) + \varepsilon\left(\frac{D\Lambda_0}{\partial t} + \frac{D\Lambda_i}{\partial a_i}\right),
\end{aligned} \tag{4.40}$$

where the functions Λ_0, Λ_i depend on t, a_i, q_i, $\partial q_i/\partial a_j$. Expanding the left-hand side to order ε, rearranging and subjecting the fields q_i to the Euler–Lagrange Equation 4.9, Noether's theorem asserts that Equation 4.40 takes the form of a continuity equation in a–t space:

$$\frac{D\mathrm{P}}{\partial t} + \frac{DJ_j}{\partial a_j} = 0, \tag{4.41}$$

where the density and current are given by

$$\mathrm{P} = \ell\xi_0 + \frac{\partial \ell}{\partial(\partial q_i/\partial t)}\left(\eta_i - \frac{\partial q_i}{\partial t}\xi_0 - \frac{\partial q_i}{\partial a_l}\xi_l\right) - \Lambda_0 \tag{4.42}$$

$$\begin{aligned}
J_j &= \ell\xi_j + \left(\frac{\partial \ell}{\partial(\partial q_i/\partial a_j)} - \frac{D}{\partial a_k}\frac{\partial \ell}{\partial(\partial^2 q_i/\partial a_j\partial a_k)}\right)\left(\eta_i - \frac{\partial q_i}{\partial t}\xi_0 - \frac{\partial q_i}{\partial a_l}\xi_l\right) \\
&\quad + \frac{\partial \ell}{\partial(\partial^2 q_i/\partial a_j\partial a_k)}\frac{D}{\partial a_k}\left(\eta_i - \frac{\partial q_i}{\partial t}\xi_0 - \frac{\partial q_i}{\partial a_l}\xi_l\right) - \Lambda_j.
\end{aligned} \tag{4.43}$$

Then, invoking Gauss's theorem and assuming the fields vanish at infinity, we obtain a conservation law:

$$\frac{d}{dt}\int_{-\infty}^{\infty} \mathrm{P}(a,t)\,d^3a = 0. \tag{4.44}$$

The continuity Equation 4.41 both generates a set of conserved quantities and determines the class of transformation functions (Equation 4.37) and associated restrictions on V that leave the action invariant, one charge being associated with each transformation function and constraint on V. To determine these functions, we note that there is no functional relationship between the derivatives of q_i beyond the Euler–Lagrange Equation 4.11. Inserting the Lagrangian density ℓ from Equation 4.4 in Equation 4.41, we must therefore set to zero the coefficients of the independent derivatives of q_i with respect to a_i and t and their products. The outcome of an arduous calculation is that the most general form of the transformation functions is the following:

$$\begin{aligned} &\xi_0 = d + \beta t + \alpha t^2, \quad \xi_i = \xi_i(a), \quad \eta_i = \left[\left(\frac{1}{2}\beta + \alpha t\right)\delta_{ij} + \omega_{ij}\right] q_j(a,t) - u_i t + c_i, \\ &\Lambda_0 = m\rho_0(a)\left(\frac{1}{2}\alpha q_i q_i - u_i q_i\right), \quad \Lambda_i = 0 \end{aligned} \tag{4.45}$$

where

$$\frac{\partial}{\partial a_i}(\rho_0 \xi_i) = 0, \tag{4.46}$$

$$\eta_i \frac{\partial V}{\partial q_i} + \xi_0 \frac{\partial V}{\partial t} + V\frac{\partial \xi_0}{\partial t} = 0, \tag{4.47}$$

and $\omega_{ij} = -\omega_{ji}$, u_i, c_i, d, α, β are constants. Unimportant functions of t and a_i in Λ_0 and Λ_i have been ignored. The constants, 12 in all, parameterize the group in addition to the function ξ_i, which is arbitrary save for the condition Equation 4.46, in which ρ_0 is a prescribed function. A series of conservation laws follow by inserting specific values of the parameters in Equations 4.42 and 4.43, as we shall see below. The absence of $\hbar$ in these transformation formulas suggests that they have a non-quantum origin. This turns out to be the case: they, in fact, define the maximal kinematical symmetry group of the classical Hamilton–Jacobi equation (obtained when V_Q is negligible in Equation 4.28).

4.4.2 Conservation in the Eulerian Picture

One of the benefits of the Lagrangian-coordinate approach to continuum mechanics is that it provides an additional means to discover Eulerian conserved charges, particularly through the application of Noether's theorem. This is especially useful in cases where the Eulerian-picture transformation corresponding to a Lagrangian-picture symmetry is trivial. We can connect the descriptions of conservation in the two pictures in two ways.

First, we can convert the Lagrangian continuity Equation 4.41 into a corresponding Eulerian one relating a density $\bar{\mathrm{P}}(x,t)$ and current $\bar{J}_i(x,t)$,

$$\frac{D\bar{\mathrm{P}}}{\partial t}+\frac{D\bar{J}_i}{\partial x_i}=0, \tag{4.48}$$

via the conversion formulas [22–24]

$$\bar{\mathrm{P}}(x,t)=\mathrm{P}(a,t)J^{-1}(a,t)\Big|_{a(x,t)}, \quad \bar{J}_i(x,t)=\left(\mathrm{P}J^{-1}\frac{\partial q_i}{\partial t}+\frac{\partial q_i}{\partial a_j}J_j\right)\Bigg|_{a(x,t)}. \tag{4.49}$$

We may thus deduce from the Lagrangian conservation law an Eulerian conservation law:

$$\frac{d}{dt}\int_{-\infty}^{\infty}\bar{\mathrm{P}}(x,t)\,d^3x=0. \tag{4.50}$$

As an example, we consider the Lagrangian density $\mathrm{P}=\rho_0$, which obeys the equation $D\mathrm{P}/\partial t = 0$ with $J_i = 0$. Then, from Equation 4.49, $\bar{\mathrm{P}}=\rho$, $\bar{J}_i=\rho v_i$, and Equation 4.48 is just Equation 4.25.

A second way to connect the Lagrangian and Eulerian accounts of conservation is to compare Equation 4.50 with the conserved charge obtained directly in the Eulerian formulation using the symmetry transformation that corresponds to Equation 4.45. With reference to the standard Lagrangian density for the Schrödinger field,

$$\hat{\ell}=\frac{i\hbar}{2}\left(\psi^*\frac{\partial\psi}{\partial t}-\psi\frac{\partial\psi^*}{\partial t}\right)-\frac{\hbar^2}{2m}\frac{\partial\psi^*}{\partial x_i}\frac{\partial\psi}{\partial x_i}-V\psi^*\psi, \tag{4.51}$$

we consider the infinitesimal transformation

$$\begin{aligned}&t'=t+\varepsilon\theta_0(x,t),\quad x_i'=x_i+\varepsilon\theta_i(x,t),\quad \psi'(x',t')=\psi(x,t)+\varepsilon\phi(x,t),\\&\psi'^*(x',t')=\psi^*(x,t)+\varepsilon\phi^*(x,t).\end{aligned} \tag{4.52}$$

The conserved density and current implied by Noether's theorem, which obey Equation 4.48, are

$$\hat{\mathrm{P}} = \hat{\ell}\theta_0 + \left[\frac{i\hbar}{2}\psi^*\left(\phi - \frac{\partial\psi}{\partial t}\theta_0 - \frac{\partial\psi}{\partial x_i}\theta_i\right) + \mathrm{cc}\right] - \hat{\Lambda}_0 \tag{4.53}$$

$$\hat{J}_j = \hat{\ell}\theta_j - \left[\frac{\hbar^2}{2m}\frac{\partial\psi^*}{\partial x_j}\left(\phi - \frac{\partial\psi}{\partial t}\theta_0 - \frac{\partial\psi}{\partial x_i}\theta_i\right) + \mathrm{cc}\right] - \hat{\Lambda}_j. \tag{4.54}$$

The transformation functions (Equation 4.52) corresponding to Equation 4.45 are

$$\begin{aligned} &\theta_0 = d + \beta t + \alpha t^2, \quad \theta_i = \left(\frac{1}{2}\beta + \alpha t\right)x_i + \omega_{ij}x_j - u_i t + c_i, \\ &\phi = \psi\left[-\frac{3}{2}\left(\frac{1}{2}\beta + \alpha t\right) + \frac{im}{\hbar}\left(\frac{1}{2}\alpha x_i x_i - u_i x_i\right)\right], \quad \hat{\Lambda}_0 = \hat{\Lambda}_i = 0, \end{aligned} \tag{4.55}$$

and the hydrodynamic variables transform as

$$\begin{aligned} &\rho' = \rho\left[1 - \varepsilon 3\left(\frac{1}{2}\beta + \alpha t\right)\right], \quad v_i' = v_i + \varepsilon\left[\omega_{ij}v_j + \alpha x_i - u_i - \left(\frac{1}{2}\beta + \alpha t\right)v_i\right], \\ &S' = S + \varepsilon m\left(\frac{1}{2}\alpha x_i x_i - u_i x_i\right). \end{aligned} \tag{4.56}$$

For a given Lagrangian-picture symmetry, the sets of functions $(\bar{\mathrm{P}}, \bar{J}_i)$ and $(\hat{\mathrm{P}}, \hat{J}_i)$ do not always coincide; an example is given in the next section. Note that the Lagrangian (Equation 4.51) is not directly connected with the Lagrangian (Equation 4.4); their relation is examined in [6].

4.5 PURE RELABELING SYMMETRY

In Equation 4.45, choose $\xi_0 = \eta_i = 0$. The nontrivial component of the transformation reduces to $a_i' = a_i + \varepsilon\xi_i(a)$, which corresponds to the infinitesimal form of the pure label transformation described in Section 4.3. To see the significance of the constraint Equation 4.46, we consider the infinitesimal transformation of the reference density, which is, in general,

$$\rho_0'(a') = \rho_0(a) + \varepsilon\delta\rho_0(a) + \varepsilon\frac{\partial\rho_0}{\partial a_i}\xi_i(a,t), \tag{4.57}$$

where $\delta\rho_0(a) = \rho_0'(a) - \rho_0(a)$ is the functional variation. The infinitesimal form of Equation 4.35 is then

$$\delta\rho_0 + \frac{\partial}{\partial a_i}(\rho_0 \xi_i) = 0. \tag{4.58}$$

Hence, comparing with Equation 4.46, we see that the latter constrains ξ_i so that ρ_0 is an invariant function ($\delta\rho_0 = 0$). Note that this constrains generally prevents ξ_i from being chosen constant.

The conserved density and current associated with the relabel symmetry are

$$\mathrm{P}(a,t) = -m\rho_0 \frac{\partial q_i}{\partial t}\frac{\partial q_i}{\partial a_j}\xi_j, \quad J_i(a,t) = \rho_0 \xi_i \left(\frac{1}{2} m \frac{\partial q_i}{\partial t}\frac{\partial q_i}{\partial t} - V(q(a.t)) - V_Q \right). \tag{4.59}$$

This conservation law is known as the analogous classical fluid theory [20]. An interesting feature of it is that the term in brackets in J_i is the Lagrangian of a particle of mass m moving in the potential $V + V_Q$. This suggests that this Lagrangian may acquire a physical significance as a component of the current.

To translate these expressions into the corresponding Eulerian quantities, we define the function $\bar{\xi}_i(x,t) = \left(\partial q_i / \partial a_j\right)\xi_j\big|_{a(x,t)}$, so that the density and current Equation 4.59 become

$$\bar{\mathrm{P}}(x,t) = -m\rho v_i \bar{\xi}_i, \quad \bar{J}_i(x,t) = \rho\bar{\xi}_j \left[-m v_i v_j + \delta_{ij}\left(\frac{1}{2} m v^2 - V - V_Q \right) \right]. \tag{4.60}$$

To obtain the conditions obeyed by $\bar{\xi}_i$, we note that the label is a constant of the motion along the trajectory it defines, so that, regarded as a function of the Eulerian independent variables, i.e., $a_i\,(x, t)$, it satisfies

$$\frac{\partial a_i}{\partial t} + v_j \frac{\partial a_i}{\partial x_j} = 0. \tag{4.61}$$

Any function ξ_i of a_i is also a constant of the motion and obeys the same equation. Translating the latter and Equation 4.46 into the Eulerian language, the function $\bar{\xi}_i$ obeys the two relations

$$\frac{\partial \bar{\xi}_i}{\partial t} + v_j \frac{\partial \bar{\xi}_i}{\partial x_j} = \bar{\xi}_j \frac{\partial v_i}{\partial x_j}, \quad \frac{\partial}{\partial x_i}\left(\rho \bar{\xi}_i\right) = 0. \tag{4.62}$$

The transformation generates an infinite class of conservation laws depending on the choice of ξ_i (subject to Equation 4.46) and encapsulates known laws as special cases. One noteworthy aspect is that this vector projects out of the linear momentum density a function that is conserved even when $\partial V/\partial q_i \neq 0$. As an example of the application of this symmetry, we shall derive from it Kelvin's theorem of the conservation of circulation for the quantum fluid, following a method employed in an analogous classical example [14]. Denote by C_0 a closed loop $a_i(s)$ in label space parameterized by the arc length s and define the transformation function to be

$$\xi_i(a) = (1/\rho_0)\oint_{C_0} \delta(a_i - a_i(s))\, da_i(s). \tag{4.63}$$

This function is easily seen to obey Equation 4.46. Substituting in Equation 4.44, we obtain

$$\frac{d}{dt}\oint_{C_0} \frac{\partial q_i(a(s),t)}{\partial t}\frac{\partial q_i(a(s),t)}{\partial a_j}\, da_j(s) = 0. \tag{4.64}$$

Changing variables from label to current position, the integration will be over a closed loop C defined by the instantaneous positions at time t occupied by the particles that originally comprised the circuit C_0 [25,26]. Then,

$$\frac{d}{dt}\oint_{C} \frac{\partial q_i(a(s),t)}{\partial t}\, dq_i(s,t) = 0, \tag{4.65}$$

which is just Equation 4.20. We conclude that Kelvin's theorem in quantum theory is a consequence of invariance with respect to the relabeling group.

The corresponding Eulerian transformation (Equation 4.52) is the identity with no useful associated conserved quantity $\hat{\mathrm{P}}$.

4.6 SPACETIME SYMMETRIES. THE SCHRÖDINGER GROUP

Here, we shall dissect the 12-parameter subgroup of Equation 4.45, which is obtained by setting $\xi_i = 0$, characterizing each case by a subset of nonzero parameters.

1) *Time translation:* $d \neq 0$. Then $\partial V/\partial t = 0$, and the conserved Lagrangian and Eulerian densities are

$$\mathrm{P} = -dH(a,t), \quad \bar{\mathrm{P}}(x,t) = \hat{\mathrm{P}}(x,t) = -d\bar{H}(x,t), \tag{4.66}$$

where

$$H = \frac{1}{2}m\rho_0\frac{\partial q_i}{\partial t}\frac{\partial q_i}{\partial t} + \rho_0 U + \rho_0 V, \quad \bar{H} = HJ^{-1}\Big|_{a(x,t)} = \frac{1}{2}m\rho v^2 + \rho U + \rho V \tag{4.67}$$

is the energy density.

2) *Space translation:* $c_i \neq 0$. Then $\partial V/\partial q_i = 0$, and

$$\mathrm{P} = c_i m\rho_0 \frac{\partial q_i}{\partial t}, \quad \bar{\mathrm{P}} = \hat{\mathrm{P}} = c_i m\rho v_i \tag{4.68}$$

is proportional to the linear momentum density.

3) *Space rotation:* $\omega_{ij} \neq 0$. Then $\varepsilon_{ijk} q_j(\partial V/\partial q_k) = 0$, and

$$\mathrm{P} = \omega_{ij} m\rho_0 q_j \frac{\partial q_i}{\partial t}, \quad \bar{\mathrm{P}} = \hat{\mathrm{P}} = \omega_{ij} m\rho q_j v_i \tag{4.69}$$

is proportional to the angular momentum density.

4) *Galilean boost:* $u_i \neq 0$. Then $\partial V/\partial q_i = 0$, and

$$\mathrm{P} = u_i m\rho_0 \left(q_i - t\frac{\partial q_i}{\partial t} \right), \quad \bar{\mathrm{P}} = \hat{\mathrm{P}} = u_i m\rho(x_i - tv_i) \tag{4.70}$$

is proportional to the Galilean momentum density.

5) *Dilation:* $\beta \neq 0$. Then

$$q_i \frac{\partial V}{\partial q_i} + 2t\frac{\partial V}{\partial t} + 2V = 0 \tag{4.71}$$

and

$$\mathrm{P} = \beta\left(\frac{1}{2} m\rho_0 q_i \frac{\partial q_i}{\partial t} - tH \right), \quad \bar{\mathrm{P}} = \hat{\mathrm{P}} = \beta\left(\frac{1}{2} m\rho x_i v_i - t\bar{H} \right). \tag{4.72}$$

6) *Extension:* $\alpha \neq 0$. Then

$$q_i \frac{\partial V}{\partial q_i} + t\frac{\partial V}{\partial t} + 2V = 0 \tag{4.73}$$

and

$$\mathrm{P} = \alpha\left[m\rho_0 q_i \left(t\frac{\partial q_i}{\partial t} - \frac{1}{2} q_i \right) - t^2 H \right], \quad \bar{\mathrm{P}} = \hat{\mathrm{P}} = \alpha\left[m\rho x_i \left(tv_i - \frac{1}{2} x_i \right) - t^2 \bar{H} \right]. \tag{4.74}$$

In all cases, the two Eulerian densities defined in Equations 4.49 and 4.53 are the same. Cases 1, 2, and 3 correspond to the usual energy, momentum, and angular momentum continuity equations for the Schrödinger field [17]. In case 4, the continuity equation reduces to (t times) the Euler–Lagrange Equations 4.11 (subject to vanishing external force), which, as can be checked, indeed have the form of Equation 4.41. These four cases together correspond to the ten-parameter Galilean group. They are supplemented by two further one-parameter kinematical transformations corresponding to cases 5 and 6. These transformations were originally discovered as symmetries of the free Schrödinger equation [27] and the complete 12-parameter set defines the "Schrödinger group." Subsequently, it was realized that dilation and extension are also symmetries for a nontrivial class of potentials [28,29]. The two classes of admissible potentials we have given in Equations 4.71 and 4.73 generalize those stated in [29], where it is assumed that the transformations are applied simultaneously. These additional symmetries were not found in the analogous prior work on a classical fluid [20] although it is known that the Schrödinger group is the maximal covariance group of classical hydrodynamics [30].

4.7 LINEAR SUPERPOSITION AS A RELABELING TRANSFORMATION

In this work, we have been concerned with transformations under which the dynamical equations of quantum hydrodynamics are covariant. These symmetries provide a method of building new solutions from known ones. This perspective suggests treating the linear superposition of wave functions, which achieves the same constructive end, as a type of symmetry, i.e., a transformation with respect to which the Schrödinger equation is covariant. In this case, the old and new solutions are, in general, not physically equivalent.

Consider two arbitrary wave functions ψ and ϕ which are both finite at the space-time point (x, t). Their linear superposition results in a new solution that is also finite at the point: $\psi' = \psi + \varepsilon\phi$. We shall regard this relation as a transformation of ψ, in which the transformation function ϕ also obeys the Schrödinger equation, restricting attention to infinitesimal transformations where the real parameter ε is chosen so that $|\varepsilon\phi| \ll |\psi|$ (we can generalize so that ε is complex but will not do so). Under this transformation, the independent and dependent variables of the Eulerian picture transform as

$$t' = t, \quad x_i' = x_i, \quad \psi'(x',t') = \psi(x,t) + \varepsilon\phi(x,t). \tag{4.75}$$

To see the value of this alternative view of superposition, we shall derive the conserved charge corresponding to this symmetry implied by Noether's theorem. Using the fact that ψ and ϕ both obey the wave equation, the Lagrangian density (Equation 4.51) transforms under Equation 4.75 as

$$\hat{\ell}\left(\psi',\partial\psi'/\partial t',\partial\psi'/\partial x'\right) = \hat{\ell}\left(\psi,\partial\psi/\partial t,\partial\psi/\partial x\right) - \varepsilon\frac{\hbar^2}{4m}\frac{\partial^2}{\partial x_i\partial x_i}(\psi^*\phi + \phi^*\psi). \tag{4.76}$$

The transformation is therefore Noetherian, i.e., it leaves the action invariant. The Noether charge is $\int (i\hbar/2)(\psi^*\phi - \phi^*\psi)d^3x$. Thus, with respect to the Lagrangian Equation 4.51, the superposition symmetry is correlated with (the real part of) the scalar product of the solutions ψ, ϕ, which is indeed conserved by the Schrödinger equation.

Choosing $\phi = -i\psi$, this approach contains as a special case an infinitesimal gauge transformation where the corresponding charge is $\int \hbar\psi^*\psi d^3x$, as expected.

The corresponding transformation of the independent and dependent variables in the Lagrangian picture is

$$t' = t, \quad a_i' = a_i'(a,t), \quad q_i'(a',t') = q_i(a,t). \tag{4.77}$$

In the Lagrangian picture, we may distinguish the wave functions under consideration at given x and t by the labels of the corresponding unique trajectories that generate them. We may then characterize the process of superposition as a transformation of label. Letting a_i label ψ, we seek the label a_i' of the unique trajectory passing point x at time t that generates ψ'. Given the initial wave function ψ_0', we can solve Equation 4.11 to find the trajectory $q_i'\left(a',t',\psi_0'\right)$. Inverting, we have $a_i'\left(q',t,\psi_0'\right)$, and substituting $q_i'\left(a',t',\psi_0'\right) = q_i(a,t,\psi_0)$ gives $a_i' = a_i'\left(a,t,\psi_0,\phi_0\right)$.

In the method just described, the new label is computed by first solving for the transformed trajectories. We now consider a special case where the transformed label may be computed from the original trajectories by supposing that the infinitesimal term in Equation 4.75 is obtained by varying a typical continuously variable parameter A (assumed dimensionless) on which the function ψ depends. For example, if ψ is a packet function, A could be proportional to its initial width. Let ψ' correspond to ψ evaluated for $A' = A + \varepsilon$. Then

$$\psi'\left(x,t,A'\right) = \psi(x,t,A) + \varepsilon\frac{\partial\psi}{\partial A}. \tag{4.78}$$

The Hamiltonian is independent of A, so the function $\partial\psi/\partial A$ is a solution if ψ is. The Eulerian fields transform as

$$\rho'\left(A'\right) = \rho(A) + \varepsilon\frac{\partial\rho}{\partial A}, \quad v_i'\left(A'\right) = v_i(A) + \varepsilon\frac{\partial v_i}{\partial A}. \tag{4.79}$$

To determine the label a_i' that generates ψ', we observe that the time dependence of $\partial\psi/\partial A$ can be computed from the trajectories in two ways: by propagating $\partial\psi_0/\partial A$ or by derivating having found ψ by propagating ψ_0. From the latter, we expect that a_i' can be expressed just in terms of a_i and ψ_0. We have

$$t' = t, \quad a_i'(a,t,A) = a_i + \varepsilon\xi_i(a,t,A), \quad q_i'\left(a',t',A'\right) = q_i(a,t,A). \tag{4.80}$$

A straightforward way to calculate ξ_i is to note first that q_i is an invariant function of A and a_i. Then,

$$q_i'\left(a',t,A'\right)=q_i\left(a',t,A'\right)$$
$$=q_i(a,t,A)+\varepsilon\frac{\partial q_i}{\partial a_j}\xi_j+\varepsilon\frac{\partial q_i}{\partial A}. \tag{4.81}$$

Combining this with the last member of Equation 4.80 gives

$$\frac{\partial q_i}{\partial a_j}\xi_j+\frac{\partial q_i}{\partial A}=0. \tag{4.82}$$

Inverting the deformation matrix using Equation 4.8 thus gives finally

$$\xi_j=-J^{-1}J_{ij}\frac{\partial q_i}{\partial A}. \tag{4.83}$$

To check that this transformation is a symmetry of the trajectory law (Equation 4.11), we only need to confirm consistency between the transformation rules of the Lagrangian velocity and density implied by the first two members of Equation 4.39 (with Equation 4.83 substituted) and the Eulerian rules (Equation 4.79). This follows straightforwardly using the following formula for conversion to the Lagrangian picture, which holds for any Eulerian function $f(x, t, A)$:

$$\frac{\partial}{\partial A}\left[f(x,t,A)\Big|_{x=q(a,t,A)}\right]=\left.\frac{\partial f(x,t,A)}{\partial A}\right|_{x=q(a,t,A)}+\left.\frac{\partial f(x,t,A)}{\partial x_i}\frac{\partial x_i}{\partial A}\right|_{x=q(a,t,A)}. \tag{4.84}$$

Inverting the procedure that led to the function Equation 4.83, we may assert that *the linear superposition of wave functions (Equation 4.78) may be generated by the following infinitesimal deformation-dependent relabeling of the trajectories, a symmetry of the law of motion (Equation 4.11):*

$$t'=t,\quad a_i'(a,t,A)=a_i-\varepsilon J^{-1}J_{ji}\frac{\partial q_j}{\partial A},\quad q_i'\left(a',t',A'\right)=q_i(a,t,A). \tag{4.85}$$

This transformation is consistent with the label transformation found in Equation 4.45 since it involves the deformation coefficients, a dependence that was excluded in the analysis of Section 4.4. It is evident that we may generalize the representation of linear superposition by a particle-relabeling transformation to the most general case, where ϕ is any solution of the Schrödinger equation.

4.8 CONCLUSION

Although it has proved fertile in computational quantum chemistry, the development of the Lagrangian picture of quantum mechanics has been uneven; key formal aspects of the picture that could have an impact in the numerical endeavor have yet to be fully investigated. Here, we have explored aspects of the Lagrangian picture that provide an alternative perspective on symmetries and conservation laws in quantum mechanics. Pursuing a variational technique, we have derived the maximal kinematical covariance group of the Schrödinger equation and the associated conserved charges and established connections between the Lagrangian method and the known Eulerian description. A novel aspect is the appearance of an infinite parameter relabeling group, which implies a wide set of conservation laws such as Kelvin's circulation theorem. Moving beyond the variational method, we have developed an alternative perspective on the linear superposition principle as a relabeling symmetry.

In our variational approach, we restricted the possible functional dependence of the transformation functions. This is sufficient to generate all the kinematical symmetries of the Schrödinger equation, together with relabelings, but future work might allow dependence of the transformation functions on the deformation coefficients. In addition, an examination of alternative Lagrangians could prove valuable. On the other hand, discovering conservation laws via the variational approach has limitations, and direct methods may be more efficient [21].

REFERENCES

1. E. Madelung. *Z. Phys.* **40**, 322 (1926).
2. S. K. Ghosh and B. M. Deb. *Phys. Rep.* **92**, 1 (1982).
3. R. E. Wyatt. *Quantum Dynamics with Trajectories* (Springer, New York, 2005).
4. D. Bohm. *Phys. Rev.* **85**, 166, 180 (1952).
5. T. Takabayasi. *Prog. Theor. Phys.* **9**, 187 (1953).
6. P. Holland. *Ann. Phys. (NY)* **315**, 503 (2005).
7. P. Holland. *Proc. R. Soc. A* **461**, 3659 (2005).
8. P. Holland. *Quantum Trajectories*. Ed. P. Chattaraj (Taylor & Francis/CRC, Boca Raton, 2010) Chap. 5.
9. P. Holland. *Found. Phys.* **36**, 369 (2006).
10. P. Holland. *J. Phys. A: Math. Theor.* **42**, 075307 (2009).
11. P. Holland. *Int. J. Theor. Phys.* **51**, 667 (2012).
12. N. Padhye and P. J. Morrison. *Phys. Lett. A* **219**, 287 (1996).
13. N. Padhye and P. J. Morrison. *Plasma Phys. Rep.* **22**, 869 (1996).
14. V. E. Zakharov and E. A. Kuznetsov. *Phys. Usp.* **40**, 1087 (1997).
15. R. Salmon. *Lectures on Geophysical Fluid Dynamics* (Oxford University Press, Oxford, 1998).
16. A. Bennett. *Lagrangian Fluid Dynamics* (Cambridge University Press, Cambridge, 2006).
17. P. R. Holland. *The Quantum Theory of Motion* (Cambridge University Press, Cambridge, 1993).
18. I. Bialynicki-Birula and Z. Bialynicki-Birula. *Phys. Rev. D* **3**, 2410 (1971).
19. P. J. Morrison. *Rev. Mod. Phys.* **70**, 467 (1998).
20. G. Caviglia and A. Morro. *J. Math. Phys.* **28**, 1056 (1987).

21. G.W. Bluman, A. F. Cheviakov, and S. C. Anco. *Applications of Symmetry Methods to Partial Differential Equations* (Springer, New York, 2010).
22. N. Padhye. *Ph.D. thesis* (University of Texas at Austin, 1998).
23. G. M. Webb, G. P. Zank, E. Kh. Kaghashvili, and R. E. Ratkiewicz. *J. Plasma Phys.* **71**, 811 (2005).
24. G. M. Webb and G. P. Zank. *J. Phys. A: Math. Theor.* **40**, 545 (2007).
25. C. Eckart. *Phys. Fluids* **3**, 421 (1960).
26. J. Casey and P. M. Naghdi. *Arch. Rational Mech. Anal.* **115**, 1 (1991).
27. U. Niederer. *Helv. Phys. Acta* **45**, 802 (1972).
28. U. Niederer. *Helv. Phys. Acta* **46**, 191 (1973).
29. B.-W. Xu. *J. Phys. A: Math. Gen.* **14**, L123 (1981).
30. L. O'Raifeartaigh and V. V. Sreedhar. *Ann. Phys. (NY)* **293**, 215 (2001).

5 Synchronization in Coupled Nonlinear Oscillators

Relevance to Neuronal Dynamics

Jane H. Sheeba, V. K. Chandrasekar, and M. Lakshmanan

CONTENTS

5.1 INTRODUCTION

Nonlinear systems often exhibit a collective phenomenon called synchronization when coupled in some fashion or under the influence of an external field. The study of the behavior of large populations of nonlinear oscillators is a topic of central research since the late 1960s, following the pioneering work of Winfree [1]. The topic of synchronization in coupled systems, including chaotic systems, is interesting because a detailed knowledge about large-scale dynamics is absolutely essential, and the collective behavior of coupled oscillator systems play a crucial role in explaining various observed phenomena in real-world systems [1–5]. In particular, biological and physiological systems usually function as collective entities. Examples include neuronal networks in the brain, predator–prey populations, flora and bacterial populations in water and soils, the spread of epidemics among the living population, etc. In particular, the study of synchronization in neuronal populations is a challenging topic from the point of view of both nonlinear dynamics and brain dynamics. This chapter gives an overview of our recent efforts in this direction as a tribute to Prof. B. M. Deb on his 70th birthday for his many-faceted scientific achievements.

In this chapter, we present our recent findings related to synchronization in coupled nonlinear oscillator systems with and without time delay as specific representation of certain neuronal behaviors. In the following section, we discuss about the ubiquitous nature of synchronization in dynamical systems. In Section 3, we explain the structure of neurons and how neurons function as a collective system. Section 4 is dedicated to the discussion of the occurrence of oscillations and synchronization in neuronal networks. In Section 5, we demonstrate the occurrence of event related desynchronization in a system of coupled nonlinear oscillators in the presence of an external field. In Section 6, we discuss the effect of time delay in coupled populations of nonlinear oscillators and demonstrate the occurrence of globally clustered chimera (GCC) states. In Section 7, we present a demand controlled delayed feedback mechanism for controlling the occurrence of mass pathological synchronization in the brain. Finally, in Section 8, we present our conclusions and summarize the chapter.

5.2 SYNCHRONIZATION IN DYNAMICAL SYSTEMS

Synchronization or concurrence between two or more oscillators is a fascinating phenomenon that is common in systems of coupled nonlinear oscillators. The phenomenon of synchronization is very common in nature and occurs in a wide variety of real world systems [1–4]. Some examples include the emulsion of light pulses by a population of fireflies in a synchronous manner, synchronization in populations of electrochemical oscillators, synchronous chirping by populations of crickets, synchronous clapping of audiences in auditoria, synchronous firing in populations of neurons, synchronized pacemaker oscillations in cardiac cells, etc. The aforementioned instances of synchronization can be explained by systems of coupled oscillators, which are represented typically as

$$\dot{X}_j = F(X_j, \varepsilon_j) + \frac{A}{N}\sum_{k=1}^{N}(X_k - X_j), \quad j = 1,2,3,\ldots,N, \quad \left(\cdot = \frac{d}{dt}\right), \tag{5.1}$$

where X_j is the state vector of the jth oscillator, A is the coupling parameter that indicates the strength of the coupling between the oscillators in the system, ε_j represents the system parameters, and $F(X_j, \varepsilon_j)$ describes intrinsic self-sustained oscillation. Other forms of couplings such as weighted coupling, nonlocal coupling, etc., are also possible and important, although we will not discuss them here. The phenomenon of phase synchronization was first reported by the famous Dutch physicist Christiaan Huygens in the 17th century [6,7]. His report was based on the observation of two pendulum clocks (see Figure 5.1) attached to a common wooden bar that continued with precise antiphase oscillation, seemingly indefinitely.

When the system of interest involves a large number of coupled oscillators, the dynamics of the system can be reduced to that of the mean field, that is, effectively of a single oscillator. When an individual oscillator in the population emits or absorbs energy, this will alter the physical states of the oscillators to which it is coupled. In particular, the periods of its neighbors are altered. If we consider a system of limit cycle oscillators (where the oscillators are evolving in a globally attracting limit cycle of constant amplitude), their phase dynamical equations can be represented by a system of coupled first-order nonlinear differential equations

$$\dot{\theta}_i = \omega_i - \frac{A}{N}\sum_{j=1}^{N} f\left(\theta_i - \theta_j\right), \quad j = 1, 2, \ldots, N. \tag{5.2}$$

This model is called the Kuramoto model (named after Yoshiki Kuramoto who first proposed the model [2]), where θ_is are the phases of the individual oscillators in the system and ω_is are the natural intrinsic frequencies of the oscillators, while f is a 2π periodic function. The model assumes weak coupling between the oscillators and that the oscillators have their own intrinsic frequencies.

The collective dynamics of the system can now be represented by the dynamics of the mean field, that is, essentially that of a single oscillator. The complex mean field parameter can be given as

$$Z = X + iY = re^{i\psi} = \frac{1}{N}\sum_{j=1}^{N} e^{i\theta_j}. \tag{5.3}$$

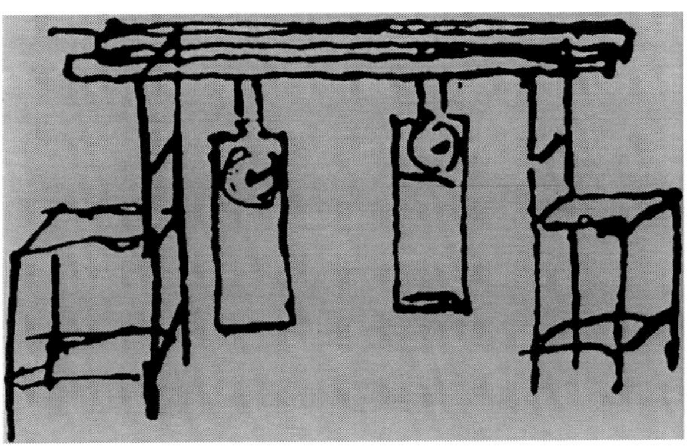

FIGURE 5.1 Original drawing of Christiaan Huygens, illustrating his observation of pendulum clock synchronization. (Data from Horologium, *The Hague*, 1658 (Oeuvres XVII); Bennett, M. et al., *Proc. R. Soc. A.*, 458(2019), 563, 2002.)

When $r = 1$, there is complete synchronization when the phases of all the oscillators are the same. When $B = 0.1$, there is no synchronization. When r takes a value between 0 and 1, there is either partial synchronization or clustering (formation of multiple small synchronized groups) in the system.

The preceding type of system of coupled nonlinear oscillators can be typically used to represent a system of neurons, where one can look for explanations for different types of synchronization that occur in the brain. In the following sections, let us discuss about the structure of a single neuron and how they are connected and function in networks, how oscillations and synchronization arise, and how these lead to significant physiological processes.

5.3 NEURON STRUCTURE AND NEURONAL NETWORKS

The brain is comprised of about 100 billion interconnected neurons. Neurons process and transmit information in the brain, and they are unique types of cells. The central nervous system is made up of neurons. Each neuron consists of a nucleus, a single axon that conveys electrical signals to other neurons, and a host of dendrites that deliver incoming signals (see Figure 5.2). A neuron's dendritic tree is connected to thousands of neighboring neurons. When one of those neurons fires, a positive or negative charge is received by one of the dendrites [8]. A process of spatial and temporal summation occurs to add the strength of all the received charges, and the resultant input is passed on to the body (soma) of the neuron. The signal does not get processed inside the soma and the enclosed nucleus. The main purpose of the existence of soma and the nucleus is to carry out the necessary maintenance to keep the neuron functional. Each terminal button of a neuron is connected to other neurons through a small gap called a synapse. Each synapse varies in its physical and

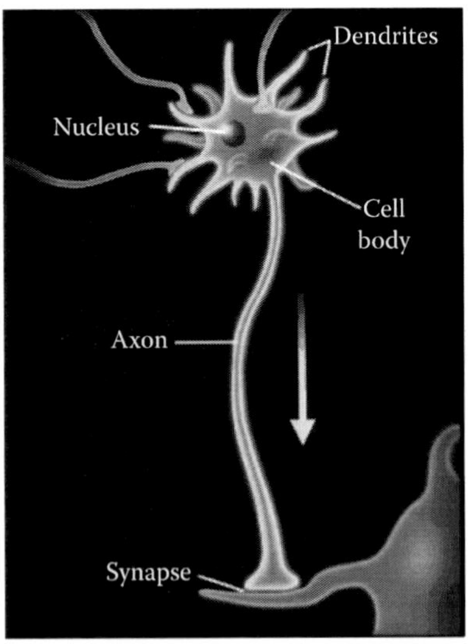

FIGURE 5.2 Structure of a single neuron. (Available at http://morphonix.com/, accessed on February 25, 2012.)

neurochemical characteristics, and this determines the strength and the polarity of the new input signal. The flexibility and vulnerability of the brain is due to this feature [9]. The impact of one neuron on a second one can be altered by altering the neurochemical composition of the synapse (done by different neurotransmitters), thus making a signal excitatory or inhibitory. Thus, interaction between the neurons takes place at the synapses.

Every neuron can have numerous inputs (as many as a thousand inputs). After firing, a neuron rests for a thousandth of a second before starting the next firing cycle. The information is encoded and carried along by the neurons through the electric signals that are generated across the outer membrane of the cell [10].

Neuronal activity occurs normally in the wave mode of the synaptic current/voltage in the populations [11]. In order for the neurons to be functional in populations, the signals should convert in the two modes, i.e., the wave mode and the pulse mode. Integration or summation of all the inputs in a neuron usually occurs in the wave mode within the neuron. Once all the input signals are summed up and processed, the transmission from one neuron to the other occurs in the pulse mode.

A neuronal population is comprised of neuronal connections, and basically, there are two types of neuronal connections: (1) anatomical and (2) functional connections [12]. Anatomical connections between neurons are due to the structure of the brain. Every anatomical connection essentially implies the possibility of interaction between two or more connected neurons but does not necessarily mean that such interactions will actually occur. On the other hand, functional connections dictate the existence of particular functions, and they usually override the anatomical connections if needed.

5.4 NEURONAL OSCILLATIONS

Any oscillatory activity is basically caused by a positive action, followed by a delayed feedback. Oscillations in the brain occur at various stages, i.e., in the membrane of the neurons (resulting in membrane potential fluctuations), between neurons (action potentials), and between neuronal populations. The functional activity of the brain is due to the neuronal oscillations and their synchronization, which involves various complex actions including information processing, awareness, motor control, sleep, and many other functions. Neuronal oscillations usually fall in a well-defined frequency band and hence are frequency specific. The frequency specific nature [11,12] of neuronal oscillations makes it possible for the brain to control various cognitive, motor, and stability processes effectively.

5.4.1 Membrane Potential Oscillations

Neurons, just like any other cell, have a cell membrane that separates the contents of the neuron from the external medium. There is a potential across this membrane due to the imbalance of ions, such as Na^+, K^+, Ca^{++}, Cl^-, etc. The imbalance is such that, at rest, the inside of the cell is at a negative potential, compared with the outside of the cell. This potential is generally –80 mV and is called the resting membrane potential [13].

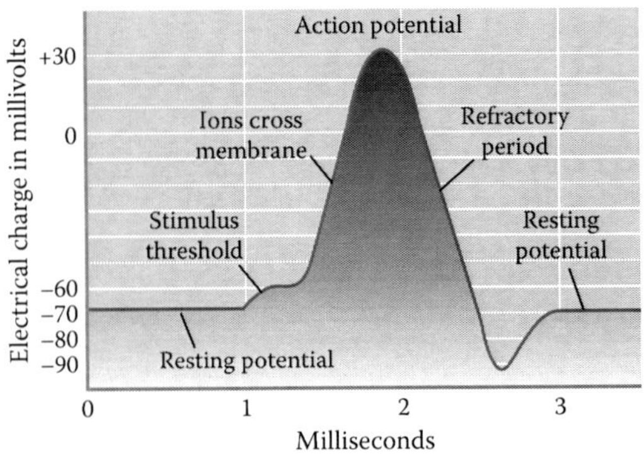

FIGURE 5.3 Schematic representation of an action potential. (Available at http://faculty.weber.edu/, accessed on February 25, 2012.)

5.4.2 Action Potential

The communication between neurons takes place through action potentials, which are the nerve impulses. The changes in the membrane potential oscillations cause action potentials [13]. Usually, neurons receive inputs from the other neurons that are functionally connected to them. The summation of the signals from those neurons causes the membrane potential of receiving neuron to oscillate. If the membrane potential reaches a particular threshold as a result of the inputs it receives, the neuron discharges (fires an action potential) and goes back to its resting state. In short, an action potential is a brief reversal of the membrane potential of the neuron such that the inside of its cell membrane becomes negative (as it would be on its resting state). A schematic representation of an action potential is shown in Figure 5.3.

5.4.3 Neuronal Oscillations and Synchronization

As we have seen so far, the intrinsic membrane potentials of the neurons is the main cause of neuronal oscillations. However, there also exist other mechanisms that can cause oscillations in the neurons. In general, every single neuron can be classified into two major types of oscillators, i.e., relaxation oscillators and conditional oscillators. Relaxation oscillators are spontaneous and oscillate in the absence of inputs, whereas conditional oscillators are those that fire as a consequence of the inputs it receives.

In the brain, synchronization and oscillation cause and affect each other. Synchronization can occur as a result of the neuronal oscillations, in which case the synchronization is called oscillation-based synchronization. On the other hand, oscillations can occur as a result of synchronization. This happens, for instance, when the occurrence of firing in a neuron predicts the occurrence of firing in the next neuron with some probability. Thus, neuronal oscillations in the brain are related to synchronization. Since oscillations and synchronization in the brain involves more than one neuron, there is no possibility that one can infer whether a single neuron is in synchrony with the others by making single cell recordings. We absolutely need to measure collective field potentials [14,15].

Synchronization or self-organization in the neuron can occur in two ways, from local to global or from global to local [16]. Here is where causality comes into the picture, and the two different ways of occurrence of synchronization reveal the direction of causality. If the oscillations of the single neuron affect the collective oscillation, then the causality is local to global. On the other hand, when the causality is global to local, the collective oscillations take control and affect the oscillatory activity of the single neurons. By being able to choose between these two types of interactions, the brain is able to perform functional activities by achieving synchronization at various levels in a self-contained manner.

5.5 ERS AND ERD

The oscillations in the brain are often perturbed by external signals that arise from any event or functional necessity. The populations of interacting neurons respond to these stimuli accordingly to either attend to or avoid any particular task or event. The effect of the event or the external stimulus is present in the changes in the level of synchronization in the neuronal populations, which are nothing but the signatures of the influence of the task or event. Such occurrences of synchronization or desynchronization due to the effect of a particular task or event are called event-related synchronization (ERS) or event-related desynchronization (ERD). The occurrence of ERS and ERD at specific frequency bands decides the acceptance or rejection of a particular task as that of a motor function, memory tasking, signals from auditory or visual cortex that need recognition, etc. Numerous experimental observations have been made in this direction, and dynamical models have also been proposed [17,18,19]. For instance, a recording was performed on male Wistar rats anaesthetized with Ketamine–Xylazine anesthetic [20]. The observations revealed a dramatic decrease in the delta band intensity (1–4 Hz) and an increase in the theta band synchronization (4–8 Hz). This change in the synchronization is reported to be the signature of the transition from deep to light anesthetic states. The results of the experimental observations are summarized in Figure 5.4.

However, there still exists a lack of a more general phenomenological explanation of the mechanism underlying the occurrence of ERS and ERD. Hence, if we model the underlying system of neurons as a system of coupled nonlinear oscillators and investigate the dynamics, it may be possible to understand the phenomena of ERS and ERD more comprehensively. For this purpose, we proceed as follows [21]:

5.5.1 Occurrence of ERD/ERS in Coupled Oscillator Models

The system of neurons in the presence of external stimulus can be represented by the following dynamical equations:

$$\dot{X}_j = F(X_j, \varepsilon_j) + \frac{A}{N}\sum_{k=1}^{N}(X_k - X_j) + BY, \quad j = 1,2,3,\ldots,N$$
$$\dot{Y} = G(Y, \varepsilon_e). \tag{5.4}$$

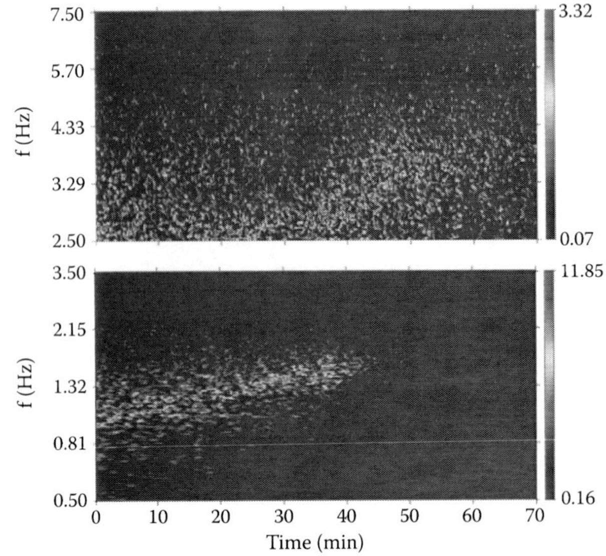

FIGURE 5.4 Time evolution of the characteristic EEG δ and θ frequencies during Ketamine–Xylazine anesthesia. (Taken from Musizza, B. et al., *J. Physiol.*, 580, 315, 2007. With permission.)

Here, N represents the size of the system. The state vector of the jth element is represented by X_j. A is the strength of the coupling between the oscillators. Y represents the state vector of the dynamic event, and B is the unidirectional coupling strength of the event. Here, ε_e represents the parameters of the external system. This model [21,22] can represent systems of populations of neurons in the presence of any external signal that represents a functional event from any extremities or other outside perturbations, stimulated microwave current driving a system of spin-torque nano-oscillators, polariton condensates in semiconductor microcavities that interact both among themselves and with the reservoir, etc. In the absence of external stimulus ($B = 0$), a sufficient coupling strength between the oscillators ($A > A_c$) causes synchronization in the system, when all the oscillators in the system behave as one. Now, if we increase the strength of the external stimulus, the expectation is that the synchronized system will enter into synchronization with the external stimulus signal. However, it need not be the case always. We have recently reported [21,22] that not all but some of the oscillators in the system that are in the synchronized state separate from the synchronized group and either synchronizes to a different frequency or becomes desynchronized, upon increasing the strength of the external stimulus signal. This is an intermediate state before the system could eventually run into synchronization with the external stimulus signal. This state is found to be extremely essential for the accomplishment of functional and behavioral tasks.

In order to demonstrate the occurrence of ERD, let us consider a system of coupled Rössler oscillators [23] in the presence of an external field, whose model equations can be represented by

$$\begin{aligned}
\dot{x}_j &= -\omega_j y_j - z_j + Bx_e, \quad j = 1,2,3,\ldots,N \\
\dot{y}_j &= \omega_j x_j + a y_j + \frac{A}{N}\sum_{k=1}^{N}\left(y_k - y_j\right) \\
\dot{z}_j &= b + z_j\left(x_j - c\right),
\end{aligned} \tag{5.5}$$

which is subject to an external stimulus x_e. Here, ω_j are the natural frequencies of the oscillator distributed uniformly between 0 and 1. We start with the state of complete synchronization and when $B = 0.1$, which is shown in Figure 5.5 (top panel), where the system is in complete phase synchronization while there is little desynchronization in the amplitude. Upon increasing the strength of the external stimulus signal to 0.18, we see that the system desynchronizes (both in amplitude and phase), and as a result, some oscillators separate themselves from the synchronized group and become desynchronized [Figure 5.5 (mid panel)]. Further increase in B to 0.23 brings back synchronization in the system, and the system of oscillators eventually becomes synchronized with the external stimulus [Figure 5.5 (bottom panel)].

In the case of the brain, the initial synchronization at a particular frequency band corresponds to a particular functional state of the body. In the particular case of the experiment on anaesthetized rats [20], this state corresponds to the state of deep anesthesia (accompanied by synchronization in the delta frequency band). In this

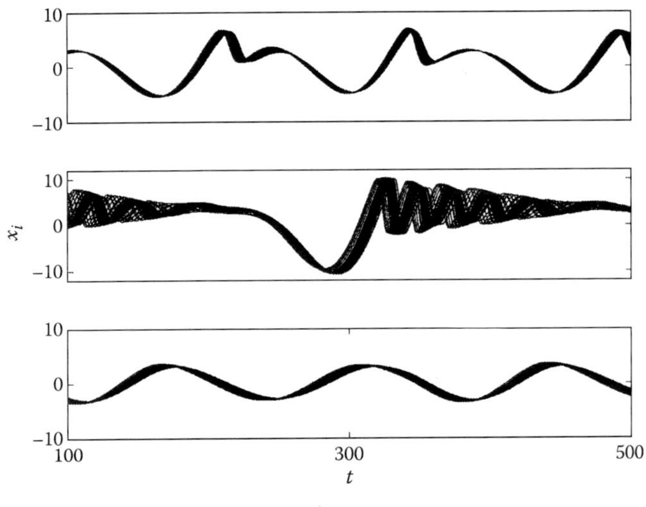

FIGURE 5.5 Occurrence of ERD in a system of coupled Rössler oscillators due to the increase in the strength of the external stimulus signal. (Top panel) Complete phase synchronization and a small amount of amplitude desynchronization. (Mid panel) Occurrence of ERD, resulting in one major synchronized group of oscillators and a small group of desynchronized oscillators. (Bottom panel) Synchronization comes back in the system. For more details, see [21].

state, the strength of the external stimulus is zero, implying that there are no incoming signals from other regions of the body that demand attention to sensory reactions (such as pain and concentration). The increase in the strength of the external stimulus represents the incoming signals from other regions of the body and the need to code information. When the strength of the incoming signal is strong enough, the synchronization in the delta band is perturbed (in order to destabilize the deep anaesthetized state), and some neurons separate themselves from the synchronized delta band and become desynchronized. As a result, there is a dramatic decrease in the intensity of delta band synchronization. This phenomenon is termed as ERD.

In the model, the separated group of oscillators can either remain desynchronized or become synchronized to a different frequency. In the results of simulation of the Rössler system, we see that the separated group of oscillators become desynchronized. In the case of the rat experiments, we see that the separated group of neurons become synchronized in the theta band for the accomplishment of information coding leading to consciousness. A simulation of this situation on a system of coupled Stuart–Landau (SL) oscillators revealed a case where the separated group of oscillators become quasi-periodically synchronized (shown in Figure 5.6). The model equation for the SL oscillator can be represented by

$$\dot{z}_j = \left(a - i\omega_j - (1+ic)\left|z_j\right|^2\right)z_j + \frac{A}{N}\sum_{k=1}^{N}\left(z_k - z_j\right) + Bz_e, \quad j = 1,2,\ldots,N. \tag{5.6}$$

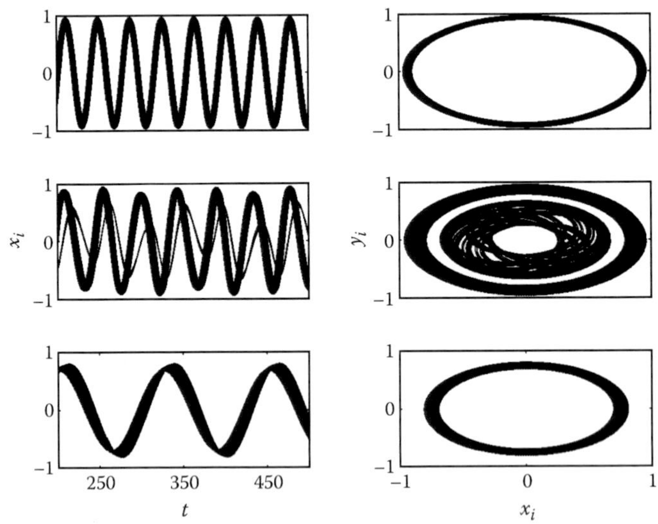

FIGURE 5.6 Occurrence of ERD in a system of SL oscillators. The panels on the left column show the time evolution of the phases of the oscillators in the system, whereas the panels on the right column show the corresponding phase portraits. ERD occurs in the mid panels, where the separated group of oscillators oscillate *quasi*periodically.

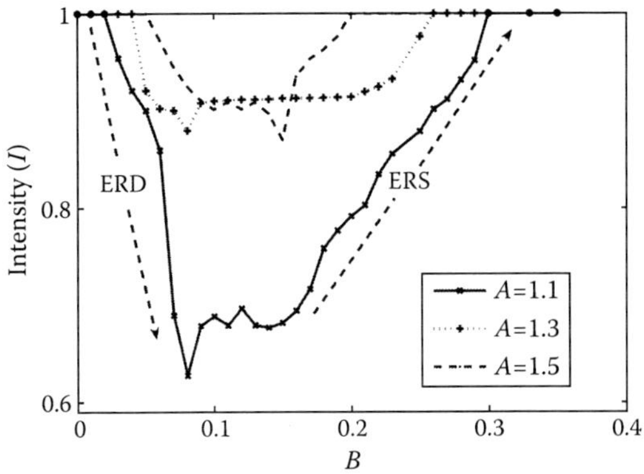

FIGURE 5.7 Change in the intensity of synchronization (I) due to the occurrence of ERD in a system of SL oscillators for various values of the coupling strength A.

Here, c is the nonisochronicity parameter, a is the Hopf bifurcation parameter, and $z_j = x_j + iy_j$ is the complex amplitude of the jth oscillator with natural frequency ω_j. The external SL oscillator is represented by the variable z_e.

The intensity of phase synchronization can be quantified using the quantity $I = \left\langle \left| \overline{e^{-i\theta_j}} \right| \right\rangle$, where $\theta_j = \tan^{-1}\left(\frac{y_j}{x_j}\right)$ is the phase of the jth oscillator, the bar represents the average over oscillators, and the angle brackets denote the average over time (see also Equation 5.3). I varies from 0 to 1. When I is 0, it represents the state of complete phase desynchronization, whereas, when $I = 1$, it represents the state of complete phase synchronization. When I takes a value between 0 and 1, it represents partial synchronization, and the value of I (the strength of the synchronization) depends on the size of the major synchronized population. There always exists a critical strength of the external stimulus B in order for ERD to occur. After a particular strength of the external stimulus, the system eventually becomes synchronized with the external stimulus signal. Hence, there exists a window of ERD, as shown in Figure 5.7, where the value of I is plotted against B. As B increases (from left to right), for a critical value, ERD occurs represented by a drop in the value of I. After a certain value of B, ERS occurs (the value of I increases with B), and eventually, the system goes back to the state when $I = 1$.

To conclude this section, systems of diffusively coupled nonlinear oscillators act as conceptual models that can explain the phenomenon of ERD. In general, the model discussed here not only applies to explaining ERD in the brain but can also explain similar occurrences of ERD in other systems such as Josephson junction arrays, Bose–Einstein condensates, system of coupled spin torque nano-oscillators, and the like.

5.6 EFFECT OF TIME DELAY: CHIMERA AND GCC STATES

Time delay is a crucial factor that occurs in all real systems. For instance, a typical example is a neuronal network where the propagation of signals between the neurons certainly involves a delay time. Similarly, a finite-time delay is involved in chemical interactions, regulation of gene transcription, and many other real-world systems. Considering instantaneous coupling in a system of interacting oscillators substantially simplifies the problem and helps us infer the dynamical properties of the system with ease. However, we should also note that considering time delay in interactions between coupled oscillators makes the model more realistic, and hence, we can get more accurate results, although we need to compromise on the ease of simulation and calculations. Since the inclusion of time delay makes the system more realistic, oftentimes, more interesting dynamical aspects are revealed when investigating systems by including time-delayed couplings rather than studying the corresponding systems with instantaneous couplings.

Upon including time-delayed coupling in the equation considered in the previous section (Equation 5.6), we have unearthed the existence of a new type of synchronized state called GCC state (stable and breathing) [24,25]. Chimera [26] literally refers to something that is composed of seemingly incompatible or incongruous parts. In a system of two populations of coupled oscillators, a chimera state represents a state when one of the populations is synchronized, whereas the other is desynchronized. A GCC state is a type of chimera state where there is one synchronized group of oscillators and one desynchronized group, while each group comprises oscillators from both the populations. We also found that this state is a characteristic of the presence of delay coupling in the system. The dynamical equations of the system are given as follows:

$$\dot{\theta}_i^{(1,2)} = \omega - \frac{A}{N}\sum_{j=1}^{N} f\left(\theta_i^{(1,2)}(t) - \theta_j^{(1,2)}\left(t-\tau_1\right)\right) - \frac{B}{N}\sum_{j=1}^{N} h\left(\theta_i^{(1,2)}(t) - \theta_j^{(2,1)}\left(t-\tau_2\right)\right). \quad (5.7)$$

Here, ω is the natural frequency of the oscillators in the populations, and it is the same for all the oscillators in both populations. Here, A and B are the coupling strengths within and between populations, and functions f and h are 2π periodic that describe the coupling. N refers to the size of the populations. The complex mean

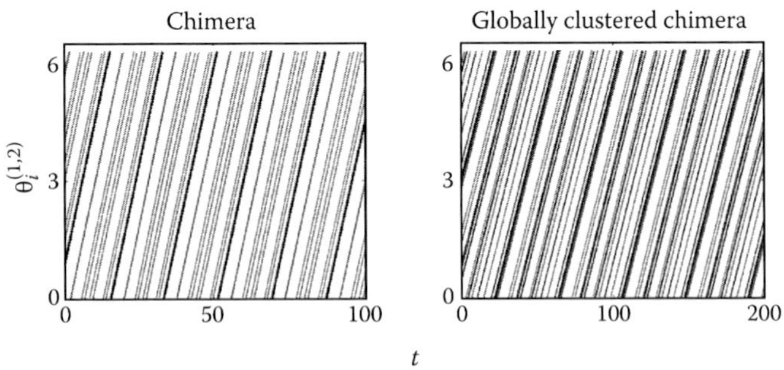

FIGURE 5.8 Occurrence of stable chimera and GCC states in Equation 5.7. The oscillators in the first and second populations are represented by black and gray lines, respectively.

field parameter $\left(r^{(1,2)} e^{i\varphi^{(1,2)}} = \frac{1}{N} \sum_{j=1}^{N} e^{i\theta_j^{(1,2)}} \right)$ can be used to quantify synchronization within a population, and here, we have considered two populations. τ_1 and τ_2 represent the coupling delay within and between the populations, respectively. The existence of different types of synchronized states in the system including the chimera and GCC states are summarized in Figure 5.8 for $\{f,h\} = \{\sin(\theta), \cos(\theta)\}$.

5.6.1 Breather and Unstable States

A breather state is one when the phases of the synchronized group remain stable whereas those of the desynchronized group fluctuate. As a result, the order parameter of the desynchronized group also fluctuates. The stability of the chimera and GCC states are affected by the time delay parameter τ. For illustration, the occurrence of stable, breathing, and unstable GCCs are shown in Figure 5.9.

The figure shows the time evolution of the phases of the oscillators in the desynchronized group. For $n = 0$, $A = 0.7$, $B = 0.4$, and $\{f,h\} = \{\sin(\theta), \sin(\theta)\}$, when $\tau = 0.85$ [in panel (a)], the GCC state is stable. This means that the desynchronized group of oscillators remain desynchronized, asymptotically. When $\tau = 1.01$, the GCC state loses its stability and ends up in a breathing state where the phases of all the oscillators switch between the synchronized (frequency suppressed) and the desynchronized states. On increasing τ further to 1.25, the GCC state becomes unstable, and the desynchronized state loses its stability and becomes synchronized. Breathers need not be periodic; they can be aperiodic, too.

The different types of synchronizations and the transition of the system from one dynamical state to the other with respect to the time delay parameter τ are summarized in Table 5.1.

To summarize, when a system of two identical coupled phase oscillator populations split into two groups in such a way that one of the populations is in synchronization and the other is in desynchronization, the dynamical state of the system is called chimera. On the other hand, if the two separated groups are such that one group is synchronized while the other is desynchronized and if each group has a fraction of

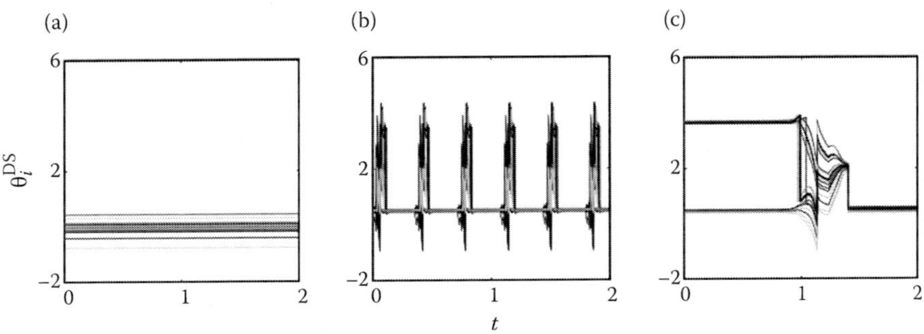

FIGURE 5.9 Occurrence of (a) stable, (b) breathing, and (c) unstable GCC states in Equation 5.7 for $\{f,h\} = \{\sin(\theta), \sin(\theta)\}$.

TABLE 5.1
Summary of the Different Types of Chimera and GCC States due to the Influence of Time Delay in the Coupled Systems (Equation 5.7)

S. No.	Value of τ	State	Description
1	2.0	Chimera	Population 1 synchronized and population 2 desynchronized.
2	3.2	Chimera	Population 1 desynchronized and population 2 synchronized.
3	4.0	GCC	One synchronized group and one desynchronized group, all containing oscillators from both the populations.
4	4.12	Unstable GCC	One synchronized group and one desynchronized group that fluctuates and becomes synchronized after a while.
5	4.13	Global synchronization	One synchronized group

oscillators from both the populations, then the dynamical state of the system is called GCC. The inclusion of time delay in the coupling reveals the existence of GCC and various other states (as summarized in Table 5.1). This indicates that the inclusion of time delay makes the system (model) more realistic and closer to real-world systems, helping us to reveal interesting dynamical states as that of GCC, which would otherwise have been missed. Chimera and GCC states are also crucial for the existence of the switching or multitasking type of functional states. For instance, some aquatic mammals engage in unihemisphere sleep (unlike most mammals with bihemispheric sleep) to mitigate the conflict between sleep and wakefulness [27].

5.7 MASS SYNCHRONIZATION AND ITS CONTROL

Although the occurrence of synchronization among the neurons at different frequency bands is highly crucial for task performance and information coding, collective synchronization is not always desirable [28–30]. Various pathological states such as Parkinson's tremors, epileptics, seizures, etc., are caused by the occurrence of strong synchronization between a large number of neuronal populations. Such synchronization is called pathological mass synchronization and also abnormal synchronization since, under normal conditions, mass synchronization does not occur. For instance, in the case of Parkinson's tremors, clusters of neurons fire in a highly synchronized manner, and the frequency of synchronization is similar to that of the tremor frequency [31]. Usually, the origin of such a strong synchronization is from a localized region of the brain. A small group of neurons in one particular region enter into strong synchronization and act as pacemakers to drive neuronal populations in the other regions into synchronization; this results in mass pathological synchronization. Essentially, there is a source population that acts as the origin of mass synchronization, which drives the target populations on to this state. A mathematical representation of the system can be of the following type [32]:

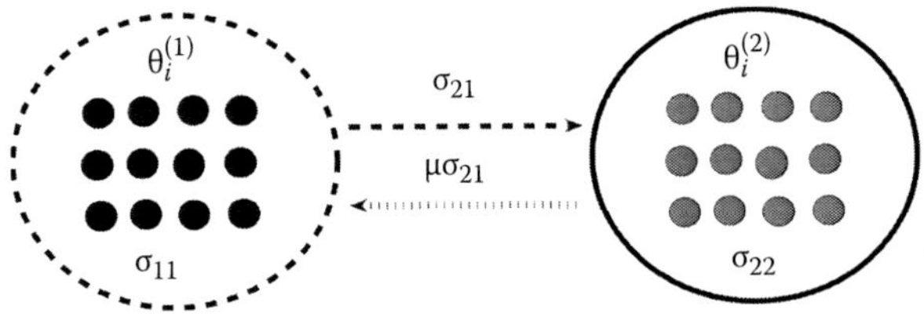

FIGURE 5.10 Schematic representation of Equation 5.8 for $N' = 2$. $\theta_i^{(1)}$ and $\theta_i^{(2)}$ are the source and target populations. Here, σ_{11} and σ_{22} are the coupling strengths within the populations. The coupling strengths from the source to the target and the target to the source are quantified by the parameters $\sigma_{12} = \mu\sigma_{21}$ and σ_{21}, respectively.

$$\dot{\theta}_i^{(\eta)} = \omega_i^{(\eta)} - \sum_{\eta'}^{N'} \frac{\sigma_{\eta\eta'}}{N} \sum_{j=1}^{N} \sin\left(\theta_i^{(\eta)}(t) - \theta_j^{(\eta')}(t) + \alpha_{\eta\eta'}\right) \tag{5.8}$$

$$i = 1,2,\ldots,N;\ \eta = 1,2\ldots,\ N'.$$

N' is the number of populations, and $\sigma_{\eta\eta'}$ is the strength of the coupling between the neurons in η' and those in η. Here, $\omega_i^{(\eta)}$ is the natural frequency of the ith neuron in population η, and $0 \le |\alpha'_{\eta\eta}| < \pi/2$ is the phase lag. For a system with one source and one target, a schematic representation is shown in Figure 5.10.

We have chosen $\sigma_{12} = \mu\sigma_{21}$ because the source drives the target on to mass synchronization and the drive of the target on the source is much lesser and negligible. The coupling strength between the oscillators within the source and within the target are represented respectively by σ_{11} and σ_{22}. Synchronization within a population can be quantified by

$$R_\eta = \langle r_\eta \rangle = \frac{1}{T}\int_0^T r_\eta\, dt, \tag{5.9}$$

where $z_\eta = r_\eta e^{i\varphi_\eta} = \left(\frac{1}{N}\right)\sum_{j=1}^{N} e^{i\theta_j^{(\eta)}}$. R_η takes a value between 0 and 1, 0 represents complete desynchronization, and 1 represents complete synchronization in the system. For partial synchronization, R_η takes a value between 0 and 1. Numerically strong synchronization can be assumed to be characterized by $R_\eta > 0.8$.

5.7.1 Occurrence of Mass Synchronization

A demonstration of the occurrence of mass pathological synchronization in a system of two populations (one source and one target) is shown in Figure 5.11.

Let us start with a state when both the source and target populations are desynchronized represented by $R_1 = 0.2$ and $R_2 = 0.3$. In this state, the coupling strength

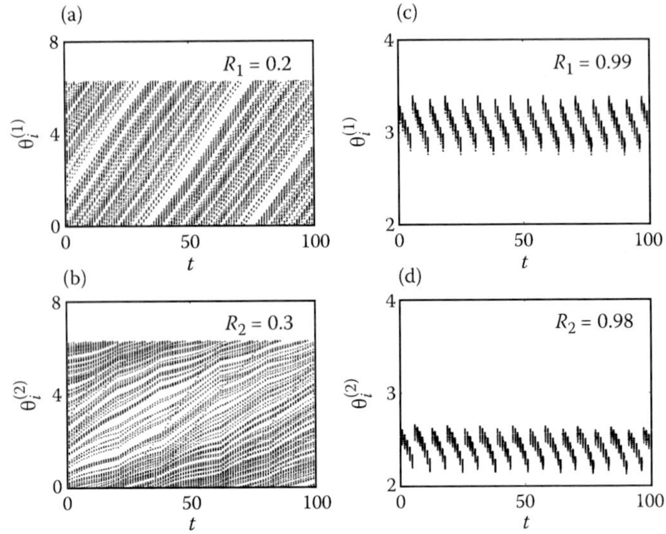

FIGURE 5.11 Plot shows the time evolution of the phases in the (a and c) source and (b and d) target populations.

within the source and target populations is maintained at 0.1, whereas there is strong coupling between the source and the target ($\sigma_{21} = 1.5$ and $\mu = 0.3$). When we increase the strength of the coupling within the source population, the source enters into strong synchronization, and, in turn, it also induces synchronization in the target population. Interestingly, the synchronization in the target population is due to the increase in the coupling strength of the source, because the strength of the coupling in the target population is still at a lower value of 0.1. In the figure, panels (a) and (b) represent the initial desynchronized state of the source and the target populations, respectively, whereas panels (c) and (d) represent the occurrence of mass pathological synchronization due to increase in the strength of coupling in the source. Similar results are observed if we increase the number of target populations to 2, 3, etc.

5.7.2 Method to Control Mass Synchronization

Since the occurrence of mass synchronization results in pathological states in the brain, a mechanism to control its occurrence is highly essential. Application of a delayed feedback can potentially force the populations to destabilize from the mass synchronization state [32]. With the inclusion of the delayed feedback, the system equations take the form

$$\dot{\theta}_i^{(\eta)} = \omega_i^{(\eta)} - \sum_{\eta'}^{N'} \frac{\sigma_{\eta\eta'}}{N} \sum_{j=1}^{N} \sin\left(\theta_i^{(\eta)} - \theta_j^{(\eta')} + \alpha_{\eta\eta'}\right) \pm FR_\tau \sin\left(\theta_i^{(\eta)} - \varphi_\tau\right), \quad (5.10)$$

where $R_\tau = R(t-\tau)$, $\varphi_\tau = \varphi(t-\tau)$, $Z(t) = \sum_{\eta'=1}^{N'} z_{\eta'} = Re^{i\varphi} =$, and $\left(\frac{1}{N'}\right)\sum_{\eta'=1}^{N'} r_{\eta'}e^{i\psi_{\eta'}}$ is the global order parameter. A schematic representation of the control mechanism is given in Figure 5.12.

If we introduce a global order parameter $R(t)$ to quantify the collective synchronization of all the populations, one can monitor the occurrence of pathological mass synchronization using the control setup. The setup includes measurement of the collective synchronization strength ($R(t)$) and the collective frequency; the occurrence of pathological synchronization can be easily recognized using these measures. Now, a delayed feedback signal with appropriate strength (F) and time delay (τ) is fed back into the system in order to destabilize the mass synchronization state.

As seen from Figure 5.13, we find that a delayed feedback signal of appropriate strength and time delay can control the occurrence of mass synchronization in the system. A rigorous mathematical analysis of the above phenomenon can be carried out in the thermodynamic limit $N \to \infty$. For details, see [29].

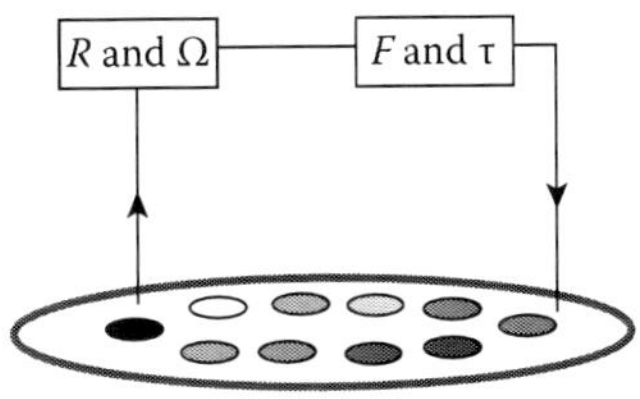

FIGURE 5.12 Schematic representation of the control setup.

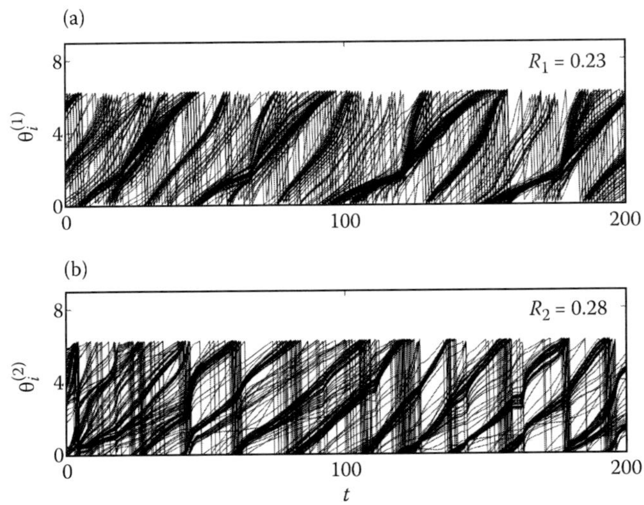

FIGURE 5.13 Occurrence of desynchronization for $\tau = 1.0$ and $F = 5.5$ in the (a) source and (b) target populations.

More importantly, this delayed feedback control of mass synchronization is demand controlled, which means that, once the mass synchronization state is destabilized, the delayed feedback signal is cut off.

To summarize, we see that synchronization is not always desirable, and in the brain, mass synchronization leads to abnormal pathological states such as epileptic seizures and Parkinson's tremors. We find that the dynamics of such states can be studied by using systems of coupled populations of phase oscillators. We have discussed a possible demand controlled delay feedback control mechanism to destabilize the mass synchronization state. We believe that this model and the accompanying results will be useful for further research in this direction and for practical implementation in deep brain stimulation therapy.

5.8 CONCLUSIONS

Systems of coupled nonlinear oscillators are used to explain the dynamics of various biological and physiological systems. This is because most of the interesting phenomena in biological systems occur as a result of the collective behavior of the entities in the system and also because we are more interested in the collective dynamics of the systems (macroscopic properties), instead of the dynamics of individual entities.

Synchronization is one such interesting collective dynamical phenomenon, which plays a crucial role in the large-scale properties of dynamical systems. In particular, in the brain, synchronization among the neurons acts as a backbone for occurrence of various functional and cognitive events. In this section, we have summarized some of the interesting physiological phenomena where synchronization plays a crucial role: (1) ERD, (2) the existence of chimera and GCC states, and (3) the occurrence of mass synchronization and its control.

Models of coupled nonlinear oscillators act as good candidates to explain the dynamical aspects in the aforementioned cases in a phenomenological manner. Even at the macroscopic level, many more details can be unearthed if we fine-tune the models to include additional factors and parameters, hence making the models closer to real-world systems. Validating the model results with the experimental results will also help fill in the gaps. Attempts are being made in this direction.

ACKNOWLEDGMENTS

The work is supported by the Department of Science and Technology (DST)—IRHPA Research Project (M. L. and V. K. C.) and a DST Ramanna Fellowship program. J. H. Sheeba is supported by a DST-FAST TRACK Young Scientist Research Project. M. Lakshmanan is supported by a DAE Raja Ramanna Fellowship Program.

REFERENCES

1. Winfree, A. T. *J. Theor. Biol.* **16**, 15 (1967).
2. Kuramoto, Y. Chemical Oscillations. *Waves and Turbulence*. Dover Books on Chemistry (2003).

3. Pikovsky, A., M. Rosenblum, and J. Kurths. *Synchronization—A Universal Concept in Nonlinear Sciences*. Cambridge University Press, Cambridge (2001).
4. Strogatz, S. H. *Physica D* **143**, 1 (2000).
5. Murray, J. D. *Mathematical Biology: I. An Introduction*. Springer-Verlag (2002).
6. Horologium. *The Hague*, 1658 (Oeuvres XVII).
7. Bennett, M., M. F. Schatz, H. Rockwood, and K. Wiesenfeld. *Proc. R. Soc. A* **458**(2019), 563 (2002).
8. Caton, R. *Brit. Med. J.* **278** (1875).
9. Scott, A. *Neuroscience*. Springer-Verlag, New York (2002).
10. Hodgkin, A. L., A. F. Huxley, and B. Katz. *Arch. Sci. Physiol.* **3**, 129 (1947).
11. Varela, F., J. P. Lachaux, E. Rodriguez, and J. Martinerie. *Nat. Rev. Neurosci.* **2**, 229 (2001).
12. Fries, P. *Trends Cogn. Sci.* **9**, 474 (2005).
13. Grundfest, H. *J. Neurophysiol.* **20** (3) (1957).
14. Leocani, L., C. Toro, P. Manganotti, P. Zhuang, and M. Hallet. *Clin. Neurophysiol.* **104**, 199 (1997).
15. Pfurtscheller, G., and F. H. Lopes da Silva. *Clin. Neurophysiol.* **110**, 1842 (1999).
16. Haken, H. *Synergetics, An Introduction*. Springer, Berlin (1983).
17. Hammond, C., H. Bergman, and P. Brown. *Trends Neurosci.* **30**, 357 (2007).
18. Kiss, I. Z., M. Quigg, S. H. C. Chun, H. Kori, and J. L. Hudson. *Biophys. J.* **94**, 1121 (2008).
19. Sheeba, J. H., A. Stefanovska, and P. V. E. McClintock. *Biophys. J.* **95**, 2722 (2008).
20. Musizza, B., A. Stefanovska, P. V. E. McClintock, M. Palus, J. Petrovcic, S. Ribaric, and F. F. Bajrovic. *J. Physiol.* **580**, 315 (2007).
21. Sheeba, J. H., V. K. Chandrasekar, and M. Lakshmanan. *Phys. Rev. Lett.* **103**, 074101 (2009).
22. J. H. Sheeba, V. K. Chandrasekar, and M. Lakshmanan. *Phys. Rev. E* **84**, 036210 (2011).
23. Rössler, O. E. *B. Math. Biol.* **39**, 275 (1977).
24. Sheeba, J. H., V. K. Chandrasekar, and M. Lakshmanan. *Phys. Rev. E* **79**, 055203 (2009).
25. Sheeba, J. H., V. K. Chandrasekar, and M. Lakshmanan. *Phys. Rev. E* **81**, 046203 (2010).
26. Abrams, D. M., and S. H. Strogatz. *Phys. Rev. Lett.* **93**, 174102 (2004).
27. Rubao, M., J. Wang, and Z. Liu. *Euro. Phys. Lett.* **91**, 40006 (2010).
28. Timmermann, L., J. Gross, M. Dirks, J. Volkmann, H. Freund, and A. Schnitzler. *Brain* **126**, 199 (2002).
29. Percha, B., R. Dzakpasu, and M. Zochowski. *Phys. Rev. E* **72**, 031909 (2005).
30. Strogatz, S. H. *Nature* (London) **410**, 268 (2001).
31. Popovych, O. V., C. Hauptmann, and P. A. Tass. *Phys. Rev. Lett.* **94**, 164102 (2005); *Biol. Cybern.* **95**, 69 (2006).
32. Sheeba, J. H., V. K. Chandrasekar, and M. Lakshmanan. *Chaos* **20**, 045106 (2010).

6 Nonperturbative Dynamics of Molecules in Intense Few-Cycle Laser Fields

Experimental and Theoretical Progress

Deepak Mathur and Ashwani K. Tiwari

CONTENTS

6.1 INTRODUCTION

A number of scientific and technological drivers are responsible for contemporary interest in studies concerned with how light and matter interact with each other, particularly with how very intense light interacts with matter. Much of the importance attributed to this area of research stems from the recognition that very intense light has very strong fields associated with it, and the dynamics that ensue when such fields irradiate atoms and molecules are of interest from both fundamental and applied viewpoints. It is useful to begin the discussion of such dynamics by first establishing what the terms "strong field" and "intense light" mean in relation to some physically established standard. One useful benchmark that may be adopted for this purpose is the Coulombic field that is experienced by an electron in the 1-s orbital of the hydrogen atom: it has a value of ~10^9 V cm^{-1}. The intensity of light that would give rise to such a field is 10^{16} W cm^{-2} (or 10 PW cm^{-2}). Pulses of light of such intensity are readily generated using commercially available Ti:sapphire laser systems incorporating an oscillator and an amplifier. Such systems emit 800-nm light, typically with pulse durations of 100 fs or less. It is not difficult to imagine that extremely high photon densities can be achieved, on the order of 10^{36} photons s^{-1} cm^{-2}, by focusing light from such femtosecond lasers. It is these remarkably high photon densities that mediate the essentially nonlinear nature of light–matter interactions in what has come to be known in recent years as the strong-field regime. In the context of what is to follow in this chapter, matter comprises individual atoms and molecules. The magnitude of the optical field that is experienced by these atoms and molecules ensures that the interaction cannot be considered perturbative. In the following, we present an overview of how the nonlinear nonperturbative dynamics of atoms and molecules in the strong-field regime is investigated using contemporary experimental and theoretical techniques.

The focus of our attention shall be on molecular dynamics in the ultrashort temporal regime. The fastest motions within molecules are those of the electrons; such motion—like orbital time periods in a Bohr-like picture of atoms—occurs on timescales of tens and hundreds of attoseconds. The nuclear degrees of freedom involve motions that are "slower": the shortest vibrational time periods in molecules are on the order of 12 fs. If intense optical fields are generated such that they last for periods shorter than vibrational time periods, the nuclear degrees of freedom can be effectively regarded as being "frozen," and the entire strong-field dynamics may be considered as being driven only by the electronic degrees of freedom in molecular systems. We present in this chapter an overview of recent progress in overcoming experimental and theoretical challenges that are encountered in studies of ultrafast molecular dynamics in the strong-field regime. Specifically, we outline details of how such ultrashort pulses may be produced from commercial laser sources supplying high-intensity pulses that are 50–100 fs long. We also provide some recent examples of studies on the ionization dynamics of triatomic and polyatomic molecules exposed to intense few-cycle pulses, and we present an overview of recent progress in theoretical methods that are proving to be of utility in rationalizing experimental observations and in predicting hitherto-unobserved processes.

In the strong-field regime, the electron and nuclear motions are strongly coupled; and therefore, under such a situation, the notion of potential energy surfaces (PESs) computed using the Born–Oppenheimer (BO) approximation becomes meaningless. In order to obtain proper insights into the dynamics under such a situation, it becomes mandatory to solve *coupled* electron–nuclear time-dependent Schrödinger equations (TDSEs). There have been a few recent attempts in this direction [1,2]. Bandrauk et al. [1] have numerically solved the exact non-BO TDSE for H_2^+ to understand the dissociative ionization of this molecule upon its irradiation by an intense laser field. In their work, the concept of PES does not arise. Gross et al. [2] have demonstrated that the full time-dependent wave function for a system of interacting electrons and nuclei, evolving under the influence of a time-varying external field, can be exactly decomposed into time-dependent electronic and nuclear wave functions. This factorization leads to an exact definition of time-dependent PESs (TDPESs). The concept of PES in this formalism is different from the PES that is conventionally developed under the BO approximation. This approach has been used to develop the TDPES for H_2^+ under the influence of a strong laser field: the TDPES has been computed for a 228-nm laser field represented by $E(t) = E_0 f(t) \sin(\omega t)$ for peak intensities of 25 and 100 TW cm^{-2}. The envelope function $f(t)$ was chosen such that the field was first linearly ramped from zero to its maximum strength at $t = 7.6$ fs and then became constant. The results of such calculations showed that the potential well that is induced by the laser field actually collapses at the peak of the pulse. These two exact methods are computationally very expensive and, therefore, can be used to carry out coupled electron–nuclear dynamical calculations only for small molecules like H_2^+. Therefore, there is a clear need for approximate methods that are not computationally expensive for bigger molecules and, at the same time, can capture the physics of coupled electron–nuclear dynamics. A recent joint experimental and theoretical work on a polyatomic molecular ion, the tetramethylsilane cation (TMS^+) [3], has shown the importance of developing TDPES for bigger molecular systems.

Kato and Kono [4] were the first to develop an approximate method for computing the TDPES in an intense laser field. In their approach, the TDPES was defined in terms of instantaneous eigenvalues of the field-dressed electronic Hamiltonian. In other words, in the Kato–Kono approach, the TDPES is basically the adiabatic potential energy that is computed by including the instantaneous value of electron-field interactions in the Hamiltonian. This method has proved to be extremely helpful in understanding many intense field phenomena in realistic molecular systems. However, this method lacks the exact *dynamical* features of the intense laser–molecule interactions, and therefore, a need still exists for a method that is not computationally as expensive as the exact methods but captures the dynamical features of electrons/molecules in strong laser fields.

Recently, Cederbaum [5] has developed a method for computing the TDPES for molecules that is based on a time-dependent BO (TDBO) approximation. This method of computing TDPES seems to be computationally manageable and, at the same time, can be used to compute the TDPES of bigger molecules. Furthermore, the TDPES computed using this method, similar to static BO PES, can be used to carry out nuclear dynamics to compute the experimental observables. However, before applying this method to the bigger system, it needs to be benchmarked against the smallest possible molecule, H_2^+. This has recently been accomplished by Garg et al. [6], who have used this method to compute the TDPES for H_2^+. The comparisons of TDPES for H_2^+ obtained from the approximate method of Cederbaum with what is obtained using the exact method of Gross et al. [2] will be presented in the following in order to provide a flavor of the progress being achieved in the development of theoretical methodologies that are becoming necessary in order to rationalize the increasingly sophisticated observations that are becoming possible due to advances in experimental methodologies. Before discussing the theoretical progress further, it is appropriate to first present an overview of recent progress in producing intense ultrashort laser pulses in the laboratory that allow direct experimental access to strong-field molecular dynamics in the few-cycle domain.

6.2 ULTRASHORT PULSES

Recent developments in techniques for producing intense laser pulses that last for only a few femtoseconds have enabled new vistas to be opened in experimental explorations of the temporal facets of molecular dynamics in the strong-field regime. Indeed, the ability to generate pulses that are long enough for only a few optical cycles has begun to yield new insights that have stimulated theoretical advances.

The generation of intense few-cycle optical pulses has become important in the contemporary pursuit of attosecond pulses, as well as in experimental probes of light–matter interactions in general and of ultrafast dynamics in particular. Currently, most practitioners utilize the hollow-fiber pulse compression technique to generate intense few-cycle pulses, although an alternate method based on filamentation in rare gases has also begun to find utility. Conventionally, linear pulse compression techniques have relied on the duration of chirped pulses being reduced by removing (or reducing) the amount of chirp. Such chirp removal or chirp reduction is usually accomplished by sending the pulses through an optical element with appropriate chromatic dispersion. Typical optical elements that may be used for such purpose include a pair of diffraction gratings (a grating compressor), a pair of prisms, an optical fiber, a chirped dielectric mirror, or a chirped-fiber or volume Bragg grating. In such cases, the shortest achievable pulse is determined by the optical bandwidth, and in ideal cases, bandwidth-limited pulses are obtained. In the high-intensity regime of interest in the present study, we have adopted a nonlinear compression technique that involves spectral broadening of unchirped pulses, followed by dispersive temporal compression. Spectral broadening is accomplished by propagating pulses through a rare-gas-filled hollow fiber, wherein most of the optical power propagates in the gas and self-phase modulation occurs [7]. Subsequent dispersive compression is achieved using a set of chirped dielectric mirrors. It is routinely possible with the hollow-fiber technique for us to compress 50-fs pulses with millijoule energies down to 5–7 fs. The shortest pulses that have been achieved using this technique are 3.2 fs long, produced with positively chirped incident pulses of 33-fs duration [8].

Upon injection of such ultrashort pulses into a gas jet, high harmonic generation can occur, and under appropriate conditions, pulse durations as short as a few hundred attoseconds may be attained [9,10]. In our laboratory, we routinely use Ne gas to fill our 1-m-long hollow fiber at a pressure of around 1 atmosphere; typical pulse durations of around 5 fs are obtained after temporal compression with a set of chirped dielectric mirrors, with the final shape of the output laser pulse being slightly elliptical. Figure 6.1 shows typical characteristics of the pulses from our hollow-fiber system. The dimensions of the 5-fs pulse are very much smaller than typical diameter values (2–3 mm) for molecular beams used in our experiments, with the number densities of the latter being kept low enough to preclude saturation effects.

We have also adopted another technique in our laboratory [11] that is based on two-stage filamentation in a rare gas. The method involves focusing 40-fs laser pulses (800-nm wavelength, 1-kHz repetition rate, and 1-mJ peak energy) into two 1-m-long tubes filled with Ar gas, typically at 1-atm pressure (Figure 6.2). White light is generated, following filamentation in Ar. This white light is temporally refocused using two sets of chirped dielectric mirrors so as to generate pulses as short as 7–10 fs (2–3 optical cycles). We note from Figure 6.1 how the laser beam profile at the output of our compression system is a marked improvement over the input beam profile. Significant improvements in the random intensity fluctuations that are inherent in high-power laser pulses have also been demonstrated [11]. Characterization

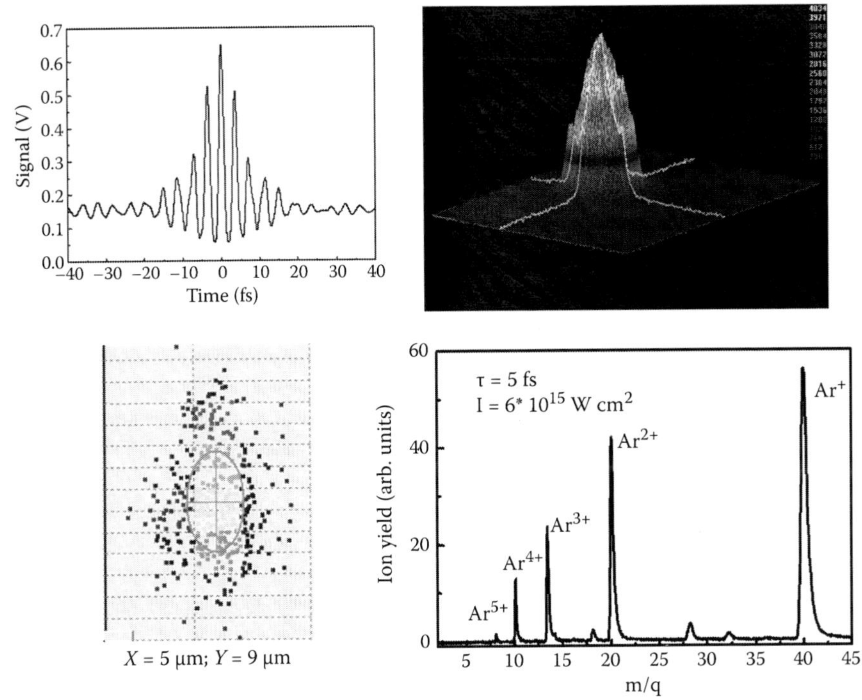

FIGURE 6.1 Characteristics of an ultrafast laser pulse delivered from the hollow-fiber system. (Top, left) Time evolution of the pulse. (Top, right) Intensity profile. (Bottom, left) Spatial dimensions of the laser beam. (Bottom, right) Typical mass spectrum of Ar that is used to calibrate the peak laser intensity. In the case that is depicted, the threshold for appearance of Ar^{5+} indicates a peak laser intensity of 6 PW cm^{-2}.

of the few-cycle pulses that we generate is by means of a spectral interferometric technique [11].

For the molecular dynamics experiments discussed in the following, the intense ultrashort pulses were steered into a high-vacuum system so as to interact with a beam of gas molecules. Ions formed as a result of the laser-molecule interaction were electrostatically extracted into a linear time-of-flight (TOF) spectrometer and detected using single-particle-counting techniques. The laser beam, molecular beam, and axis of the TOF spectrometer were mutually orthogonal to each other in an oil-free ultrahigh-vacuum environment (base pressure ~10^{-9} torr). A simple transformation of measured ion TOF spectra enabled the mapping of temporal information into mass and energy spectra.

6.3 MOLECULAR IONIZATION AND DISSOCIATION

Most experimental probes of strong-field molecular dynamics have relied on measurements of ion yields, usually employing infrared laser pulses of durations that

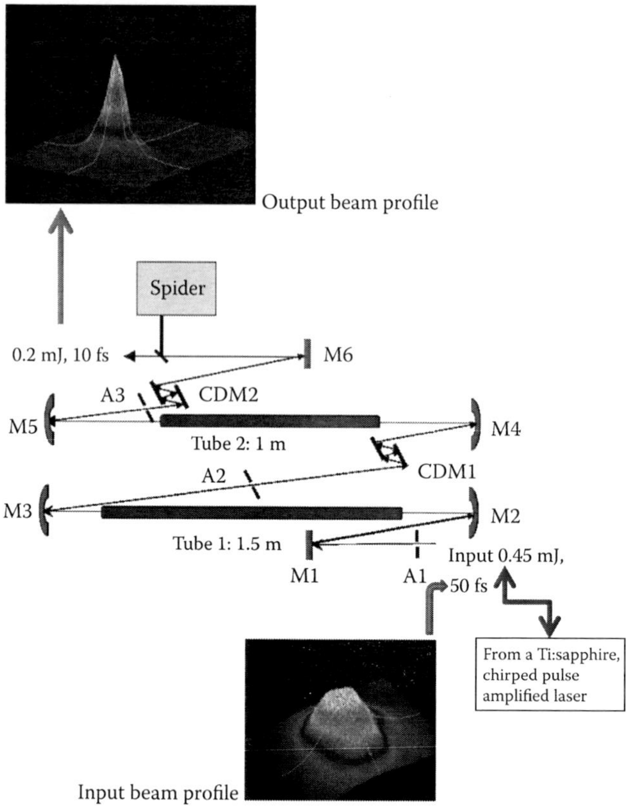

FIGURE 6.2 Generation of ultrafast laser pulses using a filamentation technique. Tube 1 and Tube 2 are filled with Ar gas at approximately 1-atm pressure in order to obtain spectral broadening. CDM1 and CDM2 are sets of chirped dielectric mirrors that are used to achieve temporal compression. M1–6 are high-reflectivity mirrors. A1–3 are apertures. The temporal characterization of the output beam is by spectral phase interferometry for direct electric-field reconstruction (SPIDER).

range from several tens to a few hundred femtoseconds. On the basis of a reasonably large body of experimental data, it seems established that the main drivers of strong-field molecular ionization and dissociation are the following: (1) enhanced ionization (EI), (2) spatial alignment, and (3) rescattering. In recent years, experiments have begun to yield indications that the dynamics that ensue when few-cycle pulses are used may be quite different to those when longer pulses are used. For one, it becomes highly improbable for dynamic alignment of molecules like O_2, N_2 to occur as the optical field is not "on" long enough to induce the necessary polarization-induced torque on the molecular axis. Second, such time periods ensure that the dynamics proceed at more or less equilibrium internuclear distances, and consequently, the EI process is expected to be effectively "switched off" when molecules are irradiated by intense few-cycle pulses; nuclei within the irradiated molecule will simply not have sufficient time to move to the critical distance at which the propensity for ionization is enhanced. Few-cycle molecular dynamics are therefore likely to be dominated only by rescattering, wherein the optical-field-ionized electron is driven by the oscillating field so as to collide with the parent molecular ion, thereby inducing further ionization. Experiments on H_2 have shown that the double ionization process is amenable to experimental control by tuning the intensity and duration of an ultrashort optical field [12]. Few-cycle pulses, therefore, offer to molecular physicists two different tantalizing prospects: first, being able to disentangle effects of different processes in strong-field ionization and dissociation dynamics and, second, being able to exercise some measure of control on the overall dynamics by tuning experimentally accessible parameters such as incident laser intensity and laser pulse duration. Indications of the former have begun to be presented [13], whereas the possibility of the latter has been established in experiments on few-cycle ionization of H_2 [12].

It is not only ionization and dissociation dynamics that offer new opportunities in the few-cycle strong-field regime. Intramolecular migration of H atoms in methanol, following strong-field double ionization, has also been recently studied [14], and differences have been discovered in the Coulomb explosion dynamics: such differences appear to depend on whether methanol is irradiated by 7- or 21-fs pulses. H atoms in H_2O have also been shown to be rearranged by strong optical fields of ~10-fs duration so as to form H_2^+ [15,16]. This atomic rearrangement is ultrafast: it has been shown to occur within a single laser pulse [15]. Experiments on the dissociative ionization of CH_4 molecules have served to highlight the importance of nonadiabatic effects: the ionization propensity in strong optical fields with durations of a few tens of femtoseconds is enhanced, compared to the situation when irradiation is by fields of longer durations [17]. Such studies have focused on the manner in which the intense optical field affects molecular ionization and dissociation. But what of the quantum mechanical properties of the molecules themselves? For instance, does the symmetry of molecular orbitals affect the strong-field dynamics?

There is now a plethora of data on *atomic* ionization in strong fields offering confirmation that the ionization rate depends only on the first ionization energy (IE) of the irradiated atom. In the tunnel ionization regime the strong optical field distorts the atom's radial potential function, making it possible for one or more valence electrons to tunnel through to the continuum. Measured energy distributions of such ionized electrons are well accounted for by the oft-used Ammosov–Delone–Krainov (ADK) theory, in which only the IE, namely the energy difference between the highest normally occupied atomic orbital and the lowest available normally unoccupied orbital, is of concern; the quantal nature of the orbitals themselves does not enter the reckoning. This is the rationalization offered for the nearly identical strong-field ionization rates that have been established for Ar and N_2, where the ratio IE(Ar)/IE(N_2) is 1.01. However, the deficiency of the ADK approach comes to the fore when taking into account the fact that, for the complementary pair O_2 and Xe, the ionization rate for the former is an order of magnitude lower than the rate for the latter, even though the ratio IE(Xe)/IE(O_2) is even closer to unity (1.005). The importance of orbital symmetry in strong-field molecular dynamics has begun to be established, following intense field *S*-matrix calculations [18] that predicted the suppression of ionization in those homonuclear molecules whose valence orbital has antibonding symmetry, like the outermost π_g orbital in O_2, but not in molecules with a bonding valence orbital, like the σ_g orbital in N_2. The conjecture was that the shape of the former orbitals results in destructive interference by the two nuclei of subwaves of the ejected electron; subsequent electron spectroscopy measurements have offered compelling support for this conjecture [19]. The quantal nature of molecular orbitals manifests itself in a simple manner in strong-field ionization of molecules like CS_2, as illustrated in the following.

Figure 6.3 shows typical ionization spectra of CS_2 obtained using intense laser pulses of different durations (10 and 40 fs) of approximately the same peak intensity (~40 TW cm^{-2}). The striking features of the 10-fs spectrum are the following: (1) the dominance of mass peaks corresponding to different charge states of the CS_2 molecule and (2) the relative absence of fragment ion peaks. The ionization pattern obtained with longer (40-fs) pulse is clearly much richer, with a gamut of atomic ions being produced up to charge state 4+. Ions like S^+ and S^{2+} are produced with large kinetic energy (~4 eV), as exemplified by the forward–backward peak splitting of their TOF peaks, indicating the formation of excited states of highly charged parent molecular ions that Coulomb explode. Long-lived CS_2^{2+} and CS_2^{3+} are also observed, but their yield is much lower, compared to the fragment ions. The rather dramatic difference between the two spectra shown in Figure 6.3 has been rationalized in the following terms [20]. The ground electronic state of CS_2 has the electronic configuration (Core)^{22} $(5\sigma_g)^2(4\sigma_u)^2(6\sigma_g)^2(5\sigma_u)^2(2\pi_u)^4(2\pi_g)^4$, yielding overall symmetry $^1\Sigma_g^+$. The outermost $2\pi_g$ orbital is mostly made up of 3p atomic orbitals of sulfur; as the equilibrium bond length is large (~1.6°A), there is relatively little π-overlap

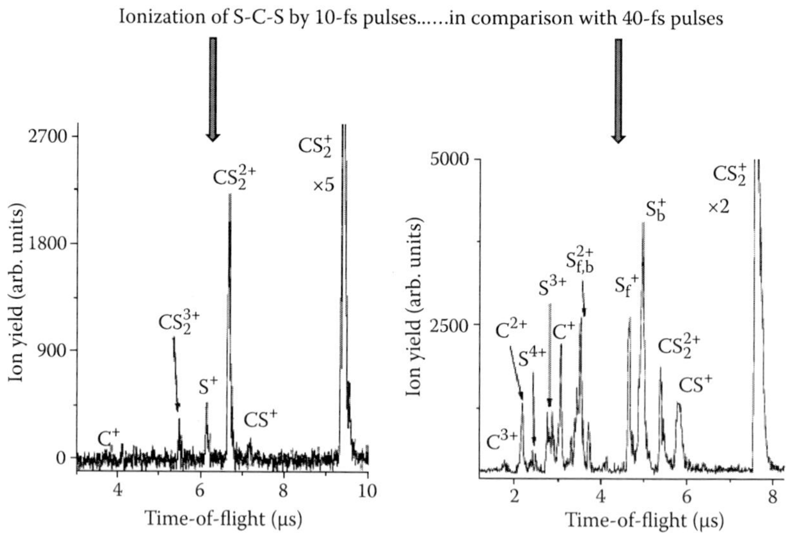

FIGURE 6.3 TOF spectra depicting ionization and fragmentation of CS_2 molecules upon irradiation with laser pulses of (left) 10-fs duration and (right) 40-fs duration. Note the relative absence of atomic fragments in the former case. The peak laser intensity was adjusted to be ~40 TW cm^{-2} in both cases.

between the two peripheral S-atoms. As a consequence, the antibonding character dominates. Upon field-induced removal of one, two, or three electrons from this antibonding orbital, single, double, and triple ionization occurs, and each electron removal effectively enhances the electronic charge density in the internuclear region of the molecule. As a result, not only the cation but also the dication and trication of CS_2 are long lived. The lifetimes of CS_2^{2+} and CS_2^{3+} have, indeed, been measured to be on the order of seconds in a storage ring experiment [21].

Another experimental pointer to the role played by the quantal description of the irradiated molecule relates to the observation of fragments like S^+ and CS^+ when 40-fs pulses are used. Such ions cannot be produced by direct field ionization of neutral CS_2 because Franck–Condon factors indicate that the dissociation continua of the X, A, and B electronic states of CS_2^+ are vertically inaccessible. The next ionic state C lies at 16.2 eV from the ground state of the neutral, well above the dissociation limits $S^+ + CS$ and $S + CS^+$, and is hence predissociative. It is the lengthening of the C–S bond that occurs in the EI process that properly accounts for the S^+ and CS^+ yields in longer pulses: population of excited electronic states of CS_2^+ then becomes likely, and it is these excited cation states that act as precursors for S^+ and CS^+. The nonappearance of these fragment ions in the case of the 10-fs spectrum is an unambiguous signature that the EI process is switched off in the ultrashort domain.

But what of rescattering that occurs on attosecond timescales? The experiments on CS_2 indicate that this is another area where quantal considerations intervene in the dynamics. Experiments on double ionization of H_2 have established [12] that, with 15-fs pulses, it is the first return recollision that dominates the rescattering dynamics. On the other hand, the third return recollision begins to be important in the case of longer pulses. Experiments on methane [13] have also shown the importance of the first return recollision when 8-fs pulses are used at intensities in the range of 100 TW cm^{-2}: no doubly charged ions were observed until longer pulses were used. In the case of CS_2, however, the molecular dication and trication dominate the 10-fs spectrum (Figure 6.3), ostensibly at the expense of the fragmentation channels, and this has been perceived [20] to be a signature of rescattering being "switched off" not because of temporal constraints but because of constraints imposed by the quantum mechanical nature of the CS_2 HOMO. The wave packet of the returning electron interferes destructively with the spatial extent of the HOMO, resulting in effective cancellation of the rescattering process. The returning electron's energy is no longer available for electronic excitation to states that would be quantum mechanically allowed to dissociate into S^+ and CS^+ fragments.

Having experimentally established that quantal considerations continue to play an important role in strong-field molecular dynamics, we present in the following a brief overview of the unfolding progress in efforts to properly incorporate into theoretical descriptions time-dependent field-dressed PESs within the TDBO approximation.

6.4 TDBO APPROXIMATION AND TDPES

The BO approximation allows one to separate fast electronic motion from slower nuclear motion. With this approximation, the total wave function can be written as a simple product of electronic and nuclear wave functions, $\phi_e(r;R)\chi_n(R)$, where

r represents all electronic coordinates and R represents all nuclear coordinates. This separation of total wave function as a simple product of electronic and nuclear wave functions is true for a time-independent Hamiltonian. However, for the case of a time-dependent Hamiltonian, as demonstrated by Cederbaum [5], a simple product of the electronic and nuclear wave functions $\phi_e(r,t;R)\chi_n(R,t)$ fails to satisfy the Schrödinger equation of the full Hamiltonian $H(t)$. In such a case, the total wave function Ψ can be written as

$$\Psi(r,R,t) = e^{i\omega(R,t)/\hbar}\phi_e(r,t;R)\chi_n(R,t), \quad (6.1)$$

where $\omega(R,t)$ is the time-dependent topological phase factor. ϕ_e is obtained for each value of R by solving the following equation:

$$i\hbar\frac{\partial}{\partial t}\phi_e(r,t;R) = H_e\phi_e(r,t;R). \quad (6.2)$$

Inserting $\Psi(r,R,t)$ into the Schrödinger equation of the full Hamiltonian H and projecting on the nuclear space gives us

$$i\hbar\frac{\partial\chi_n}{\partial t} = \left[\tilde{T}_n + \frac{\partial\omega}{\partial t}\right]\chi_n, \quad (6.3)$$

$T_n = -\frac{\hbar^2}{2M}\Delta$, with Δ denoting the Laplacian operator and M being the average nuclear mass. $\tilde{T}_n$ is given by

$$\tilde{T}_n = T_n - \frac{\hbar^2}{2M}\left[b + \frac{i}{\hbar}\Delta\omega + \frac{2i}{\hbar}(\nabla\omega).a - \frac{1}{\hbar^2}(\nabla\omega)^2 + 2\left[\frac{i}{\hbar}(\nabla\omega) + a\right].\nabla\right], \quad (6.4)$$

where a and b are time-dependent electron–nuclei coupling terms defined as

$$a(R,t) = \int \phi_e^*(r,t;R)\nabla\phi_e(r,t;R)\,dr \quad (6.5)$$

and

$$b(R,t) = \int \phi_e^*(r,t;R)\Delta\phi_e(r,t;R)\,dr. \quad (6.6)$$

The gradient and Laplacian operator are derivatives in the nuclear space. We rearrange Equation 6.4 in a simpler form as

$$\tilde{T}_n = -\frac{\hbar^2}{2M}\left[\nabla + a + \frac{i}{\hbar}(\nabla\omega)\right]^2 + \frac{\hbar^2}{2M}[\nabla.a + a.a - b]. \quad (6.7)$$

$\tilde{T}_n$ can be further simplified by eliminating the nuclear momentum coupling operator by using the condition

$$\nabla\omega = i\hbar a. \tag{6.8}$$

After determining the time-dependent topological phase from the above condition, the TDPES can be written as

$$\tilde{V}(R,t) = \frac{\partial\omega}{\partial t} + \frac{\hbar^2}{2M}[\nabla.a + a.a - b] + \int \phi_e^*(r,t;R)\frac{1}{R}\phi_e(r,t;R)\,dr. \tag{6.9}$$

This approach is used to compute the TDPES for the ground electronic state $1s\sigma_g$ of H_2^+ in the presence of linearly polarized, ultrashort, and intense laser pulses. Restricting the motion of the single electron and the two nuclei to the direction of the polarization axis of the laser field, we solved the problem using a 1-D Hamiltonian featuring soft-core Coulombic interactions [2]:

$$H_e(t) = -\frac{1}{2m_e}\frac{\partial^2}{\partial r^2} + V_l(r,t) - \frac{1}{\sqrt{1+(r-R/2)^2}} - \frac{1}{\sqrt{1+(r+R/2)^2}}, \tag{6.10}$$

where $m_e = (2M_p)/(2M_p + 1)$ is the electronic reduced mass and M_p is the proton mass. The interaction of the electron with the intense laser field is given by $V_l(r,t) = q_e rE(t)$, where $E(t)$ is the electric field amplitude and q_e is the electronic charge.

In order to generate the exact electronic wave function at each nuclear configuration, the time relaxation method can be used [22]. An arbitrarily chosen Gaussian wave packet is propagated in imaginary time by using the following equation:

$$\phi(\tau) = e^{-H_e\tau}\phi(0) = \sum_n c_n e^{-E_n\tau}\phi_n(0), \tag{6.11}$$

where $\tau = it$. In Equation 6.11, each eigenfunction relaxes to zero at a rate that is proportional to its eigenvalue. It is the ground-state wave function, which relaxes slowly, that persists. Higher states are obtained by numerically subtracting the ground-state wave function from the initial wave packet. Another round of propagation gives the first excited state. Propagation for a long time can mix the eigenfunctions, forcing the initial wave packet to relax to the ground state. This propagation scheme is applied at each nuclear configuration so as to generate the exact electronic wave function. We have computed the field-free molecular potential in this soft core potential by using

$$V(R) = \left\langle \phi_e(r;R)\middle|T_e\middle|\phi_e(r;R)\right\rangle + \left\langle \phi_e(r;R)\,|\,V_{soft}\,|\,\phi_e(r;R)\right\rangle + \left\langle \phi_e(r;R)\,|\,\frac{1}{R}\,|\,\phi_e(r;R)\right\rangle.$$

We discuss in the following the TDPES for H_2^+ [6] computed under the influence of a laser field with $\lambda = 800$ nm that is represented by $E(t) = E_0 e^{-2\ln 2(t-t_0)^2/\tau^2}\cos(\omega(t-t_0))$. Two values of peak intensity of the applied laser pulse are used: $I = |E_0|^2 = 1$ and 100 TW cm^{-2}.

The ground-state TDPESs [6] computed using the TDBO approximation, for pulses with durations of 2 and 5 fs, are depicted in Figures 6.4 and 6.5, respectively. The energy reference for the TDPES in all the cases is the energy value in the asymptotic region. It is clear from these figures that the well depth in the TDPES in both cases starts decreasing during the rising edge of the laser pulse, and the well depth collapses around the peak of the pulse. It is also clear from these figures that the TDPES tries to return to field-free PES form during the trailing edge of the pulse. However, even after the pulse is over, the TDPES does not completely regain its original form. In fact, during the trailing edge of the pulses, the PESs appear to acquire some wells at large values of internuclear distance R. These results are consistent with the exact calculations of Gross et al. [2]. Additionally, the calculations of Garg et al. [6] also show that the TDPES for the $2p\sigma_u$ state always remains repulsive, with only slight changes in the energy values; the TDPES for the $1s\sigma_g$ state for weaker laser pulses (of intensity 10^{12} W/cm^2) does not feature any collapse of the potential well as the field peaks.

Garg et al. [6] used these TDPESs to carry out nuclear dynamics on the $1s\sigma_g$ state of H_2^+. They found that part of the nuclear wave packet leaked out toward the asymptotic region in the case of the 5-fs pulse, whereas, for the shorter 1- and 2-fs pulses, it remained trapped within the well of the TDPES. This implies that the

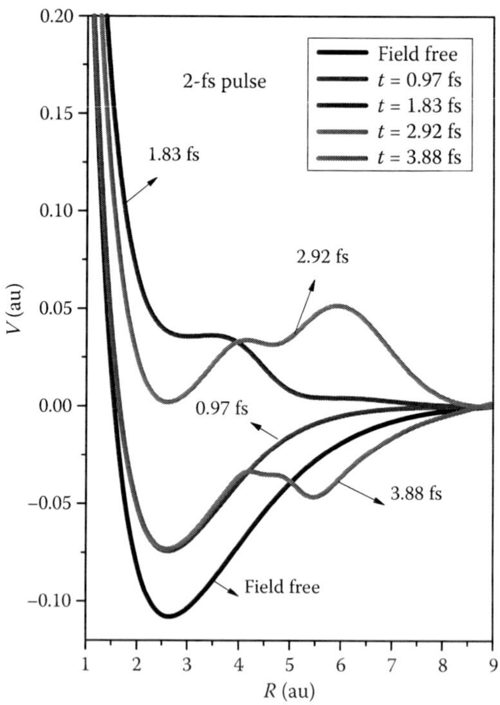

FIGURE 6.4 Ground-state TDPESs of H_2^+ computed using the time-dependent BO approximation for pulses of 2-fs duration. The energy scale is with reference to the energy value in the asymptotic region.

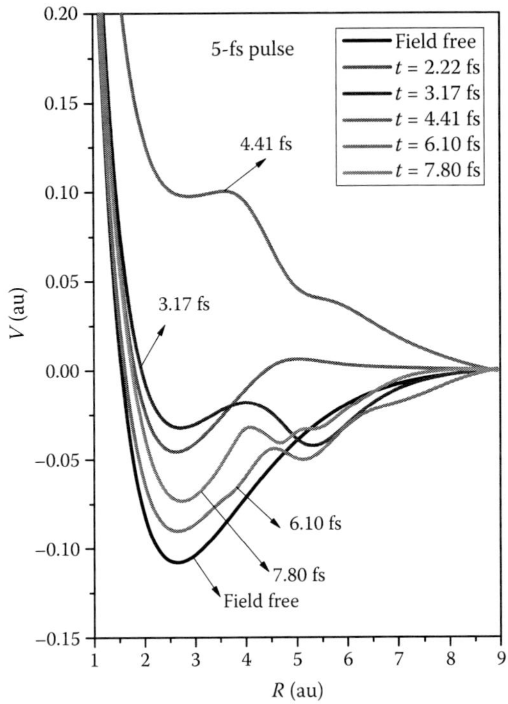

FIGURE 6.5 Ground-state TDPES of H_2^+ computed using the time-dependent BO approximation for pulses of 5-fs duration. The energy scale is with reference to the energy value in the asymptotic region.

longer rising and trailing edges of adiabatic laser pulses will eventually lead to the leaking of the initially trapped nuclear wave packet. These results support the time-dependent bond hardening in the TMS^+ under the influence of intense laser pulses that have recently been experimentally observed [3].

ACKNOWLEDGMENTS

The work outlined in this chapter was made possible through the contributions of our talented colleagues, A. Dharmadhikari, J. Dharmadhikari, K. Dota, and M. Garg, and we are proud to express our gratitude to each of them.

REFERENCES

1. Chelkowski, S., C. Foisy, and A. D. Bandrauk. 1998. Electron–nuclear dynamics of multiphoton H_2^+ dissociative ionization in intense laser fields. *Phys. Rev. A* **57:** 1176–1185.
2. Abedi, A., N. T. Maitra, and E. K. U. Gross. 2010. Exact factorization of the time-dependent electron–nuclear wave function. *Phys. Rev. Lett.* **105:** 123002-1–4.
3. Dota, K., M. Garg, A. K. Tiwari, J. A. Dharmadhikari, A. K. Dharmadhikari, and D. Mathur. 2012. Intense two-cycle laser pulses induce time-dependent bond hardening in a polyatomic molecule. *Phys. Rev. Lett.* **108:** 073602-1–4.

4. Kato, T. and H. Kono. 2004. Time-dependent multiconfiguration theory for electronic dynamics of molecules in an intense laser field. *Chem. Phys. Lett.* **392:** 533–540.
5. Cederbaum, L. S. 2008. Born–Oppenheimer approximation and beyond for time-dependent electronic processes. *J. Chem. Phys.* **128:** 124101–124109.
6. Garg, M., A. K. Tiwari, and D. Mathur. 2012. Quantum Dynamics of H_2^+ in intense laser fields on time-dependent potential energy surfaces. *J. Phys. Chem.* A **116**: 8762–8767.
7. Nisoli, M., S. De Silvestri, and O. Svelto. 1996. Generation of high energy 10 fs pulses by a new pulse compression technique. *Appl. Phys. Lett.* **68:** 2793–2795.
8. Sung, J. H., J. Y. Park, T. Imran, Y. S. Lee, and C. H. Nam. 2006. Generation of 0.2-TW 5.5-fs optical pulses at 1 kHz using a differentially pumped hollow-fiber chirped-mirror compressor. *Appl. Phys. B* **82:** 5–8.
9. Drescher, M., M. Hentschel, R. Kienberger, G. Tempea, Ch. Spielmann, G. A. Reider, P. B. Corkum, and F. Krausz. 2001. X-ray pulses approaching the attosecond frontier. *Science* **291:** 1923–1927.
10. Paul, P. M., E. S. Toma, P. Breger, G. Mullot, F. Augé, Ph. Balcou, H. G. Muller, and P. Agostini. 2001. Observation of a train of attosecond pulses from high harmonic generation. *Science* **292:** 1689–1692.
11. Dharmadhikari, A. K., J. A. Dharmadhikari, F. A. Rajgara, and D. Mathur. 2008. Polarization and energy stability of filamentation-generated few-cycle pulses. *Opt. Express* **16:** 7083–7089.
12. Alnaser, A. S., X. M. Tong, T. Osipov, S. Voss, C. M. Maharjan, P. Ranitovic, B. Ulrich, B. Shan, Z. Chang, C. D. Lin, and C. L. Cocke. 2004. Routes to control of H_2 Coulomb explosion in few-cycle laser pulses. *Phys. Rev. Lett.* **93:** 183202-1–4.
13. Mathur, D. and F. A. Rajgara. 2006. Nonadiabatic response of molecules to strong fields of picosecond, femtosecond, and subfemtosecond duration: an experimental study of the methane dication. *J. Chem. Phys.* **124:** 194308–194316.
14. Itakura, R., P. Liu, Y. Furukawa, T. Okino, K. Yamanouchi, and H. Nakano. 2007. Two-body Coulomb explosion and hydrogen migration in methanol induced by intense 7 and 21 fs laser pulses. *J. Chem. Phys.* **127:** 104306-1–5.
15. Rajgara, F. A., A. K. Dharmadhikari, D. Mathur, and C. P. Safvan. 2009. Strong fields induce ultrafast rearrangement of H atoms in H_2O. *J. Chem. Phys.* **130:** 231104-1–4.
16. Garg, M., A. K. Tiwari, and D. Mathur. 2012. Quantum dynamics of proton migration in H_2O dications: formation of H_2^+ on ultrafast timescales. *J. Chem. Phys.* **136:** 024320-1–5.
17. Mathur, D. and F. A. Rajgara. 2006. Nonadiabatic response of molecules to strong fields of picosecond, femtosecond and sub-femtosecond duration: an experimental study of the methane dication. *J. Chem. Phys.* **124:** 194308–19313.
18. Muth-Bohm, J., A. Becker, and F. H. M. Faisal. 2000. Suppressed molecular ionization for a class of diatomics in intense femtosecond laser fields. *Phys. Rev. Lett.* **85:** 2280–2083.
19. Okunishi, M., K. Shimada, G. Prümper, D. Mathur, and K. Ueda. 2007. Probing molecular symmetry effects in the ionization of N_2 and O_2 by intense laser fields. *J. Chem. Phys.* **127:** 064310–064313.
20. Mathur, D., A. K. Dharmadhikari, F. A. Rajgara, and J. A. Dharmadhikari. 2008. Molecular symmetry effects in the ionization of CS_2 by intense few-cycle laser pulses. *Phys. Rev. A* **78:** 013405-1–4.
21. Mathur, D., L. H. Andersen, P. Hvelplund, D. Kella, and C. P. Safvan. 1995. Long-lived, doubly charged diatomic and triatomic molecular ions. *J. Phys. B* **28:** 3415–3423.
22. Kosloff, R. and H. Tal-Ezer. 1986. A direct relaxation method for calculating eigenfunctions and eigenvalues of the Schrödinger equation on a grid. *Chem. Phys. Lett.* **127:** 223–230.

7 Selective Photodynamic Control of Bond Dissociation Using Optimal Initial Vibrational States

Bhavesh K. Shandilya, Manabendra Sarma, Vandana Kurkal-Siebert, Satrajit Adhikari, and Manoj K. Mishra

CONTENTS

7.1 INTRODUCTION

The formation and dissociation of chemical bonds is of prime concern in chemistry, and with the advent of ultrafast high-intensity lasers, selective bond dissociation using appropriate laser pulses as molecular tweezers and scissors has received extensive and intense attention.[1–90] Theoretical approaches to laser-assisted control of chemical reactions have kept pace and demonstrated remarkable success,[24,28] with experimental results[17,18,20,26] reinforcing the theoretical ideas. The development of theory and experiments has largely focused on the design of appropriate laser pulses to obtain the desired product from photodissociation reactions, often requiring field attributes that may not be realized on the basis of chemical considerations and may also be difficult to reproduce under normal laboratory conditions.[6–10,21,24,28] The photodissociation yield has, however, also been found to be extremely sensitive to the initial vibrational state from which photolysis is induced, and results for H_2^+[25], HI,[22,29] HCl,[19] and HOD[3,4,23,46,49,55–61] reveal a crucial role for the initial state of the system in product selectivity and enhancement. This critical dependence on the initial vibrational state indicates that a suitable choice of a single optimal initial vibrational state or an optimized linear superposition of the field-free vibrational states may be another route to photodynamic control of selective bond dissociation. The different established theoretical methods that use lasers to control chemical reactions have been reviewed recently[6–13,15,23,24,27,28] and are based on field design. It is our purpose in this contribution to present results from our work that provide encouraging alternatives for selective control of bond dissociation using simple laser pulses, in conjunction with the use of an appropriately chosen single field-free vibrational state[3,4,34,57–61] or an optimal linear combination[4,14,30–33,57,58] thereof as the initial state to be subjected to this simple laser pulse.

Along this line, a scheme for establishing the optimal linear mix of the field-free vibrational eigenstates for the given photolysis pulse and the chosen photodissociation objective has been investigated in our group,[4,14,30–33,57,58] whereby the emphasis is shifted from control through the design of an appropriate field to the design of an optimal linear combination of the field-free vibrational eigenstates for the chosen photolysis pulse and the preferred photodissociation objective. The field-optimized initial state (FOIST)–based approach, where control of bond dissociation shifts from control via field design to the design of an optimal initial state for the chosen field and which is computationally simple to implement, has found great success in selective control of product yields from dissociation of diatomic bonds.[14,30–33] Based on the mechanistic insights obtained from FOIST-based selective photodynamic control of products in the photodissociation of diatomic molecules,[14,30–33] the use of appropriately chosen infrared (IR) pulses to obtain a desirable linear combination of excited bond-specific modes, followed by the use of an apt ultraviolet (UV) pulse, has been found to provide near-total[60] selective dissociation of only the selected bond in HOD. Some results from a preliminary test of these ideas in the case of $O^{18}O^{16}O^{16}$ are also included in this presentation.

An outline of the computational approaches and field attributes used in our investigations are presented in the next section. Discussion of some key results is provided in Section 3, and a summary of the main observations is highlighted as concluding remarks.

7.2 METHOD

7.2.1 Diatomic Systems: IBr and HI

The effect of radiation field $\vec{\varepsilon}(t)$ on the molecule having dipole moment $\vec{\mu}$ may be realized by solving time-dependent Schrödinger equation

$$i\hbar\frac{\partial}{\partial t}\psi = \hat{H}(t)\psi,\ \hat{H}(t) = \hat{H}_0 - \vec{\mu}\cdot\vec{\varepsilon}(t), \tag{7.1}$$

where $\hat{H}_0$ is the field-free Hamiltonian.[14] The wave function ψ at any time t can be evaluated as $\psi(t) = \hat{U}(t,0)\psi(0)$, where $\hat{U}(t,0) \simeq e^{-i\hat{H}t/\hbar} = e^{-i(\hat{H}_0 - \vec{\mu}\cdot\vec{\varepsilon}(t))t/\hbar}$ is a propagator, which is not necessarily unitary, and $\psi(0)$ is the initial state upon which field $\vec{\varepsilon}(t)$ is applied.

The product yield in the desired channel is related to the time-integrated flux

$$\begin{aligned} f &= \int_0^T dt\left\langle \psi(t)\middle|\hat{j}_i\middle|\psi(t)\right\rangle \\ &= \int_0^T dt\left\langle \psi(0)\middle|\hat{U}^\dagger(t,0)\hat{j}_i\hat{U}(t,0)\middle|\psi(0)\right\rangle = \left\langle \psi(0)\middle|\hat{F}\middle|\psi(0)\right\rangle, \end{aligned} \tag{7.2}$$

with $\psi(t) = \hat{U}(t,0)\psi(0), \hat{F} = \int_0^T dt\hat{U}^\dagger(t,0)\hat{j}_i\hat{U}(t,0)$ and $\hat{j}_i = \frac{1}{2\mu_i}\left[\hat{p}_i\delta\left(r_i - r_i^d\right) + \delta\left(r_i - r_i^d\right)\hat{p}_i\right]$, where $\hat{U}$ is the time evolution operator, $\hat{j}_i$ is the flux operator in the ith channel, and μ_i, $\hat{p}_i$, and r_i^d are the reduced mass, the momentum operator, and a grid point in the asymptotic region of the ith channel. As we can see from Equation 7.2, the product yield $f = \left\langle \psi(0)\middle|\hat{F}\middle|\psi(0)\right\rangle$ in a desired channel may be altered by altering the field-dependent part $\hat{F}$ or $\psi(0)$.[4,14]

In the FOIST scheme,[4,14,30–33,57,58] the product yield is maximized through the preparation of the initial wave function $\psi(0)$ as a superposition of the field-free vibrational wave function $\{\phi_m\}$ of the ground electronic state

$$\psi(0) = \sum_{m=0}^{M} C_m\phi_m. \tag{7.3}$$

With the time-integrated flux operator $\hat{F}$ being Hermitian, the optimization of the channel- and field-specific flux functional $\langle\psi(0)|\hat{F}|\psi(0)\rangle$ with respect to the coefficients C_m employed in Equation 7.3 leads to the Rayleigh–Ritz eigenvalue problem[4,14]

$$\mathbf{FC} = \mathbf{Cf}, \tag{7.4}$$

where $\mathbf{f}$ is the diagonal matrix comprising the eigenvalues of the time-integrated flux matrix $\mathbf{F}$. The matrix elements of $\mathbf{F}$ in the ith channel are given by[4,14,30]

$$F_{kl}^{i} \approx \Delta t \sum_{n=0}^{N_t} \left\langle \psi_k(n\Delta t) \middle| \hat{j}_i \middle| \psi_l(n\Delta t) \right\rangle, \tag{7.5}$$

where Δt is the step size for the numerical time propagation such that $N_t \Delta t = T$ is the total propagation time chosen to ensure almost full dissociation. We propagate the $M + 1$ initial states shown in Equation 7.3 using the appropriate field and calculate the accumulated flux matrices (F_{kl}^{i}) in the ith channel. The accumulated (F_{kl}^{i}) matrices for the ith channel are diagonalized and eigenvector ($C_m^{\max}$) corresponding to the highest eigenvalue $f_{\max}$ indicates the maximum possible dissociation yield available for the chosen field and the manifold of vibrational eigenstates in that calculation. $C_m^{\max}$ defines the initial wave function $\psi^{\max}(0) = \sum_m C_m^{\max} \phi_m$, which will provide ($f_{\max} \times 100\%$) dissociation in the chosen channel for the field used and the expansion manifold of field-free vibrational states ($M + 1$) utilized in the calculation.[4,14]

To evaluate the product branching ratio for both the IBr and the HI molecules, we have carried out time-dependent wave packet (TDWP) analysis,[91] where the Schrödinger equation

$$H_{ex} \chi_l(0) = i\hbar \frac{\partial \chi_l(0)}{\partial t} \tag{7.6}$$

is solved with $\chi_l = \mu_{0l}\phi(0)$ as the initial wave function on the lth excited state, where μ_{0l} is the transition dipole moment between the ground (0) and the lth excited state, $\phi(0)$ is a free vibrational ground state [which is the optimal linear combination $\psi^{\max}(0)$]. The time evolution of the above defined wave function $\chi_l(0)$ is governed by $\chi_l(t) = \exp\left[-iH_{ex}t/\hbar\right]\chi_l(0)$, and the overlap $\left\langle \chi_l(0) \middle| \chi_l(t) \right\rangle$ of the time-evolving wave function $\chi_l(t)$ with the initial $\chi_l(0)$ is called the autocorrelation function. To ensure the correct branching ratio,[92] the autocorrelation function $\left\langle \chi_l(0) \middle| \chi_l(t) \right\rangle$ is evaluated after a sufficiently large time interval T such that the norm of the wave function on different curves has stabilized and the system population is completely out of the curve-crossing region. The Fourier transform of the autocorrelation functions yields a frequency-dependent partial absorption cross section:

$$\sigma(\omega) = C\omega \int_{-\infty}^{\infty} e^{-i(\omega+E_0)t} \left\langle \chi(T) \middle| \chi(T+t) \right\rangle dt, \tag{7.7}$$

where C is a constant,[91] ω is the frequency of the incident radiation, and E_0 is the energy corresponding to the initial state. The branching ratio is given by the ratio of the sum of partial photoabsorption cross sections for the two channels.

The photodissociation of IBr[14,30–34] and HI[31–34] has been studied extensively and are our representative systems of interest. The result to be presented are for the photolysis of IBr and HI employing a multicolor continuous-wave (cw) field of the form $\varepsilon(t) = A\sum_{p=0}^{2}\cos(\omega - \omega_{p,0})t$, where A is the amplitude, ω is the photodissociation frequency, and $\omega_{p,0} = (E_p - E_0)/\hbar$ is the Bohr frequency for transition between the pth and the ground (zeroth) vibrational energy levels. The vibrational eigenvalues E_p and the corresponding eigenfunctions ϕ_p of the electronic ground state are computed using the Fourier grid Hamiltonian (FGH) method.[93] The split-operator fast Fourier transform (FFT)[94,95] with Pauli matrix propagation[96,97] was utilized to integrate the time-dependent Schrödinger equation.

7.2.2 Triatomic System: HOD

In HOD calculations, we have considered only the ground and the first excited states of this molecule. The bending mode is inactive in the first absorption band $\left(\tilde{A}\,{}^1B_1 \leftarrow \tilde{X}\,{}^1A_1\right)$ of HOD; thereby, excitation from the ground state $\left(\tilde{X}\,{}^1A_1\right)$ to the first excited state $\left(\tilde{X}\,{}^1B_1\right)$ induces a negligible change in the bending angle.[36,37] Hence, the internal kinetic energy operator in terms of the conjugate momenta $\hat{p}_1$ and $\hat{p}_2$ associated with the O–H (r_1) and O–D (r_2) stretching coordinates, respectively, is taken as[4,38,40,41,46,53,54]

$$\hat{T} = \frac{\hat{p}_1^2}{2\mu_1} + \frac{\hat{p}_2^2}{2\mu_2} + \frac{\hat{p}_1\hat{p}_2}{m_O}\cos\theta, \tag{7.8}$$

where

$$\hat{p}_j = \frac{\hbar}{i}\frac{\partial}{\partial r_j},\ j = 1,2, \mu_1 = \frac{m_H m_O}{m_H + m_O},\ \mu_2 = \frac{m_D m_O}{m_D + m_O} \tag{7.9}$$

and θ is the equilibrium bond angle (104.52°). The potential energy surfaces (PESs),[38,40,98–100] the transition dipole moment surfaces, and the IR– and UV–molecule interaction Hamiltonians are taken from earlier investigations.[40,46] The time evolution of the corresponding nuclear motion can then be performed using the time-dependent Schrödinger equation

$$i\hbar\frac{\partial}{\partial t}\begin{pmatrix}\Psi_g \\ \Psi_e\end{pmatrix} = \begin{pmatrix}\hat{H}_g + \hat{H}_{ir} & \hat{H}_{uv}(t) \\ \hat{H}_{uv}(t) & \hat{H}_e(t)\end{pmatrix}\begin{pmatrix}\Psi_g \\ \Psi_e\end{pmatrix}, \tag{7.10}$$

where $\Psi_g = \Psi_g(r_1,r_2,t)$ and $\Psi_e = \Psi_e(r_1,r_2,t)$ are the wave functions associated with nuclear motion in the ground and the first excited states, respectively. $\hat{H}_g = \hat{T} + \hat{V}_g$ and $\hat{H}_e = \hat{T} + \hat{V}_e$ are the Hamiltonians for the two electronic states, and $\hat{H}_{uv}$ couples and perturbs both the electronic states. The IR–HOD interaction Hamiltonian is the same as that used in an earlier investigation.[46] We solve Equation 7.10 with the initial condition that the ground state wave function Ψ_g is a single field-free vibrational state of the HOD electronic ground state, and the excited state wave function $\Psi_e = 0$ at $t = 0$. The vibrational eigenfunctions of the ground electronic state of the HOD molecule were obtained using the FGH method[93] modified for two dimensions.[101]

The propagation of the wave functions $\{\Psi_g(t), \Psi_e(t)\}$ has been performed using Equation 7.10, where the effect of the kinetic energy operator on the wave function is evaluated with a 2-D FFT[102] and the time propagation is carried out using the Lanczos scheme.[103] The wave function is represented on a spatial grid spanning $r_{\text{O–H}}/r_{\text{O–D}}$ bond lengths in the range of 1 a_0–10 a_0 in 128 steps with $\Delta r_{\text{O–H}} = \Delta r_{\text{O–D}} \approx 0.0703 a_0$ (≈ 0.0372 Å), and the propagation of fields under the influence of a generic Gaussian UV pulse[3,4,57–61] of the form $E(\text{t}) = 0.09a(t)(\cos\omega t)$, where $a(t) = \exp[-\gamma(t - t_{\text{UV}})^2]$, with FWHM $= \sqrt{\dfrac{4\ln 2}{\gamma}}$, is done in time steps of $\Delta t = 1$ a.u. of time ≈0.0242 fs. On the other hand, for IR dynamics,[3,4,60] we have used $E(t) = A_0 a(t)\exp[-\gamma(t - t_0)^2](\cos\omega t)$, where $A_0 = E_0\cos\phi$, $\phi = (\pi - \theta)/2$, $\theta = 104.52°$, and $a(t) = \left(8\gamma t_l^2/\pi\right)^{1/4}$ with FWHM as defined above.

The product yield can be maximized through the FOIST scheme, as described in the earlier section. However, it should be noted that the flux operator corresponding to the ith channel $\hat{j}_i$ is to be replaced by $\hat{J}_i$, which, in this case, is expressed as $\hat{J}_1 = \left(\hat{j}_1 + \dfrac{\mu_2\cos\theta}{m_o}\hat{j}_2\right)$ and $\hat{J}_2 = \left(\hat{j}_2 + \dfrac{\mu_1\cos\theta}{m_o}\hat{j}_1\right)$, with the H + O–D channel labeled as 1 and the H–O + D channel as 2.

The expressions for the total flux J in the H + O–D and H–O + D channels are then given by[3,4]

$$J_{\text{H+O–D}} = \int_0^{r_{2d}}\int_0^{T}\Psi^*(r_1, r_2, t)\times\left(\hat{j}_1 + \frac{\mu_2\cos\theta}{m_O}\hat{j}_2\right)\Psi(r_1, r_2, t)\,dr_2\,dt \tag{7.11}$$

$$J_{\text{H–O+D}} = \int_0^{r_{1d}}\int_0^{T}\Psi^*(r_1, r_2, t)\times\left(\hat{j}_2 + \frac{\mu_1\cos\theta}{m_O}\hat{j}_1\right)\Psi(r_1, r_2, t)\,dr_1\,dt. \tag{7.12}$$

As mentioned earlier, in the present contribution, we provide some of the main results[4,57–59] obtained from our application of the FOIST scheme for different chosen fields, as well as from (IR + UV)-based calculations[60] for this prototypical molecule.

7.2.3 Triatomic System: $^{18}O^{16}O^{16}O$

In case of the $^{18}O^{16}O^{16}O$ molecule, we have included the ground and the second excited PES[63] for dynamical calculations on this system. The internal kinetic energy operator in terms of the conjugate momenta $\hat{p}_1$ and $\hat{p}_2$ associated with the $^{18}O–^{16}O$ (r_1) and $^{16}O–^{16}O$ (r_2) stretching coordinates, respectively, is taken as[72]

$$\hat{T} = \frac{\hat{p}_1^2}{2\mu_1} + \frac{\hat{p}_2^2}{2\mu_2} + \frac{\hat{p}_1\hat{p}_2}{2m_{^{16}O}}\cos\theta, \tag{7.13}$$

where

$$\hat{p}_j = \frac{\hbar}{i}\frac{\partial}{\partial r_j},\ j = 1,2, \mu_1 = \frac{m_{^{18}O}m_{^{16}O}}{m_{^{18}O} + m_{^{16}O}},\ \mu_2 = \frac{m_{^{16}O}m_{^{16}O}}{m_{^{16}O} + m_{^{16}O}} \tag{7.14}$$

and θ is the equilibrium bond angle (116.8°). The transition dipole moment surfaces and the UV–molecule interaction Hamiltonian are those utilized in earlier investigations.[64,69,72]

Like HOD, here, the propagation of the wave functions $\{\Psi_g(t), \Psi_e(t)\}$ has also been performed using Equation 7.10, where the effect of the kinetic energy operator on the wave function is once again evaluated with a 2-D FFT[102] and the time propagation is carried out using the Lanczos scheme.[103] The wave function is represented on a spatial grid spanning r_1/r_2 bond lengths between 2a$_0$ and 11 a$_0$ in 128 steps, with $\Delta r_1 = \Delta r_2 \approx 0.0703a_0$ ($\approx$ 0.0372 Å), and propagation is done with the same Gaussian UV pulse form that we used in case of the HOD molecule.[3,4,57–61] An absorbing ramp potential is placed at asymptotic cuts to avoid unphysical reflection from the edges.

The flux expressions for the total flux J in the $^{18}O + {}^{16}O{-}^{16}O$ and $^{18}O{-}^{16}O + {}^{16}O$ channels labeled as subscripts 1 and 2 are similar to the H + O–D and H–O + D channels seen in Equations 7.11 and 7.12 for the HOD molecule. These are

$$J_1 = \int_0^{r_{2d}}\int_0^{T} \Psi^*(r_1, r_2, t) \times \left(\hat{j}_1 + \frac{\mu_2\cos\theta}{2m_{^{16}O}}\hat{j}_2\right)\Psi(r_1, r_2, t)\,dr_2\,dt \tag{7.15}$$

and

$$J_2 = \int_0^{r_{1d}}\int_0^{T} \Psi^*(r_1, r_2, t) \times \left(\hat{j}_2 + \frac{\mu_1\cos\theta}{2m_{^{16}O}}\hat{j}_1\right)\Psi(r_1, r_2, t)\,dr_1\,dt. \tag{7.16}$$

In this contribution, we will present some initial results from our calculations using UV-induced dynamics in the $^{18}O^{16}O^{16}O$ molecule.

7.3 RESULTS AND DISCUSSION

7.3.1 Selective Control of IBr Photodissociation

The potential energy curves are portrayed in Figure 7.1, with labels 1 and 2 denoting 1: B (O^+) and 2: $B\left({}^3\Pi_0^+\right)$, which are nonadiabatically coupled excited states with curve crossing at R = 6.08 a_0. These two excited states are optically coupled with the ground state (0) having the transition dipole moment $\mu_{01} = 0.25\mu_{02}$. The results

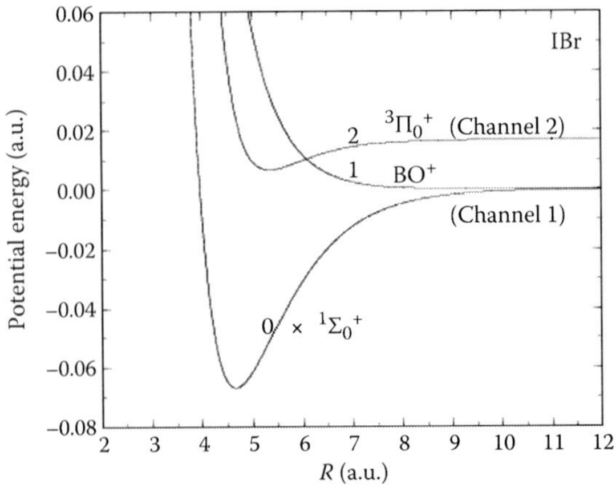

FIGURE 7.1 Ground (0), first excited (1), and second excited (2) potential energy curves of the IBr molecule.

to be presented are for the cw field $\varepsilon(t) = A\sum_{p=0}^{2} \cos(\omega - \omega_{p,0})t$, with details in an earlier review.[14]

Using optimal combinations ψ_1^{max} and ψ_2^{max} of the ground ($\upsilon = 0$), first ($\upsilon = 1$), and second ($\upsilon = 2$) vibrational states (filled symbols), the variation of the maximized flux out of channels 1 (I + Br) and 2 (I + Br*) as a function of the field amplitude at a laser frequency of $\omega = 0.087$ a.u. (19,094 cm^{-1}) is compared with the maximum from any one of the $\upsilon = 0$, $\upsilon = 1$, and $\upsilon = 2$ vibrational eigenstates (open symbols) in Figure 7.2. Since the optimal FOIST combination comprises the $\upsilon = 0$, $\upsilon = 1$, and $\upsilon = 2$ states, the optimized flux is compared with the maximum achievable flux if the total initial population was to be in any one of these and not just yielded from $\upsilon = 0$.

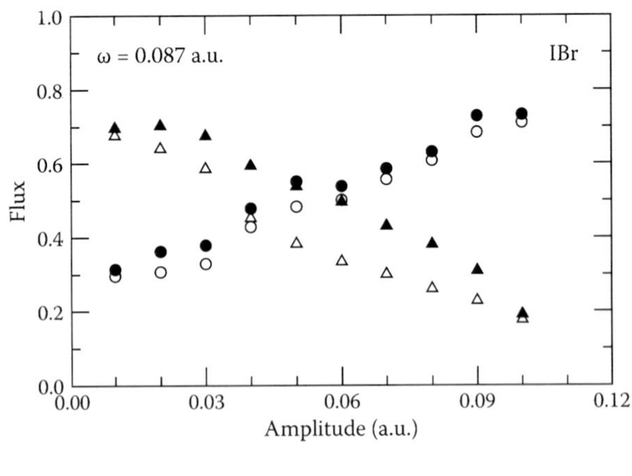

FIGURE 7.2 Flux out of channels 1 (I + Br) and 2 (I + Br*) as a function of the field amplitude for the IBr molecule.

It can be seen from Figure 7.2 that, for a broad range of field strengths, the sum of flux I + Br (represented by circles) and I + Br* (represented by triangle) adds to 1, implying the total dissociation of the molecule. Moreover, the flux out of I + Br* is greater at lower amplitude, and at larger amplitude, the flux is predominantly from the I + Br channel.

For the chosen frequency, almost-total dissociation is attainable for a range of amplitudes, and the desired yield, either I + Br or I + Br*, can be obtained with a simple cw field employed in the FOIST scheme. It is therefore inferred that almost any branching ratio may be achieved using the scheme[14,31,33] presented. The extent of optimization, however, varies with the field amplitude, and the FOIST results do not provide much improvement over the maximum available result from one of the pure vibrational states in the optimization manifold. It should, however, be noted that FOIST alone can provide whichever of $\upsilon = 0$, $\upsilon = 1$, or $\upsilon = 2$ is optimal for the chosen objective.

The selective flux maximization, as shown in Figure 7.2, in this FOIST scheme is achieved by altering the spatial profile of the initial state to be subjected to photolysis pulse, and since changes in flux are due to flowing of probability density, it is useful to examine the attributes of the probability density profiles from the FOISTs. The probability density plots of the first three vibrational states ϕ_0, ϕ_1, and ϕ_2 of the IBr molecule are plotted in Figure 7.3, whereas the probability density profiles from the optimal superpositions ψ_1^{max} and ψ_2^{max}, which maximize flux out of the I + Br and I + Br* channels, respectively, for a field frequency of $\omega = 0.087$ a.u. (19,094 cm^{-1}) and an amplitude of $A = 0.03$ a.u., which is the lowest amplitude for which almost 100% dissociation[31] occurs using a pulse length of 480 fs, followed by further propagation without field for another 387 fs, are depicted in Figure 7.4.

From Figures 7.3 and 7.4, it is clear that the optimal wave function ψ_2^{max}, which peaked at 4.58 a_0, seems to maximize flux out of channel 2 (I + Br*) by localizing the probability density left of the ϕ_0 wave function (which peaked at 4.66 a_0) and ψ_1^{max},

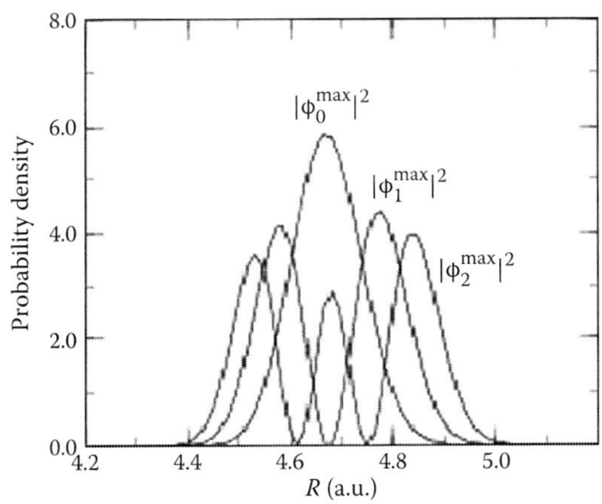

FIGURE 7.3 Probability density profiles for the vibrational eigenstates ϕ_0, ϕ_1, and ϕ_2 of the IBr molecule.

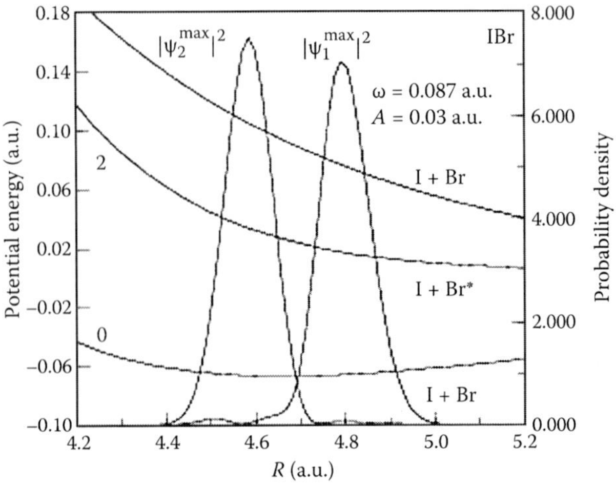

FIGURE 7.4 Probability density profiles for the optimal superpositions ψ_1^{max} and ψ_2^{max} of the IBr molecule.

which peaked at 4.66 a_0, maximizes flux out of channel 1 (I + Br) by localizing the probability density right of the ϕ_0 wave function. The peaking of ψ_2^{max} to the left of ϕ_0 and the peaking of ψ_1^{max} to the right of ϕ_0 are persistent throughout the frequency range specified earlier, but the results presented here are only for a representative frequency. It can also be seen that the probability density profiles for ψ_1^{max} and ψ_2^{max} are mutually exclusive and more compact as compared to those from ϕ_0, ϕ_1, and ϕ_2. Moreover, the probability density profiles from pure eigenstates ϕ_0, ϕ_1, and ϕ_2 (Figure 7.3) subsume the spatial attributes of both ψ_1^{max} and ψ_2^{max} (Figure 7.4), which explains why selective photodissociation cannot ensue from the use of only one of these molecular eigenstates as the initial state. This is achieved in the cases studied by us through the dominant component coming from $\upsilon = 0$ for ψ_2^{max} and ψ_1^{max} and from $\upsilon = 1$ at lower amplitude and $\upsilon = 2$ at higher amplitudes for ψ_1^{max}.

This need for a suitable mixing of vibrational states for selective control of photodissociation is also seen in the optimal control theory–based calculation[104] on IBr. The additional frequency components of the optimal field separated from each other by the IBr ground state vibrational spacings and large expectation value for the internuclear distance on the electronic ground state corresponding to vibrational stretch for highly excited vibrational levels of IBr and extremely intense fields required to achieve this in the very beginning of the control procedure point to the same central role of the initial mixing of vibrational states in achieving selective control. In our FOIST scheme, ψ_1^{max} and ψ_2^{max} represent the premixing of vibrational states required for selective control, with an additional advantage that the photolysis pulse may be chosen beforehand for practical convenience.

The $B\left({}^3\Pi_0^+\right)$ state of the IBr is four times more strongly coupled with the ground state $\mu_{01} = 0.25\ \mu_{02}$, compared to the B($O^+$) state. State B($O^+$) is far off resonance within the frequency band considered here. Therefore, the transfer of amplitude of ψ_1^{max} or ψ_2^{max} to be transported to the excited state will be much larger in the $B\left({}^3\Pi_0^+\right)$ state and will dominate the photodissociation outcome. The broad mechanistic details may therefore be inferred from arguments employing the dynamics ensuing from the evolution of the wave packet on the $B\left({}^3\Pi_0^+\right)$ curve alone. We may therefore deduce from the structural features of $\left|\psi_2^{max}\right|^2$ (Figure 7.4) that, for any given frequency, transition from the initial state represented by ψ_2^{max} will occur in the energetically higher or steeper and more repulsive region of the excited $B\left({}^3\Pi_0^+\right)$ potential energy curve. The excited molecule described by ψ_2^{max} will therefore traverse $B\left({}^3\Pi_0^+\right)$ and B(O^+), crossing with greater velocity, and exit out of the excited I + Br* channel, compared with the molecule represented by ψ_1^{max}, which transfers to a relatively smoother region of the $B\left({}^3\Pi_0^+\right)$ state potential energy curve, thereby facilitating a slower adiabatic exit out of the lower I + Br channel. Selective maximization is effected through the localization of the probability density at internuclear distances, which enable Frank–Condon transitions to the appropriate region of the excited states. Transfer of the wave function to the steeper repulsive part of the excited potential energy curves favor high-velocity diabatic exit into the higher channel. Localization away from the repulsive wall favors slow adiabatic exit into the lower channel. This interpretation of the control mechanism utilizing ψ_1^{max} and ψ_2^{max} as the initial states is consistent with the analysis of the frequency dependence of IBr photodissociation as a function of the molecular radial velocity using the Landau–Zener theory presented by De Vries et al.,[105] where the increase in photodissociation yield out of the I + Br* channel with the increase in frequency is well correlated with an increase in radial velocity at the crossing point. From Figure 7.4, it is obvious that $\left|\psi_1^{max}\right|^2$ and $\left|\psi_2^{max}\right|^2$ are localized in a mutually exclusive manner and that the optimization leads to significant alteration of the spatial profiles to excite the molecule to the region most suited for directing flux out of the desired channel.

The mechanism of FOIST-based selective control of IBr photo dissociation has been further probed using ψ_1^{max} and ψ_2^{max} in the TDWP calculation of the IBr absorption spectrum (Figure 7.5) and the branching ratio (Figure 7.6). The absorption cross section obtained using ψ_2^{max} as the initial condition is smooth, with only a negligible interference pattern around 600 nm. The absorption spectrum from ψ_1^{max} as the initial condition contains a series of sharp peaks that are characteristic of the predissociation dynamics[106] in the higher wavelength region, is dominated by predissociation in the vicinity of the energy values around the $B\left({}^3\Pi_0^+\right)$ to B(O^+) crossing, and leads to the complicated interference pattern seen in the ψ_1^{max} absorption profile. In contrast, the smoothness of the absorption spectrum from ψ_2^{max} as the initial condition stems from the initial placement of the ψ_2^{max} wave packet on the steeper part of the excited curve, which facilitates faster diabatic exit with little time for interference. The absorption spectrum from ϕ_0 (Figure 7.5) as the initial condition has also been plotted and compares well with other calculated and experimental[107,108] absorption spectra for IBr. The absorption spectra peak at the wavelengths corresponding to the vertical Frank–Condon transition energies, with ψ_1^{max}, ψ_2^{max}, and ϕ_0 as the initial states.

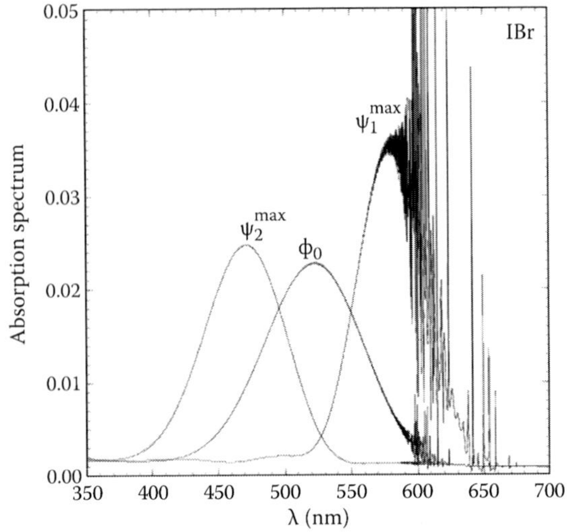

FIGURE 7.5 Total absorption spectrum for IBr with ϕ_0, ψ_1^{max}, and ψ_2^{max} as the initial conditions.

The branching ratio $\Gamma(Br^*/Br)$ with ϕ_0, ψ_1^{max}, and ψ_2^{max} as the initial conditions are plotted in Figure 7.6. At all energy values, $\Gamma(Br^*/Br)$ is much larger in magnitude with ψ_2^{max} as the initial condition, compared to that with ϕ_0 or ψ_1^{max} as the initial condition, and $\Gamma(Br^*/Br)$ is uniformly smaller with ψ_1^{max} as the initial condition, compared to that with ψ_2^{max} or ϕ_0 as the initial condition. This does seem to suggest that

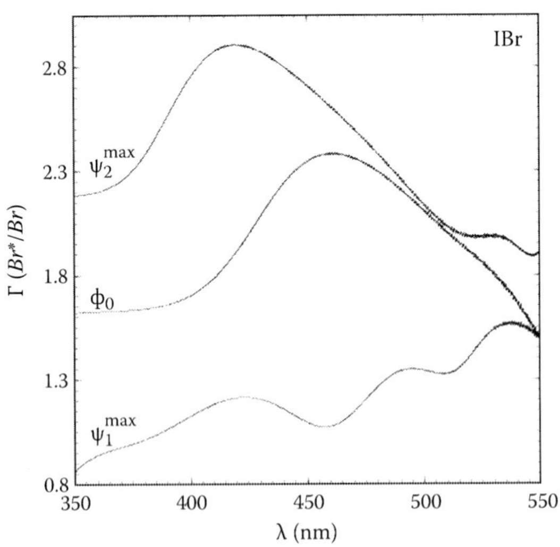

FIGURE 7.6 Branching ratio $\Gamma(Br^*/Br)$ with ϕ_0, ψ_1^{max}, and ψ_2^{max} as the initial conditions.

the preparation of the initial state for a suitable photolysis pulse using our FOIST scheme can provide selective control by constructing an appropriate linear combination of vibrational states.

7.3.2 Selective Control of HI Photodissociation

The potential energy curves (Figure 7.7a) and nonadiabatic optical coupling elements are taken from earlier investigations[22,29] of HI photodissociation. The probability density profiles employed in optimization manifold are displayed in Figure 7.7b. The $^1\Sigma_0$ ground state (labeled as 0) is optically coupled with the four excited states $^3\Pi_1$, $^1\Pi_1$, $^3\Pi_0$, and $^3\Sigma_1$ (which are labeled 1, 2, 3, and 4, respectively). Furthermore, states $^3\Pi_1(1)$ – $^1\Pi_1(2)$, $^3\Pi_1(1)$ – $^3\Sigma_1(4)$, and $^1\Pi_1(2)$ – $^3\Sigma_1(4)$ are nonadiabatically mutually coupled, whereas state $^3\Pi_0(3)$ has no coupling with any of the states. The crossing between states 2 and 4, which is the outermost, will control the final flux redistribution occurring at $R = 3.83\ a_0$. The altered spatial profile for ψ_1^{max} and ψ_2^{max} from our investigation spanning the frequency range of 0.16–0.26 a.u. (36,000–50,000 cm^{-1}) are similar, and we offer a representative probability density plot of $\left|\psi_2^{max}\right|^2$ for $\omega = 0.20$ a.u. (44,000 cm^{-1}) and $A = 0.01$ a.u. in Figure 7.7c, where ψ_1^{max} and ψ_2^{max} once again represent the FOISTs that maximize flux out of the lower H + I and the higher H + I* channels, respectively. The probability density of the ground vibrational level peaks at 3.80 a_0, whereas ψ_1^{max} and ψ_2^{max} peak at 3.29 a_0 and 3.01 a_0, respectively; as earlier discussed for the IBr molecule, ψ_2^{max} peaked once again to the left of the ϕ_0 peak, and ψ_1^{max} peaked to the right.

In Figure 7.8a and b, we present a plot of the variation in the individual norm $\langle\chi_i(t)|\chi_i(t)\rangle$ on each excited state and also in the total norm on the H + I/H + I* channels resulting from the use of ψ_1^{max} and ψ_2^{max} as initial states in the TDWP calculation. As seen in Figure 7.8a, with ψ_1^{max} as the initial state, there is a net transfer of amplitude from the H + I* channels (states 1 and 2) to the H + I channel (state 4). $^3\Pi_0(3)$ has adiabatic coupling to any other state and therefore experiences no change in the norm transferred to it. There is a steady depletion from both states 1 and 4 into state 2 initially. Within 5 fs, the norm on all these repulsive curves for this extremely light ($\mu \approx m_H$) system is well past the outermost crossing (Figure 7.8c). Unlike in the case of the heavy and slow-moving IBr, the norms stabilize much more quickly, and the system covers an average distance of 12.5 a_0 within 25 fs. Results from the TDWP analysis with ψ_2^{max} as the initial condition are plotted in Figure 7.8b, where a fast depletion from state 2 into both state 4 (H + I*) and state 1 (H + I) is clearly seen. The buildup in the H + I* channel is therefore entirely due to the depletion from the more repulsive $^1\Pi_1$ state, and selectivity will be assisted by tuning the laser to the frequency corresponding to the Frank–Condon transitions to the $^1\Pi_1$ state. The molecule, when described by ψ_2^{max}, has to travel a greater distance but still traverses the outer crossing between states 2 and 4 faster than when described by ψ_1^{max}. However, due to the low and reduced mass of HI and, consequently, very fast motion, which is a considerable overlap between the spatial profiles of ψ_1^{max} and ψ_2^{max} for HI, the net change in the $\langle r\rangle_t$ at 25 fs is only marginally higher with ψ_2^{max} as the initial state. At

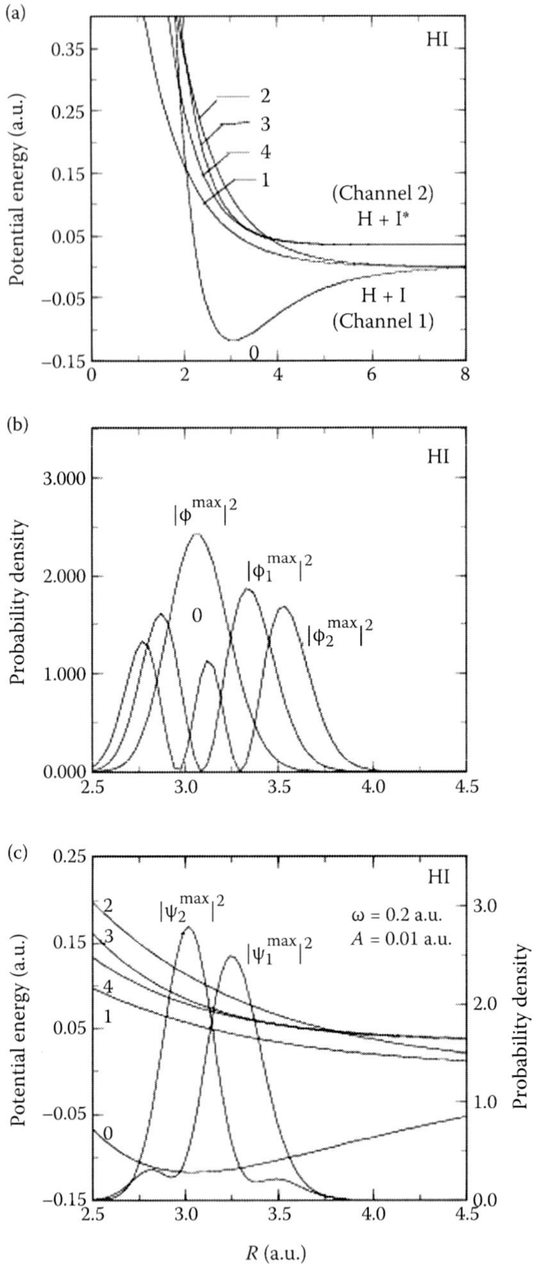

FIGURE 7.7 (a) Potential energy curve for the electronic states 0:$^1\Sigma_0$, 1: $^3\Pi_1$, 2: $^1\Pi_1$, 3: $^3\Pi_0^+$, and 4: $^3\Sigma_1$ of HI. (b) Probability density plots for the ϕ_0, ϕ_1, and ϕ_2 ($\upsilon = 0, 1, 2$) vibrational eigenstates of the HI electronic ground state. (c) Probability density plots for the optimized superpositions ψ_1^{max} and ψ_2^{max} for flux maximization through the H + I and H + I* channels for the field $\varepsilon(t) = A\sum_{p=0}^{2} \cos(\omega - \omega_{p,0})t$ with $A = 0.01$ a.u. and $\omega = 0.02$ a.u. for HI.

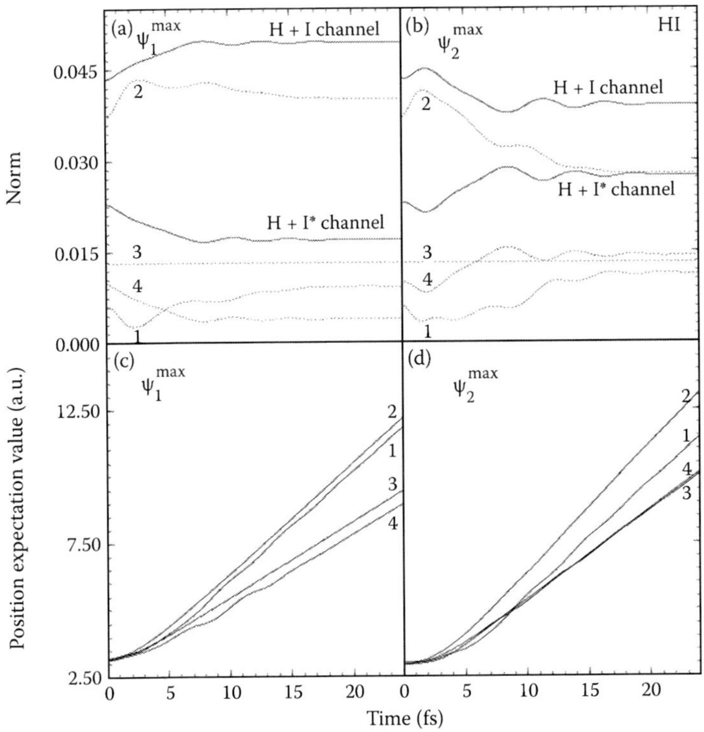

FIGURE 7.8 Norm evolution on the different potential energy curves of HI with (a) ψ_1^{max} and (b) ψ_2^{max} as the initial conditions and position expectation values $\langle r \rangle_t$ with (c) ψ_1^{max} and (d) ψ_2^{max} as the initial conditions.

around 25 fs, the norms are well stabilized, and as seen from the position expectation value plots of Figure 7.8c and d, the wave function is indeed well past the outer crossing. The autocorrelation functions $\sum_i \langle \chi_i(\tau) | \chi_i(\tau + t) \rangle$, with both ψ_1^{max} and ψ_2^{max} as the initial conditions, have therefore been calculated with $\tau = 25$ fs and are displayed in Figure 7.9.

In all cases of HI, since all the four excited curves are repulsive, the possibility for recurrences is negligibly small, and the autocorrelation plots in Figure 7.9 fall to zero much faster due to a quick dephasing of the wave packet in the coordinate space. Figure 7.10 contains the total absorption cross sections obtained from the full Fourier transform of the autocorrelation function, where the absorption spectrum with ϕ_0 as the initial state is once again in excellent agreement with the absorption profile for HI.[29] Figure 7.11 portrays the branching ratio $\Gamma(I^*/I)$ with ϕ_0, ψ_1^{max}, and ψ_2^{max} as the initial conditions, and the $\Gamma(I^*/I)$ ratio with ψ_2^{max} as the initial condition is much larger than that obtained with ϕ_0 and ψ_1^{max} as the initial conditions. Hence, the linear combination of vibrational eigenfunctions leading to ψ_1^{max} favors the formation of H + I products, whereas the linear combination ψ_2^{max} favors the H + I* expected by IBr analysis.

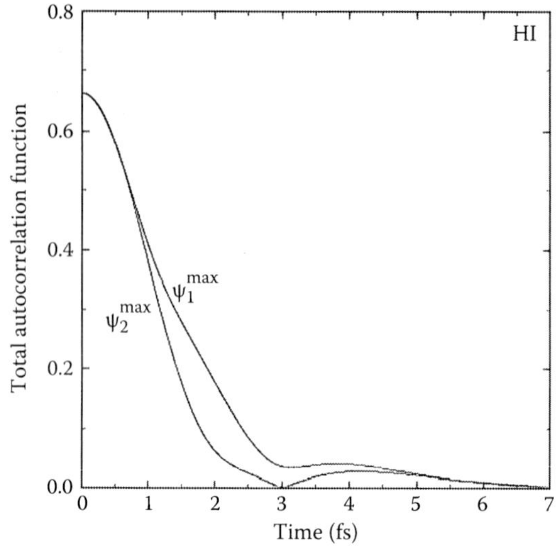

FIGURE 7.9 Total autocorrelation functions for HI with ψ_1^{max} and ψ_2^{max} as the initial conditions.

The use of FOISTs, even for this more complicated system, leads to requisite alterations in the spatial profiles, so that they peak at the internuclear distances required to facilitate Frank–Condon transitions to the appropriate portion of the excited state potential energy curves, enhancing photodissociation out of the desired channel.

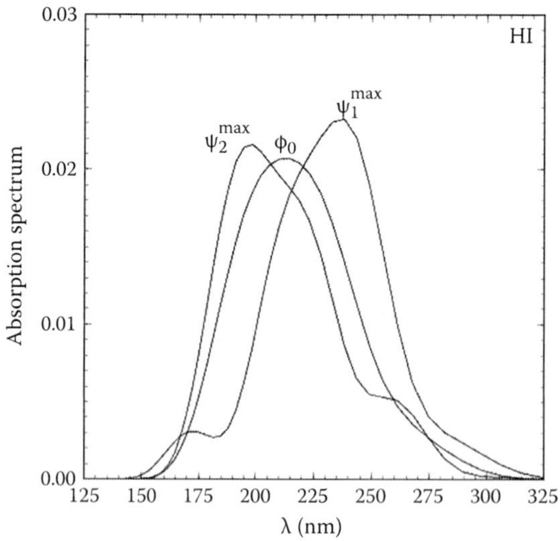

FIGURE 7.10 Total absorption spectrum for HI with ϕ_0, ψ_1^{max}, and ψ_2^{max} as the initial conditions.

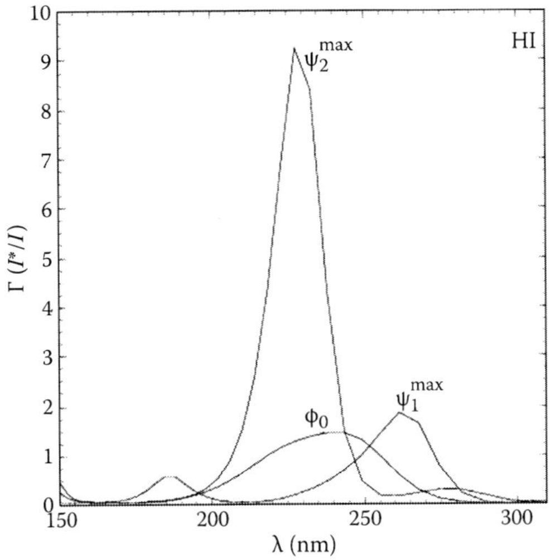

FIGURE 7.11 Branching ratio $\Gamma(I^*/I)$ with ϕ_0, ψ_1^{max}, and ψ_2^{max} as the initial conditions.

7.3.3 Selective Control of HOD Photodissociation

Following the more recent work on HOD,[46,53] we too have utilized a Gaussian UV pulse for photolysis, except that the temporal width of our pulse is larger (FWHM = 50 fs) to permit easy separation of frequencies involved in photodynamics. The simple Gaussian pulse used by us and its power spectrum are depicted in Figure 7.12. The frequency dependence of flux of the dissociative channels H + O–D (J_{H+O-D}) and H–O + D (J_{H+O-D}) of the HOD molecule using the field profile of Figure 7.12 and the frequency range covering the first absorption band for photolysis using |0,0>, |0,1>, and |0,2> (where the first integer corresponds to the quantum of excitation in the O–H mode and the second integer corresponds to the quantum of excitation in the O–D mode) as initial states are presented in Figure 7.13.

It can be seen from Figure 7.13 that, with the ground vibrational state |0,0> as the initial state, due to the lower mass of the H atom and the consequent ease for large amplitude vibrations in O–H vis-à-vis O–D, O–H dissociation dominates over O–D dissociation, as expected[36–38,40] for the entire range of frequencies. However, unlike in the previous studies,[42–45,53] a large variety in the H + O–D/H–O + D product yield may be achieved without having to provide additional quanta of excitation in the O–H bond.

The |0,1> vibrational state with one quantum of excitation in the O–D mode is 2727 cm^{-1} higher than the |0,0> level and comes in resonance with the excited surface at lower UV frequencies. Hence, the dissociation in both the O–H and O–D modes picks up at frequencies lower than that for the |0,0> level. The |0,1> state has larger O–D stretch, with the probability density peaking in the H–O + D channel. The natural preference for H + O–D dissociation therefore is reversed and the H–O + D flux is much more than the H + O–D flux in a broad range of frequencies. At higher

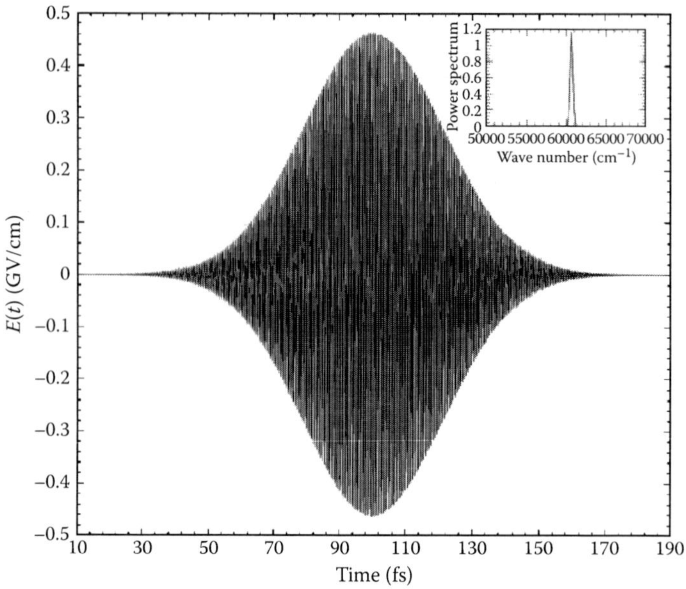

FIGURE 7.12 UV laser pulse $E(t) = 0.09a(t)\cos \omega t$, where $a(t) = \exp[-\gamma(t - t_{uv})^2]$, with $\text{FWHM} = \sqrt{4\ln 2/\gamma} = 50\text{fs}$, $t_{uv} = 100\text{fs}$, and $\omega = 60{,}777$ cm^{-1}. The maximum field amplitude is 0.46 GV/cm, and the maximum field intensity is 178 TW/cm^2. The corresponding power spectrum is shown in the inset.

frequencies, the |0,1> state begins to go off-resonance with the excited surface, and the H + O–D flux predominates. Finally, at very high frequencies, the |0,1> level is completely off-resonance vis-à-vis the repulsive excited surface, and both the H + O–D and the H–O + D flux values drop down to negligible levels.[4,57,58]

Similarly, with |0,2> as the initial state with two quanta of vibrational excitations in the O–D mode, the O–D bond is stretched much more, and as a result, the H–O + D flux values increase to a maximum of about 83%. The |0,2> level being 5369 cm^{-1} above |0,0> comes in resonance in the repulsive excited state at lower frequencies. The dissociation begins at frequencies lower than even those for |0,1>, and with much more pronounced bias in the probability distributions favoring the H–O + D channel, high dissociation in H–O + D is understandable.[4,57,58]

To summarize, as shown in Figure 7.13, |0,0> favors cleaving of the O–H bond and gives the maximum H–O + D flux of about 33% at 60,777 cm^{-1} and the maximum H + O–D flux of 81.9% at 67,169 cm^{-1}. |0,1> gives the maximum H–O + D flux of about 62.5% at 59,703 cm^{-1}, and for |0,2>, it is the 54,372 cm^{-1} pulse that gives the maximum H–O + D flux of 82.8%. These values are collected in Table 7.1.

To understand the mechanistic features that may assist in maximal selectivity and yield, we have also examined the time evolution of $|\Psi_g(r_1, r_2, t)|^2$ and $|\Psi_e(r_1, r_2, t)|^2$ on the ground and excited PESs. Results from the time evolution of |0,0> on the ground and the first excited electronic states, under the influence of a UV field frequency of

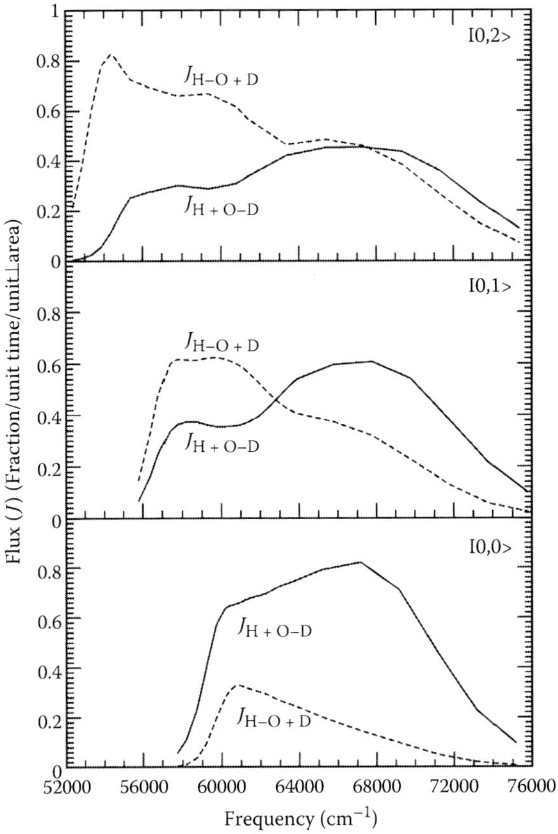

FIGURE 7.13 Plot of flux versus frequency for different initial states.

67,169 cm^{-1} and with the profile of Figure 7.12, are presented in Figures 7.14 and 7.15, along with the corresponding flux in both channels. In Figures 7.16 and 7.17, we have provided the time-evolution plots and flux profiles for the |0,1> and |0,2> states, with 59,703 and 54,372 cm^{-1} as the carrier frequencies, respectively. All these frequencies were chosen for having provided the maximum H + O–D or H–O + D flux for the corresponding initial states as noted previously.[4]

As mentioned above, Figure 7.15 shows the time evolution of the probability density of |0,0> on both the ground and the excited surfaces. In the ground electronic state, the wave function profile of |0,0> changes considerably in its spatial extent, which influences the transfer to different regions of the excited surface. As can be seen from the contour plots of the probability density amplitude on the excited surface, there is considerable variation in the spatial displacement pattern as the field gains strength at 75 fs, and corner cutting by the probability density in the H + O–D valley of the excited surface, leading to a highly oscillating structure in the H–O + D valley, is a clear pointer to the field-induced changes that may be tapped for selective control. The field-induced displacement of the wave function in the ground state

TABLE 7.1
Flux Obtained Using a Single Initial State and Laser Pulse $E(t) = 0.09a(t)\cos \omega t$, Where $a(t) = \exp[-\gamma(t - t_{uv})^2]$, with $\text{FWHM} = \sqrt{4\ln 2/\gamma} = 50$ fs and $t_{uv} = 100$ fs

Initial State	Frequency ω(cm⁻¹)	H + O–D Flux (%)	H–O + D Flux (%)
|0,0>	60,777	65.6	33.0
|0,0>	67,169	81.9	14.5
|0,1>	59,703	35.4	62.5
|0,2>	54,372	12.2	82.8

Note: The maximum field amplitude is 0.46 GV/cm, and the maximum field intensity is 178 TW/cm^2.

pushes the probability density flow in the excited state into the H + O–D channel with a marked enhancement at 75 and 150 fs. At 75 fs, the field starts impacting the molecule in a more pronounced manner, whereas, at 150 fs, the field effect ebbs, and the amplitude remaining in the excited state flows through the H + O–D channel with a sharp pickup in the H + O–D flux as seen from the flux-versus-time plot of Figure 7.14. The cross talk between the two surfaces seems to have an important preparatory role, and the beat structure (7072 cm^{-1}) of the population transfer oscillations (inset of Figure 7.14b) closely equals the vibrational frequency of the |0,2> state (7250 cm^{-1}). As a result, the HOD molecule in resonance with this beat frequency may be mimicking the O–H bond oscillations with two quanta of excitation, inducing a more favorable dissociation of the O–H bond, as observed.[4,59]

A few snapshots from the time evolution of the |0,1> state at $t = 50$, 75, 100, 125, and 150 fs on the ground and the repulsive excited surfaces are presented in Figure 7.16a and b, respectively. As can be seen from the time-evolution plots on the ground surface (Figure 7.16a), the nodal topology of the |0,1> state at 50 fs undergoes considerable distortions in the 75–150-fs plots, and these signal an active manipulation of the spatial attributes of the |0,1> probability density profile through field-induced mixing with other vibrational states, most probably by dumping from the excited electronic surface to different vibrational levels of the ground surface. This premise is buttressed by the population and flux plots of Figure 7.16c, where synchronized population transfer between the ground and the excited states is clearly seen. We surmise that this change in the probability density profile on the ground surface leads to Franck–Condon transitions to different regions of the excited surface, and the initial bias of greater amplitude in the H + O–D channel at 50 fs (Figure 7.16b and c) is altered in favor of the greater flux in the H–O + D channel at 75, 100, and 125 fs (Figure 7.16c) until the amplitude values are negligibly small, a surge in the H + O–D channel at 150 fs has little effect on the overall bias of much greater flux in the H–O + D channel, and the final flux in the H–O + D channel (Figure 7.16c) is approximately twice as large as that in the H + O–D channel. Furthermore, the

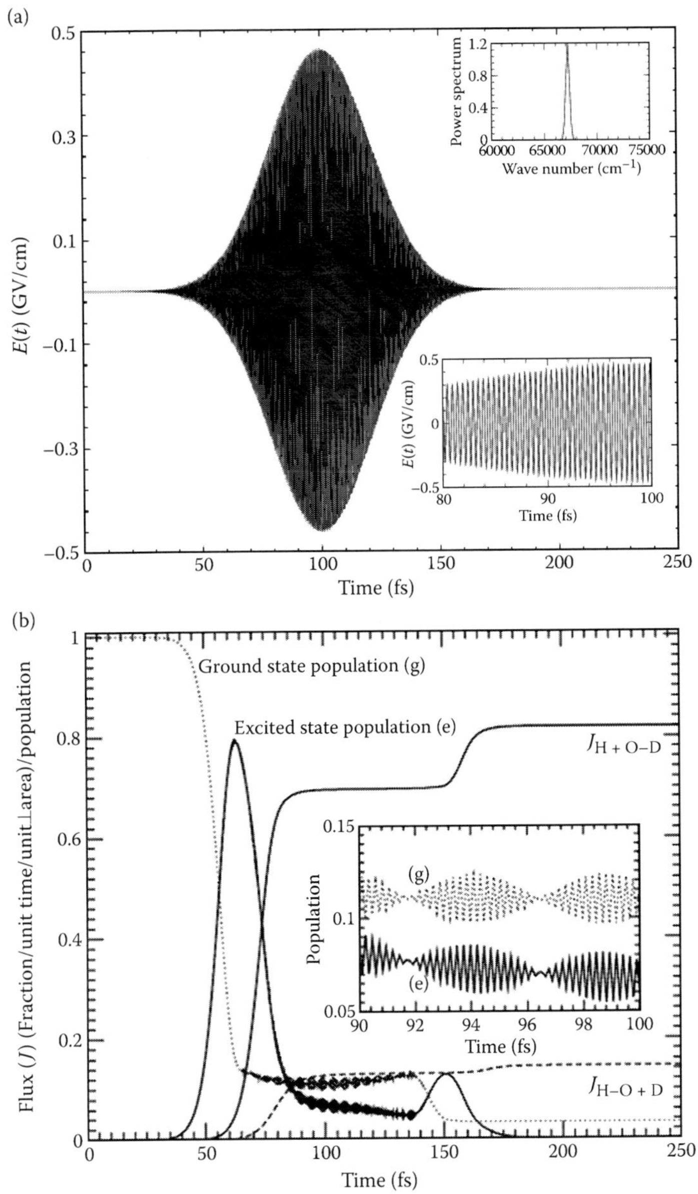

FIGURE 7.14 (a) Single-color UV laser pulse with the same attributes as that shown in Figure 7.12, except with frequency $\omega = 67{,}169\ \text{cm}^{-1}$. The power spectrum and greater resolution are presented in the insets. (b) Ground and excited state populations and accumulated H + O–D and H–O + D flux from |0,0> as the initial state under the influence of the field shown in (a).

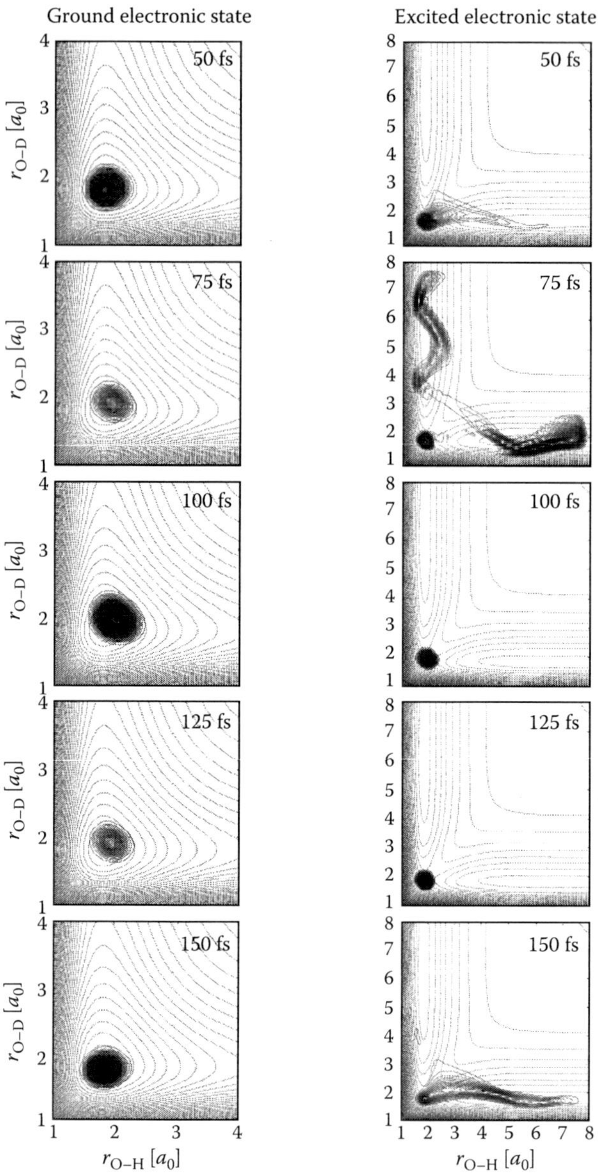

FIGURE 7.15 Time evolution of the |0,0> state on the ground and excited surfaces at $t = 50$, 75, 100, 125, and 150 fs for the field profile shown in Figure 7.14a.

dynamic changes in the lobal topology of the |0,1> state in Figure 7.16a show that appealing insights based on a fixed static Franck–Condon window may not always be correct.[4,57]

Some snapshots from the time evolution of the |0,2> state at 50, 75, 100, 125, and 150 fs on the ground and excited surfaces are given in Figure 7.17a and b,

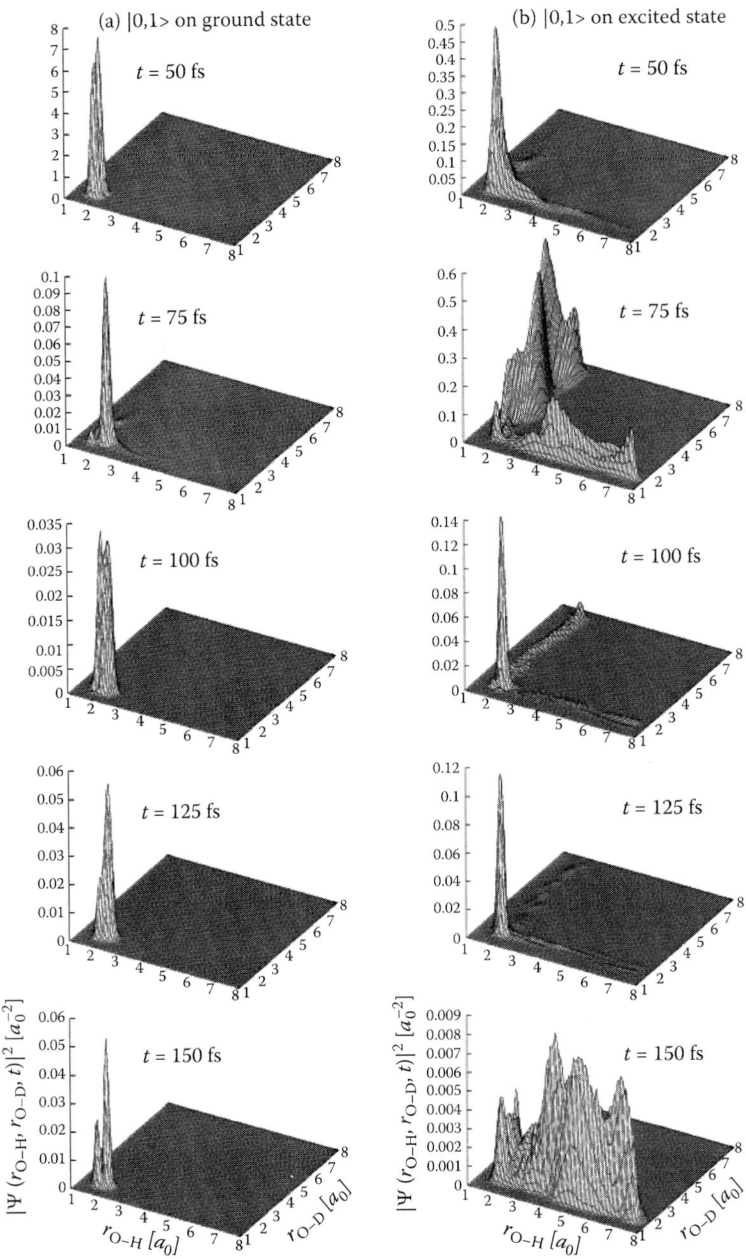

FIGURE 7.16 Time evolution of the |0,1> state on the (a) ground and (b) excited electronic states.

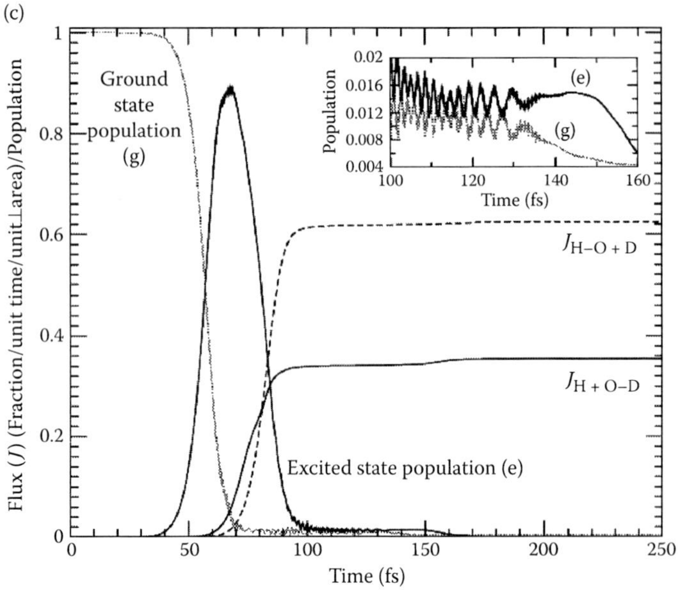

FIGURE 7.16 (Continued) (c) ground and excited state populations and the accumulated H + O–D and H–O + D flux under the influence of the Gaussian UV pulse of Figure 7.12, except with carrier frequency $\omega = 59{,}703\ \text{cm}^{-1}$.

respectively. It can be seen from the time-evolution plots on the ground and excited surfaces (Figure 7.17a) that the nodal topology characteristic of the |0,2> state at 50 fs undergoes some distortion in the 75–150-fs plots, but these are not as large as that for the |0,1> state. The spatial profiles on the excited state in Figure 7.17b have much greater diversity and a marked flow in the H–O + D channel. The more dominant theme as seen from the population and flux profiles of Figure 7.17c is the nearly similar magnitude of the amplitude on the ground and excited surfaces once major depletion has taken place between the 50–75-fs interval. As can be seen from Figure 7.17c, there is near similarity of the total amplitude on the ground and excited surfaces, from 75 fs onward with continuous exchange of field-mediated probability flow between the two surfaces. Also, due to the greater spatial bias of the |0,2> probability distribution toward the H–O + D channel, the H–O + D flux dominates the H + O–D flux from the very beginning, and as can be seen in Figure 7.17c, there is a surge of population in the excited state once the field is cut off at 150 fs. In the absence of field-induced dumping, the depletion from the excited state stops around 150 fs, and in Figure 7.17b and c, we do see a marked pickup in the H + O–D and H–O + D fluxes at the time.[4,57]

We have also sampled H + O–D and H–O + D fluxes for a few different combinations of colors/initial states, and some prominent results[4,57,58] are collected in Table 7.2. Tracking of the time evolution of even a single vibrational state on both surfaces is extremely demanding of computational resources, and a very comprehensive

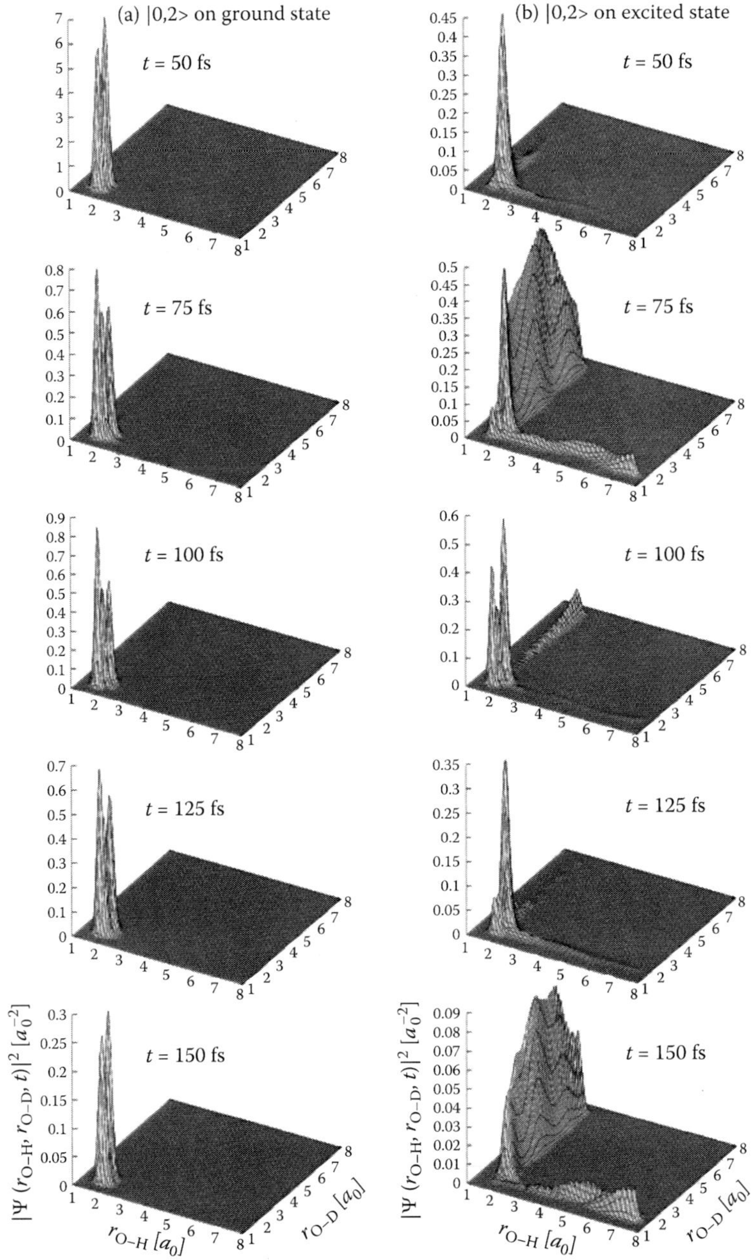

FIGURE 7.17 Time evolution of |0,2> on the (a) ground and (b) first excited states.

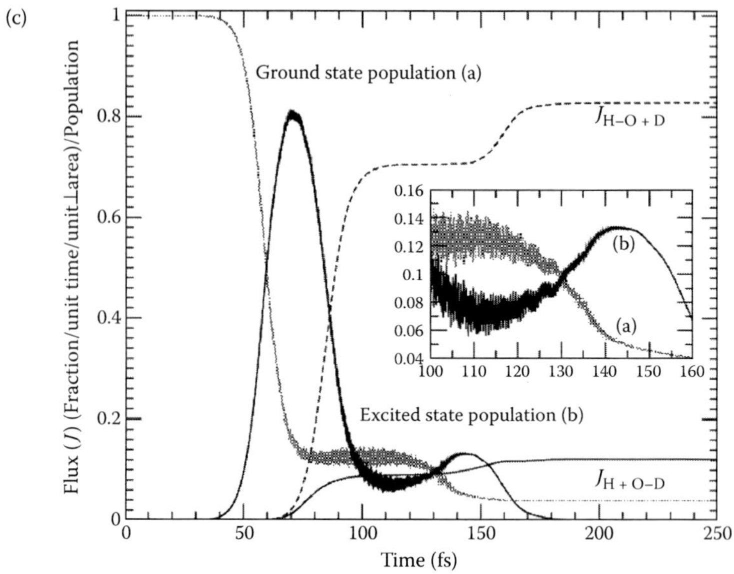

FIGURE 7.17 (Continued) (c) ground and excited state populations and the accumulated H + O–D and H–O + D flux from |0,2> as the initial state under the influence of the field with the same attributes as in Figure 7.12, except with a carrier frequency of $\omega = 54{,}372$ cm^{-1}.

investigation of FOIST-based mixing of many vibrational states has, therefore, not been attempted in our investigation. For the frequencies sampled here, mixing of |0,0> + |0,1> using a single-color photolysis pulse (rows 1 and 2 of Table 7.2) does not offer any improvement over those obtained using only |0,1> as the initial state. The use of a two-color photolysis pulse in combination with FOIST-based mixing of |0,0> and |0,1> (rows 3 and 4 of Table 7.2) however does offer 6%–8% more output in the H–O + D channel than that achieved without FOIST.[4]

We have shown[14,34] that mixing of states may be replaced by mixing of colors, and we have therefore extended our approach to examine if two-color lasers may be used for preferential dissociation of the O–D bond. Using |0,0> as the initial state and two lasers with frequencies of 54,920 and 52,193 cm^{-1}, where the frequency difference between these two lasers corresponds to the energy gap between the |0,0> and |0,1> vibrational levels, the H–O + D flux is much larger than the H + O–D flux. It is therefore a welcome surprise to report that the mixing of states may be supplemented by mixing of colors (Table 7.2, rows 5 through 7), and for the two laser setups, the H–O + D flux is approximately thrice (Table 7.2, row 7) the H + O–D flux, even with the |0,0> as the initial state.[4]

The two-color field employed here for selective dissociation of the O–D bond and its power spectrum is shown in Figure 7.18a. The beat structure of the resulting pulse displays a frequency of 2752 cm^{-1}, which resonates with the vibrational frequency of the |0,1> state with one quantum of excitation in the O–D mode (2727 cm^{-1}) and is nearly half of the vibrational frequency for the |0,2> state (5369 cm^{-1}). Moreover, the combinational UV frequency of this 50-fs two-color pulse is ~55 160 cm^{-1} (lower

TABLE 7.2
Flux Obtained Using Combinations of Color(s)/Initial State(s) Using Laser Pulse $\varepsilon(t) = 0.09 * a(t)\sum_{i=1}^{2}\cos\left(\omega_{UV}^{i}\right)t$, Where $a(t)$ Is the Same as That Described in Table 7.1

Initial State(s)	Frequencies (cm^{-1})	H + O–D Flux (%)	H–O + D Flux (%)
|0,0> + |0,1>	59,703	35.5	62.4
|0,0> + |0,1>	60,777	35.9	59.9
|0,0> + |0,1>	60,777 and 59,703	31.8	66.2
|0,0> + |0,1>	60,777 and 59,203	30.3	67.9
|0,0>	54,920 and 52,193	29.0	44.3
|0,0>	54,920 and 52,203	26.4	47.2
|0,0>	54,920 and 52,303	18.5	52.2

Note: ω_{UV}^{1} and ω_{UV}^{2} are the carrier frequencies that provide maximum flux from the |0,0> and |0,1> states selected from a large sampling of individual and two-color combinations.

inset of Figure 7.18a), which is very close to the resonance frequency for population transfer from the |0,2> state on the ground surface to the excited electronic state. From this choice of the two-color laser pulse, we would therefore expect the dissociation pattern for the H–O–D characteristic of H–O–D with excited vibration in the O–D bond, which has been shown to favor O–D dissociation,[59] and as can be seen from the flux-versus-time plot of Figure 7.18b, the final flux in the H–O + D channel is indeed thrice as large as that in the H + O–D channel. The kinematic bias favoring the dissociation of the O–H bond can therefore be reversed from the ground vibrational state of the ground electronic state of HOD with a suitably chosen two-color UV pulse.[4,59]

To obtain mechanistic insight, some snapshots of the time evolution from the |0,0> vibrational state of the ground electronic state on both the ground and the excited surfaces at 50, 75, 100, 125, and 150 fs for the two-color laser pulse discussed above are provided in Figure 7.19. The cross talk characterizing the population transfer between the ground and the excited surfaces is quite intense (Figures 7.18b and 7.19). The initial |0,0> probability density profile on the ground electronic state (Figure 7.19) undergoes considerable distortions in the 100–150-fs interval, and the nodal topology of the |0,0> state begins to change from 100 fs onward. The change of this nodal topology indicates an active manipulation by the two-color laser field employed here, and with two nodes clearly visible at 125 fs in Figure 7.19, it seems that the excited local O–D modes are being prepared by the two-color laser field since, at 125 fs, the probability density in the ground state is like that of the |0,2> state.[59]

The field-induced synchronized population transfer between the ground and the excited surfaces seen in Figure 7.18b, and its correlation with the pulse profile is

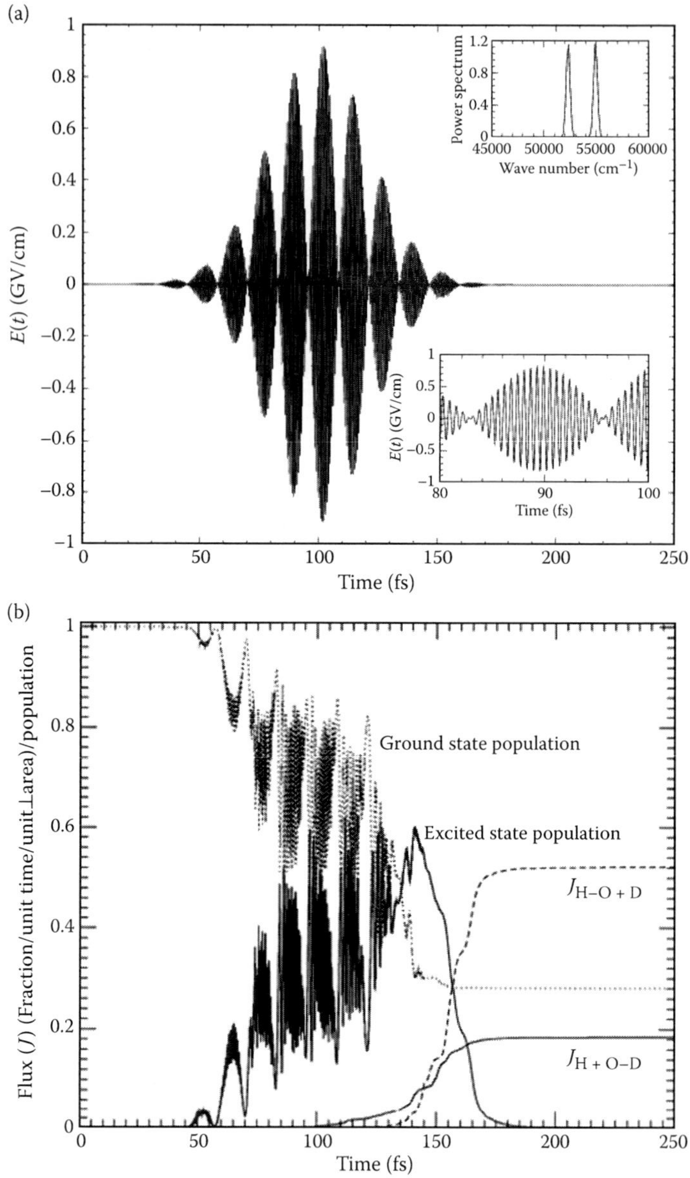

FIGURE 7.18 (a) Two-color laser field $E(t) = 0.09a(t)(\cos \omega_1 t + \cos \omega_2 t)$, with $\omega_1 = 54{,}920$ cm^{-1} and $\omega_2 = 52{,}303$ cm^{-1} with the same attributes for $a(t)$ as in Figures 7.12 and 7.14a. The power spectrum and higher resolution are presented in the insets. (b) Ground and excited state populations and accumulated H + O–D and H–O + D flux from |0,0> as the initial state under the influence of the two-color laser pulse shown in (a).

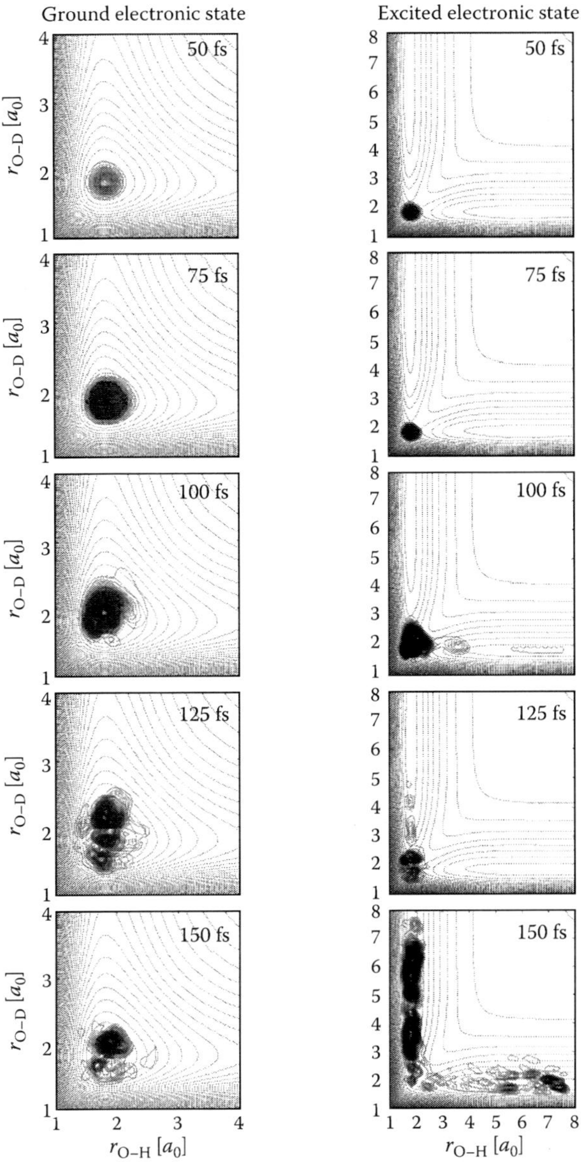

FIGURE 7.19 Time evolution of the |0,0> state on the ground and excited surfaces at t = 50, 75, 100, 125, and 150 fs with the field profile shown in Figure 7.18a.

similar to that discussed for Figure 7.14b earlier. The time-evolution plots on the excited surface at 50–75 fs (Figure 7.19) show a simple transfer of probability amplitudes from the ground surface. The flow of probability density into competing channels starts well after the field has peaked from 100 fs, with the buildup in the H–O + D channel starting only after 140 fs, when the field intensity is low enough to interrupt the pronounced population exchange and the flushing of populations can proceed with a small kick in H–O + D flux at 155 fs as expected. Initially (100 fs), the flow of amplitude is toward the H + O–D channel, which is altered from 125 fs onward, and this alteration in the flow of probability amplitude continues until the end of the dissociation process. The final flux in the H–O + D channel (~52%) is approximately three times that of the flux in the H + O–D channel (~18%).

As mentioned earlier, there is a kinematic bias favoring preferential dissociation of the O–H bond, and successful selective dissociation of this bond has been the original precursor for the prototypical status of HOD in various approaches to selective control.[36–46,48,50–54,56–59] A fully quantal two-surface examination of this established route to selective control, where an IR pulse is used for selective vibrational churn in the desired bond before HOD is exposed to an appropriate UV pulse for transfer to the repulsive excited surface, is therefore an obvious and desirable extension of our mechanistic examination.[60] A careful analysis of the microdynamical details arising from the exposure of HOD to IR pulses, which produce localized O–H excitations, and the role of the photolysing UV pulse frequency in selective dissociation of this bond, is detailed in Figures 7.20 through 7.22.

The IR and UV pulses with their power spectra are depicted in Figure 7.20a, and the resulting vibrational mix is shown in Figure 7.20b. As can be seen in Figure 7.20b, there is a stable combination of pure O–H modes with maximum population in the |3,0> mode. This combination has sufficient prior stretch in the O–H bond, and results from subjecting this combination to UV pulses with different frequencies are displayed in Figures 7.21 and 7.22 and collected in Table 7.3. The flux values of Table 7.3 isolate the UV pulse with a frequency of 51,090 cm^{-1} to be most effective, which incidentally will provide sufficient energy to place the HOD molecule with 3 quanta of excitation in the O–H mode—the dominant component of the linear combination resulting from the IR churn—just above the saddle point of the upper repulsive surface. Details of the population transfer dynamics and dissociative flux in the H + O–D and H–O + D channels are shown in Figures 7.21 and 7.22. The results in Figure 7.21 show that, as soon as the UV pulse begins to build sufficient power around 300 fs, there is a quick transfer of population to the excited surface, which reaches a maximum by 325 fs. With the excited upper surface being repulsive, the population in the excited state begins to flow almost simultaneously, and due to the prior stretching of the O–H bond on the ground surface, flux builds up in the H + O–D channel, with 72% dissociation of the O–H bond being achieved within the next 25 fs. The final flux of 82% in the H + O–D channel is achieved once the UV field switches off at 420 fs, shutting off the cross talk between the ground and the excited surfaces.[60] There is no further dumping from the excited surface, and this allows flushing of population on the excited surface into the competing channels, as seen from a kick in both the H + O–D and H–O + D flux values around 400 fs.

The localization of the vibrational excitation in the O–H bond is clearly seen in the 75–175-fs probability density plots of Figure 7.22, where the gradual transformation of the |0,0> ground vibrational state into the linear combination of pure O–H modes is easily seen. This linear combination resulting from the vibrational churn retains its shape until the UV field comes into play with sufficient power at 300 fs, and as can be seen in the probability density plots on the excited surface, the role of

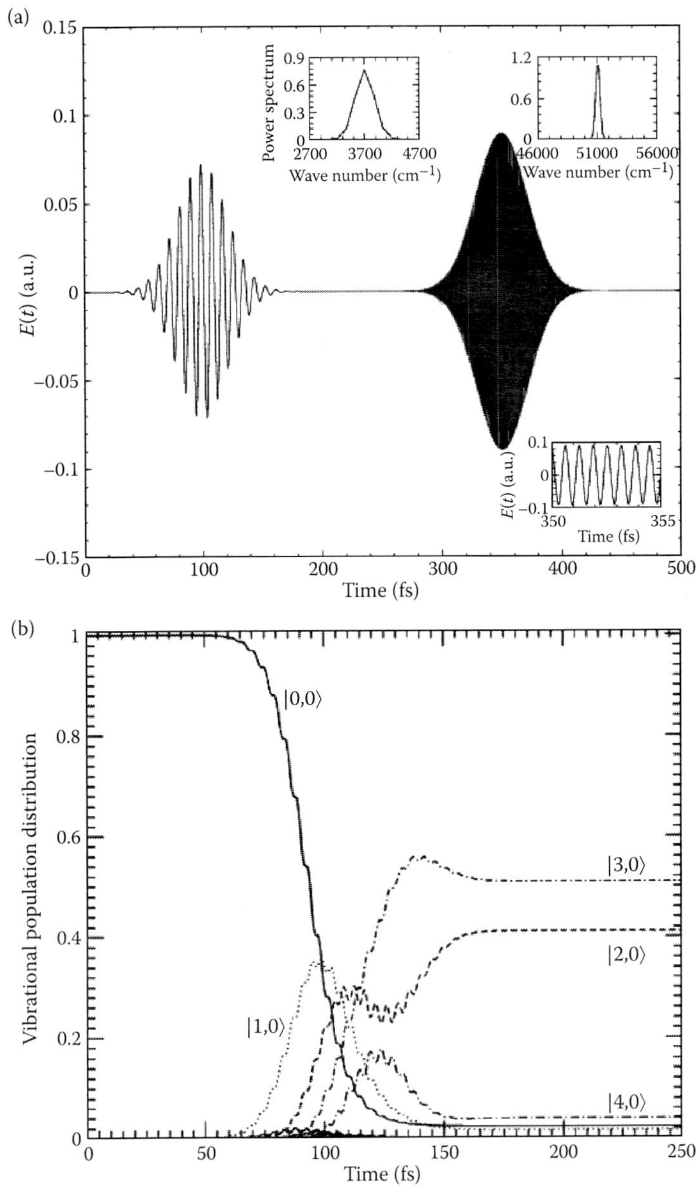

FIGURE 7.20 (a) IR and UV laser pulses with pulse profile $E(t) = A_0 a(t)\exp[-\gamma(t - t_0)]^2$ $(\cos \omega t)$. For the IR pulse $A_0 = E_0 \cos\phi$ with $\phi = (\pi - \theta)/2$, $\theta = 104.52^0$, $E_0 = 0.025$, and $a(t) = \left(8\gamma t_l^2/\pi\right)^{1/4}$, with FWHM = $(4\ln 2/\gamma)^{1/2}$ = 50 fs, t_l = 250 fs, t_0^{ir} = 100 fs, and ω_{ir} = 3706 cm^{-1}. For the 50-fs UV laser pulse $A_0 = 0.09$, $a(t) = 1$, t_0^{uv} = 350 fs, and ω_{uv} = 51 090 cm^{-1}. The power spectra for the IR and UV pulses and a more resolved UV field profile are presented in the insets. (b) Vibrational population distribution from |0,0> as the initial state under the influence of the IR pulse depicted in (a).

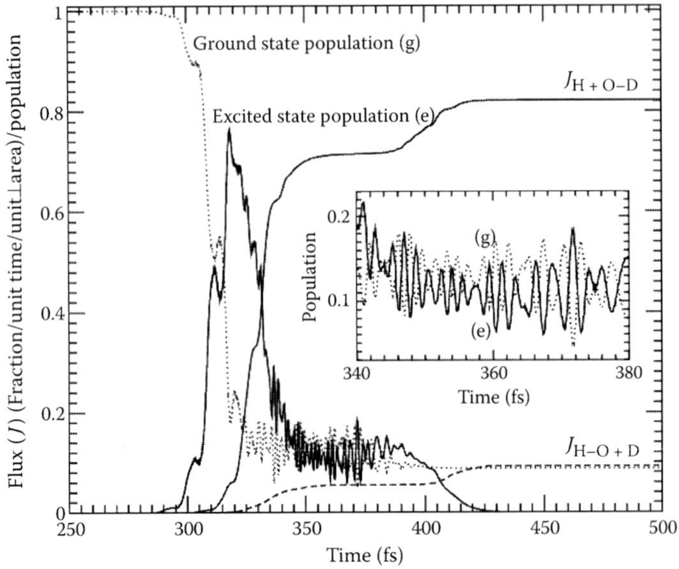

FIGURE 7.21 Ground and excited state populations and accumulated H + O–D and H–O + D flux from the vibrational population distribution shown in Figure 7.20b under the influence of the IR + UV pulses shown in Figure 7.20a.

this prior vibrational excitation in engineering favored a flow into the H + O–D channel of the excited surface is quite effective indeed. It is also seen that, unlike what has been assumed in some earlier analyses,[36,38,40,55] it is not only a small wedge of the ground state probability density accessible at the pulse frequency that is transferred to the upper surface, but there is a more or less complete transfer of the ground state population to the upper surface, which ensures selectivity with substantial yield, as seen in Table 7.3. The transmodal flow into the H–O + D channel is seen to be initiated by the cross talk between the two surfaces induced by the UV pulse from 325 fs onward with nonnegligible amplitude in the O–D mode, as well as on both the ground and the excited surfaces. The flow in the H–O + D channel of the excited surface is seen from 350 fs onward with a kick at 400 fs, as remarked earlier. The transmodal flow into the H–O + D channel from 350 fs onward, even in this simple case with well-separated bond frequencies, is difficult to miss.[60]

The IR and UV pulses employed for selective control of O–D dissociation[60] with their power spectrum are depicted in Figure 7.23a, and the resulting vibrational mix is shown in Figure 7.23b. As can be seen in Figure 7.23b, there is a stable combination of pure O–D modes with maximum population in the |0,5> mode. This combination has sufficient prior stretch in the O–D bond, and results from subjecting this combination to UV pulses with different frequencies are displayed in Figures 7.24 and 7.25 and collected in Table 7.4. The flux values of Table 7.4 isolate the UV pulse with a frequency of 46,062 cm^{-1} to be most effective, which incidentally will provide sufficient energy to place the HOD molecule with 5 quanta of excitation in the

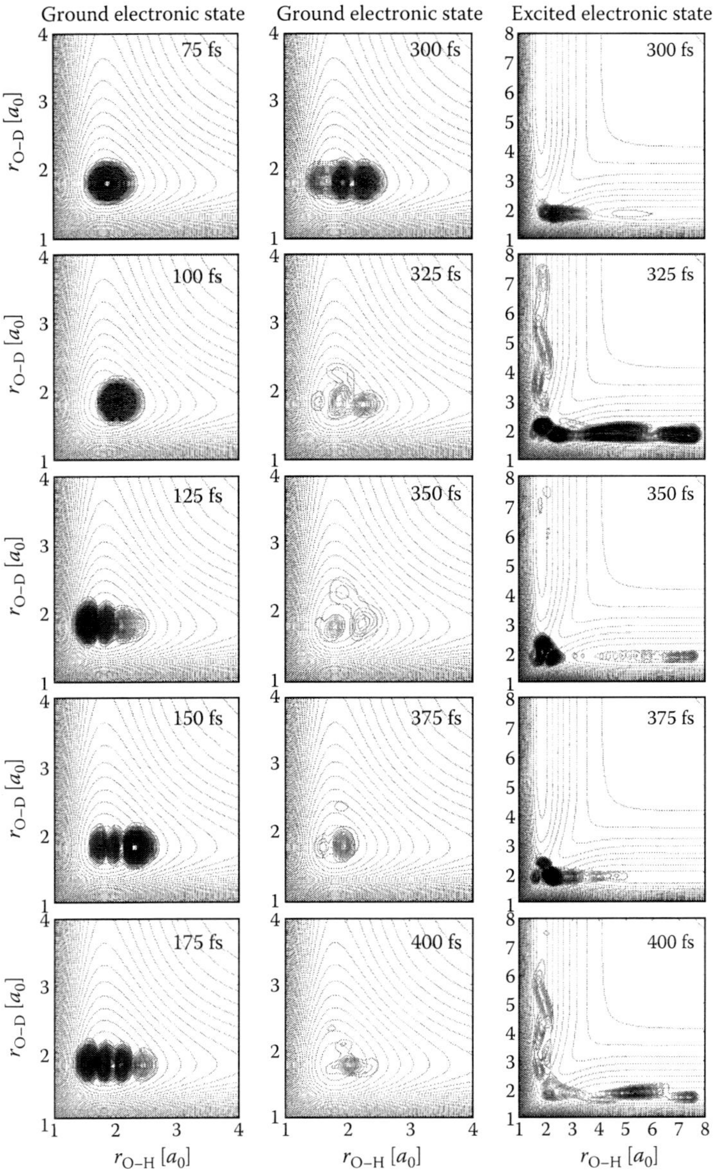

FIGURE 7.22 Time evolution of the |0,0> state on the ground and excited surfaces for the IR + UV fields profiled in Figure 7.20a.

O–D mode—the dominant component of the linear combination resulting from the IR churn—just above the saddle point of the upper repulsive surface. Details of the population transfer dynamics and dissociative flux in the H–O + D and H + O–D channels are shown in Figures 7.24 and 7.25. Results in Figure 7.24 show that, as soon as the UV pulse begins to build sufficient power around 300 fs, there is a quick

TABLE 7.3
Photodissociation Flux and Branching Ratio for Local O–H Modes under the Influence of the Field Shown in Figure 7.20a

UV Frequency, ω_{uv} (cm^{-1})	H + O–D Flux	H–O + D Flux	H + O–D/H–O + D Flux Ratio
59,090	0.73	0.17	4.29
57,090	0.73	0.14	5.21
55,090	0.71	0.17	4.18
53,090	0.79	0.11	7.19
51,090	0.82	9.2×10^{-2}	8.90
49,090	0.70	1.86×10^{-2}	37.63
47,090	0.55	5.88×10^{-3}	94.82
45,090	0.42	3.10×10^{-4}	1354.84
43,090	0.12	2.00×10^{-5}	6000.00

transfer of population to the excited surface, which reaches a maximum by 350 fs. With the excited upper surface being repulsive, the population in the excited state begins to flow almost simultaneously, and due to the prior stretching of the O–D bond on the ground surface, flux builds up in the H–O + D channel, with ~72% dissociation of the O–D bond being achieved within the next 10 fs. The final flux of 79% in the H–O + D channel is achieved once the UV field switches off at 420 fs, shutting off the cross talk between the ground and the excited surfaces. There is no further dumping from the excited surface, and this allows flushing of population on the excited surface into the competing channels as seen once again, from a kick in both the H–O + D and H + O–D flux values around 400 fs.

The gradual buildup of localized excitations in the O–D bond is depicted in the 75–175-fs probability density plots of Figure 7.25, where gradual transformation of the |0,0> ground vibrational state into the linear combination of pure O–D modes is easily seen. This linear combination resulting from the vibrational churn retains its shape until the UV field comes into play with sufficient power at 300 fs, and as can be seen in the probability density plots on the excited surface, the role of this vibrational excitation in engineering a favored flow into the H–O + D channel of the excited surface is quite effective and in line with the results discussed earlier for selective control of O–H bond dissociation.[60] It is also seen that, unlike what has been assumed in some earlier analyses,[36,38,40,55] once again, it is not only a small wedge of the ground state probability density accessible at the pulse frequency that is transferred to the upper surface, but there is a more or less complete transfer of the ground state population to the upper surface, which is required for ensuring selectivity with substantial yield, as seen in Table 7.4. On the ground surface, the cross talk between the two surfaces induced by the UV pulse leads to a small amplitude in the O–H mode, as well at 350 fs, and a small flow in the H + O–D channel of the excited surface is seen from 350 fs onward with a kick at 400 fs as remarked earlier.[60]

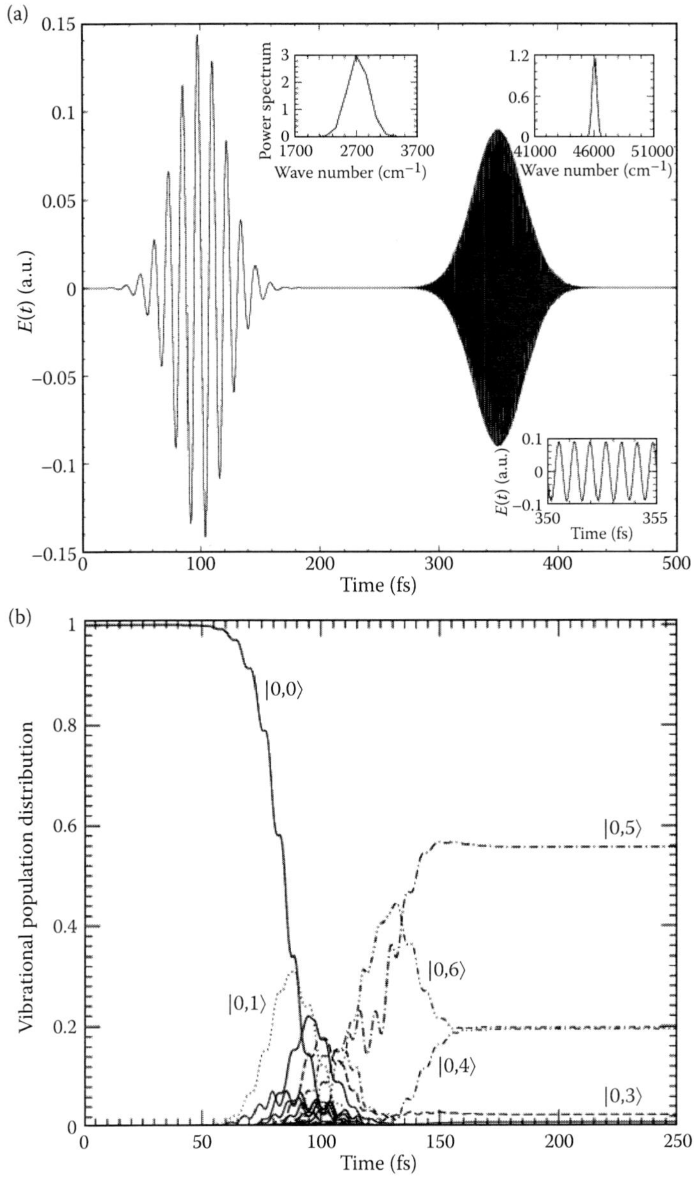

FIGURE 7.23 Same as Figure 7.20, except that $E_0 = 0.05$, $\omega_{ir} = 2727\ \text{cm}^{-1}$, and $\omega_{uv} = 46062\ \text{cm}^{-1}$.

Transmodal flow from 350 fs onward, even in this simple case with well-separated bond frequencies, is similar to that seen earlier.

Finally, deciphering the mechanistic basis of selective dissociation in HOD is also substantiated using the expectation values of stretch and momentum in the two bonds on both surfaces, and a selection of our results from this approach[61] is offered

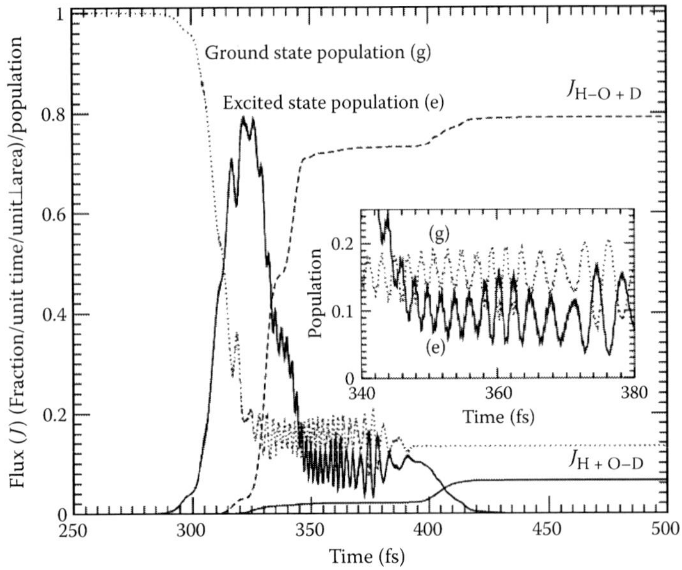

FIGURE 7.24 Same as Figure 7.21, except that the IR + UV combination is that presented in Figure 7.23.

in Figures 7.26 and 7.27. The carrier frequency for the Gaussian UV pulse profiles of Figures 7.26a and 7.27a are chosen to maximize selective photodissociation, depending on the initial vibrational state of the HOD molecule on the ground electronic state, and the results using a UV field of 67,169 cm^{-1} frequency with |0,0> as the initial state are presented in Figure 7.26. Figure 7.26b provides the pattern for population transfer from the ground to the excited surface and the resulting flux in the H + O–D and H–O + D channels on the upper surface. It can be seen from Figure 7.26b that, as the field begins to gain sufficient strength from 40 fs onward, there is a rapid population transfer from the ground to the excited electronic state, and the population buildup in the excited state goes on until approximately 60 fs, at which time the flux in the H + O–D channel begins to pick up. This is because the upper excited surface is entirely repulsive,[98,99] and any population deposited in the totally repulsive H + O–D or H–O + D channels of the excited surface is bound to lead to the dissociative downhill motion in both channels and a buildup of flux at the cost of diminution of population.[61]

What is seen only faintly in Figure 7.26b but much more so in Figure 7.27b is a field-induced cross talk between the ground and the excited electronic state populations, and as the flux (J_{H+O-D}) builds up from 65 to 80 fs in Figure 7.26b, there is, as expected, a sharp decrease in the excited state population, which is being flushed out as dissociative flux in the H + O–D and H–O + D channels. Beginning at 65 fs, the cross talk between two surfaces is quite significant. By 90 fs, the flux in H + O–D has built up to 72%, and there is more or less stable population in the ground and excited states as stable flux in the two channels, which does not change until the field

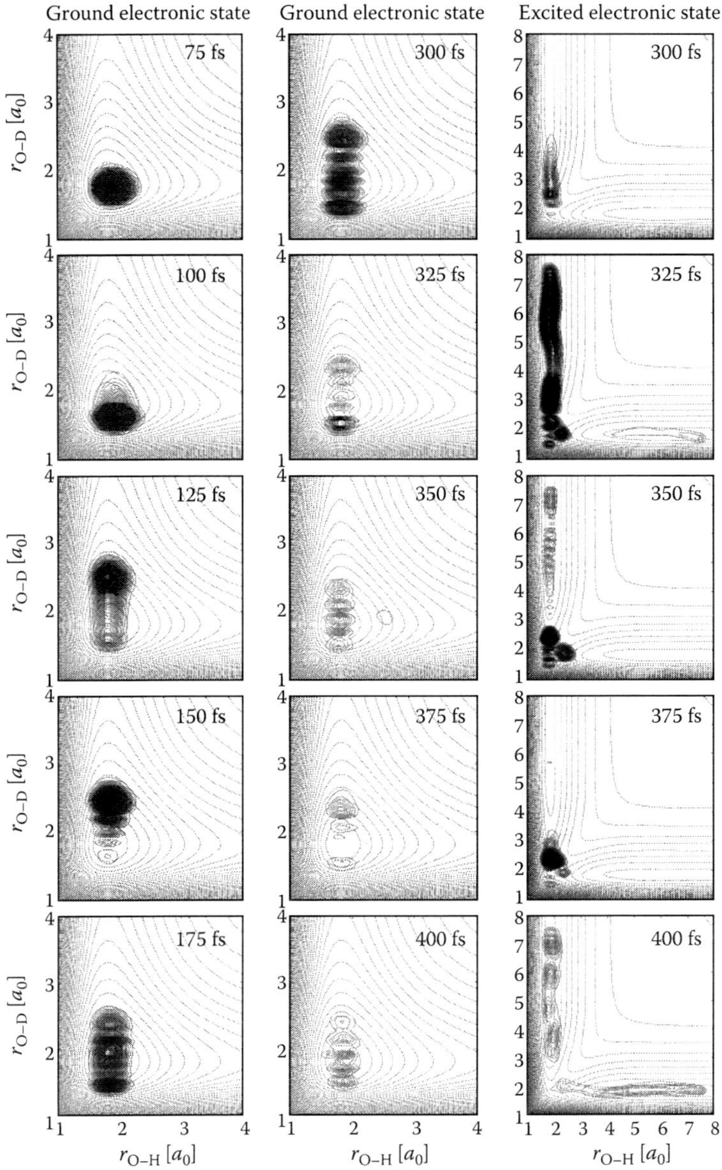

FIGURE 7.25 Same as Figure 7.22, except that the IR + UV combination is that presented in Figure 7.23.

is more or less switched off at 150 fs. As the field is switched off, the field-induced cross talk stops, and the population in the excited state comes down quickly to near-zero with a kickup in the H + O–D flux and a near-total dissociation of the H–O–D molecule with ~82% flux (J_{H+O-D}) in the H + O–D channel and ~15% (J_{H-O+D}) in H–O + D channel.

TABLE 7.4
Photodissociation Flux and Branching Ratio for Local O–D Modes under the Influence of the Field Profile Shown in Figure 7.23a

UV Frequency, ω_{uv} (cm^{-1})	H + O–D Flux	H–O + D Flux	H + O–D/H–O + D Flux Ratio
54,062	0.25	0.64	2.56
52,062	0.26	0.64	2.46
50,062	0.19	0.70	3.68
48,062	0.15	0.76	5.07
46,062	6.57×10^{-2}	0.79	12.02
44,062	8.97×10^{-3}	0.75	83.61
42,062	3.09×10^{-3}	0.41	132.62
40,062	4.10×10^{-4}	0.20	487.80
38,062	1.20×10^{-4}	4.50×10^{-2}	375.00

Between 80 and 140 fs, there is almost stable flux in both the H + O–D and the H–O + D channels, and as a result, the population in the ground and excited states is also nearly stable but with persistent cross talk. The stability of flux implies that, on the excited state, the bond stretching must be in the near vicinity of the equilibrium $\langle r_{O-H} \rangle$ and $\langle r_{O-D} \rangle$ values, as seen in the excited state stretch expectation value profile of Figure 7.26e. The near-constant stability of features between 100 and 150 fs in the excited-state expectation values of Figure 7.26e and f is therefore linked to an absence of flux buildup and the stable cross talk between the ground and the excited state populations in this interval, where the probability density profile on the excited state surface remains anchored near equilibrium $\langle r_{O-H} \rangle$ and $\langle r_{O-D} \rangle$ values on the excited state surface (as seen in Figure 7.15). However, as also seen in 100–150-fs plots of Figure 7.15, there is considerable sloshing around in the ground state probability density profiles. This sloshing around in the ground state probability density profile induced by the cross talk between the ground and the excited state populations indicates that the dumping from the excited surface is not always to the ground vibrational state of the ground electronic state, and the field-induced dumping from the upper excited electronic surface may also be to the excited vibrational levels of the ground surface, which leads to the mixing of higher vibrational excited states in the ground state populations, which we believe is responsible for the additional field-induced stretching in the 80–140-fs interval of Figure 7.26c, where other expectation values are stable. Of course, as the field is switched off around 150 fs, the kickup in the flux seen in Figure 7.26b is through a complicated downhill sloshing in repulsive H + O–D and H–O + D channels on the excited surface, giving rise to the sampling of larger $\langle r_{O-H} \rangle$ and $\langle r_{O-D} \rangle$ values in Figure 7.26e. On the ground electronic surface, the stretch in the O–H and O–D bonds reverts to oscillations around the equilibrium values once the field has been switched off. The ground state stretching expectation value profile after the field is switched off is oscillatory in Figure 7.26c, and there is a classical complementarity between the stretch/momentum expectation value profiles

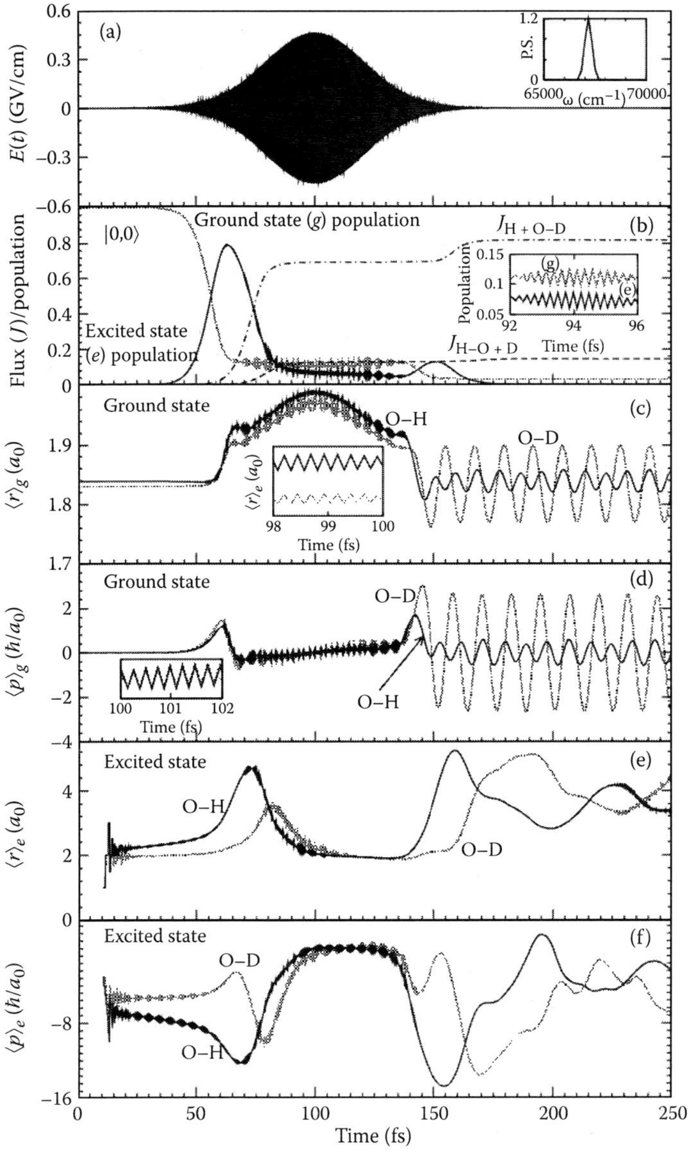

FIGURE 7.26 Results from the exposure of the ground vibrational state |0,0> of HOD on the ground electronic state to the (a) Gaussian UV pulse with same attributes as that depicted in Figure 7.14a. The power spectrum (P.S.) is provided as an inset. (b) Ground and excited state populations and accumulated H + O–D and H–O + D flux. (c) Average bond lengths $\langle r \rangle_g$ on ground surface for the O–H and O–D bond modes. (d) Average momenta $\langle p \rangle_g$ on the ground surface for the O–H and O–D modes. (e) Average bond lengths $\langle r \rangle_e$ on the excited surface. (f) Average momenta $\langle p \rangle_e$ on the excited surface.

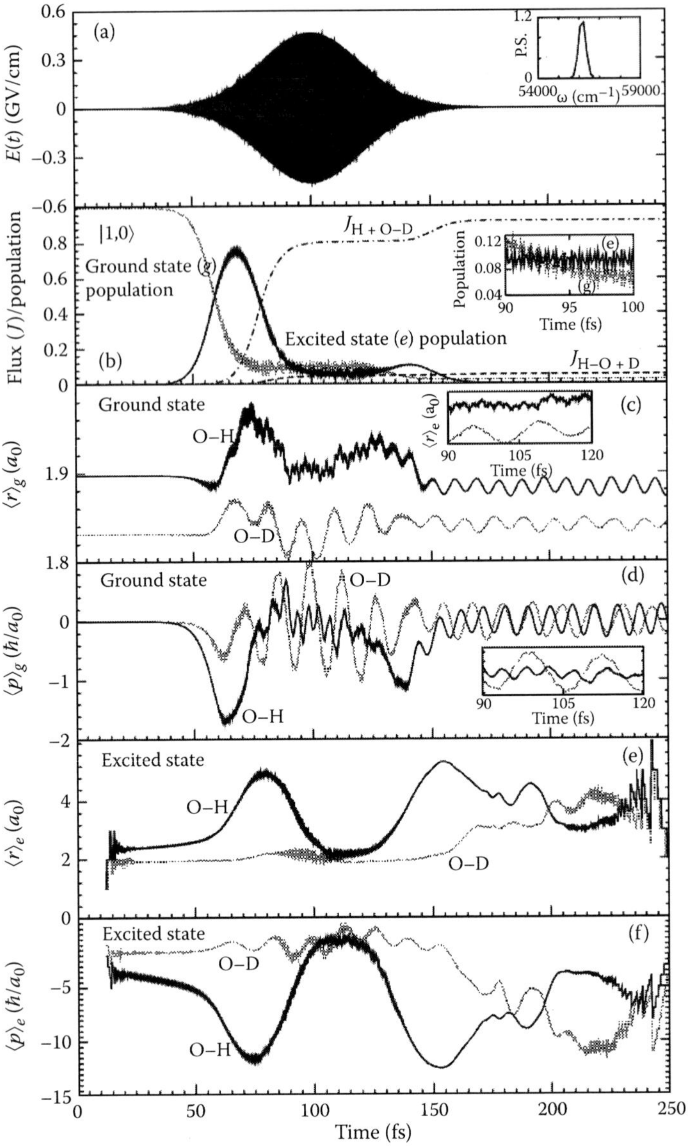

FIGURE 7.27 Results from the exposure of the |1,0> vibrational state of HOD on the ground electronic state to the same Gaussian UV pulse as in Figure 7.26 but with $\omega = 56{,}155\ \text{cm}^{-1}$. (a)–(f) Same as in Figure 7.26.

of Figure 7.26c and d and Figure 7.26e and f. The features discussed for Figure 7.26 become more pronounced and easily discernible in Figure 7.27.

As can be seen from Figure 7.26a and c, the onset of buildup in field strength leads to a concerted buildup of stretch in both the O–H and the O–D bonds, which normalize to expected low amplitude oscillations as the field is switched off, with lower energy O–D bond showing larger amplitude motion. The inset makes it clear that, for the ground vibrational state |0,0>, O–H and O–D stretch or contraction is synchronized with the field-induced stretch being slightly more in the O–H bond, and as long as the field is on, the finer undulations in the inset are also larger for the O–H bond, as expected. The corresponding average momentum profile in the ground vibrational state for the two bonds are plotted in Figure 7.26d, and the field-induced average momentum distribution in the two bonds on the ground surface are coupled and comparable. Although the field-induced flux in Figure 7.26b is clearly higher for the breakup of the O–H bond, this does not seem to ensue from any buildup of much larger momentum in this bond in comparison to the O–D bond. The longtime field-free average momentum profile for both bonds mirrors the average stretch profile of Figure 7.26c.

The real reason for the higher H + O–D flux is made clear by Figure 7.26e and f, where a small bias toward greater initial stretch in the O–H bond on the ground surface (Figure 7.26c) leads to a quick downhill motion (larger negative momentum) of Figure 7.26f in the H + O–D valley on the upper surface, sampling dissociation limit average stretch in the excited state (Figure 7.26e), which is much larger and quicker for the O–H bond, and our results seem to clearly favor a critical role for an initial bias in bond stretch as an effective facilitator for selective dissociation of the O–H bond in this case. The variations in the average stretch and momentum profiles are a confirmation of the complexities in the downhill motion in the H + O–D and H–O + D valleys of the repulsive upper surface. Sloshing around of the wave function in these valleys is well established,[37,38,40,41,44,46,51–53,55,57–60] and the excited state average stretch and momentum profiles are a manifestation of this complex probability distribution in the dissociative downhill motion in the H + O–D and H–O + D valleys, on the upper repulsive surface.[57–60]

The dominant role of initial stretch in controlling the selective outcome of photodissociation is further examined by using the initial vibrational state |1,0> with one quantum of vibrational excitation in the O–H stretch. The flux in H + O–D rises from ~82%, with |0,0> as the initial state and 67,169 cm^{-1} UV pulse, to ~93%, with |1,0> as the initial state and 56,155 cm^{-1} UV pulse (Figure 7.27a) used to transfer the |1,0> population in the ground state to the repulsive $^1B_1$ surface.[61] The population transfer and flux profiles are presented in Figure 7.27b, and it is easily seen that the cross talk between the ground and the excited surface sets in as soon as the field strength builds up and is quite pronounced thereafter. This back-and-forth population exchange as the molecule samples different regions of the upper surface leads to a mixing of vibrational states on the ground surface, which is easy to see in the average stretch profile for the O–H bond on the ground surface plotted in Figure 7.27c, which oscillate considerably. The large initial O–H stretch built into the |1,0> state gives rise to a small initial contraction (Figure 7.27c) and a negative initial average momentum $\langle p_{O\text{-}H}\rangle$ in the beginning (Figure 7.27d). The $\langle r_{O\text{-}H}\rangle$ and $\langle p_{O\text{-}H}\rangle$, as well as the $\langle r_{O\text{-}D}\rangle$ and $\langle p_{O\text{-}D}\rangle$, profiles seem to show a near-classical interlocking of the momentum maxima with stretch minima and the momentum minima with stretch maxima.[61]

The large initial stretch in the O–H bond accentuates the effects seen earlier in Figure 7.26e and f, where quick sampling of dissociatively large stretch regions of the excited state with bond lengths that are large enough to mimic dissociation (Figure 7.27e) and even larger average negative momentum (Figure 7.27f) than that seen for the |0,0> state, quickening the downhill motion in the H + O–D valley of the upper surface, facilitates larger flux in the H + O–D channel. The extended features of the excited state average stretch and momentum profiles (Figure 7.27e and f) once again mirror the complex sloshing of the probability density flow on the upper surface.[57]

7.3.4 Selective Control of $^{18}O^{16}O^{16}O$ Photodissociation

Finally, we present some initial results from our computational attempts to achieve selective photodynamic control of bond dissociation in the $^{18}O^{16}O^{16}O$ molecule. In our investigation, our treatment of the isotopic ozone ($^{18}O^{16}O^{16}O$) molecule is similar to that outlined for the HOD molecule, where, again, there are two distinct dissociation channels, viz., ^{18}O + ^{16}O–^{16}O and ^{18}O–^{16}O + ^{16}O. Results from our calculations show that the ^{16}O–^{16}O bond may be preferentially dissociated compared to the ^{18}O–^{16}O bond.

The strategy employed for selective bond dissociation in HOD is repeated for $^{18}O^{16}O^{16}O$ as well, and we begin by examining the UV dynamics for |0,0>, |1,0>, and |2,0> (where the first integer corresponds to the quantum of vibrational excitation in the ^{18}O–^{16}O mode and the second integer corresponds to the quantum of vibrational excitation in the ^{16}O–^{16}O mode) states of the isotopic ozone ($^{18}O^{16}O^{16}O$) molecule. In this investigation, we present some results from initial calculations on $^{18}O^{16}O^{16}O$ photodissociation, where only one initial vibrational state is involved. In $^{18}O^{16}O^{16}O$, ^{18}O and ^{16}O are connected through the ^{16}O atom, giving rise to very narrowly spaced eigenstates, and these eigenmodes are strongly coupled. The difference between |1,0> and |0,1> is only ≈170 cm^{-1} in our calculations, whereas the energy difference between these two states calculated in an earlier study[72] was only 53 cm^{-1}. Similar trends are observed in other vibrational states as well. It is, therefore, clear that preparing pure ^{16}O–^{16}O or ^{18}O–^{16}O mode is a difficult task unlike in the case of HOD, where the difference between two consecutive modes is more than ≈1000 cm^{-1}.

Employing the Gaussian UV pulse of FWHM in 50 fs, we have carried out the UV dynamics in the Hartley absorption band (32,000–55,000 cm^{-1}) for |0,0>, |1,0>, and |2,0> vibrational states, and the results are collected in Tables 7.5 through 7.7, respectively. However, in our calculation, we have used different field amplitudes (E_0) for the three chosen initial states in order to get maximum flux in the ^{18}O + ^{16}O–^{16}O and ^{18}O–^{16}O + ^{16}O channels. A value of 0.02 a.u. as E_0 was used for the |0,0> and |2,0> states, where, as for the |1,0> state, we have used E_0 = 0.04 a.u. It can be seen from Table 7.5 that, at lower UV frequencies, the flux in both channels are negligible due to the insufficient energy for transferring the probability amplitude to the excited surface. It is also remarkable that there is little or insignificant population that is transferred from the ground to the excited surface, even with sufficient energy pulse, which is a signature of the Franck–Condon window. The transfer of population from ground to excited state is feasible for an optimal choice of frequencies, and flux appears in both channels. The maximum flux in the ^{16}O–^{16}O mode is 39.5% at

TABLE 7.5
Mapping of UV Frequencies with |0,0> as the Initial State and 0.02 a.u. as E_0 in the Hartley Absorption Band (32,000–55,000 cm^{-1})

UV Frequency ω_{uv} (cm^{-1})	J_1 (^{18}O + ^{16}O–^{16}O Channel) (%)	J_2(^{18}O–^{16}O + ^{16}O Channel) (%)	Ground State Population (%)	Excited State Population (%)	Total Population after 250 fs (%)
35,000	0.28	0.3	98.8	0.6	100
38,000	20.9	23.5	28.6	21.6	93.6
38,919	22.7	38.4	6.2	21.6	88.8
39,000	22.3	39.3	5.9	21.6	88.8
39,019	21.9	39.4	5.9	21.7	88.8
39,069	21.8	39.5	5.7	21.8	88.7
39,119	21.9	39.3	5.4	22.0	88.6
39,169	22.2	38.9	5.1	22.3	88.5
39,219	22.7	38.2	4.6	22.7	88.3
39,319	24.1	36.5	3.5	23.5	87.7
40,000	25.3	32.4	1.9	22.5	82.1
41,000	19.5	29.8	10.0	20.8	80.0
45,000	7.0	9.4	59.4	11.5	87.2
50,000	2.6	4.39	73.1	10.7	90.6
55,000	1.4	1.6	91.0	2.7	96.6

Note: The time-integrated flux in the ^{18}O + ^{16}O–^{16}O (J_1) and ^{18}O–^{16}O + ^{16}O (J_2) channels and population distribution in the ground and excited states after 250 fs are also provided.

TABLE 7.6
Mapping of UV Frequencies with |1,0> as the Initial State and E_0 = 0.04 a.u. in the Hartley Absorption Band (32,000–55,000 cm^{-1})

UV Frequency ω_{uv} (cm^{-1})	J_1(^{18}O + ^{16}O–^{16}O Channel) (%)	J_2 (^{18}O–^{16}O + ^{16}O Channel) (%)	Ground State Population (%)	Excited State Population (%)	Total Population after 250 fs (%)
33,062	0.01	0.09	99.95	0.01	99.99
37,062	7.3	9.46	78.12	2.39	97.35
38,062	17.09	29.71	35.42	7.65	89.88
38,162	17.17	31.54	32.37	8.03	89.12
38,562	16.75	34.39	27.31	8.52	86.98
39,662	12.83	32.94	25.69	9.84	81.31
40,062	17.79	29.37	17.72	15.27	80.17
45,062	61.35	8.46	69.03	6.25	89.88
50,062	5.27	5.96	67.13	7.79	86.16

Note: The time-integrated flux in the ^{18}O + ^{16}O–^{16}O (J_1) and ^{18}O–^{16}O + ^{16}O (J_2) channels and population distribution in the ground and excited states after 250 fs are also provided.

TABLE 7.7
Mapping of UV Frequencies with |2,0> as the Initial State and 0.02 a.u. as E_0 in the Hartley Absorption Band (32,000–55,000 cm^{-1})

UV Frequency ω_{uv} (cm^{-1})	J_1 (^{18}O + ^{16}O–^{16}O Channel) (%)	J_2 (^{18}O–^{16}O + ^{16}O Channel) (%)	Ground State Population (%)	Excited State Population (%)	Total Population after 250 fs (%)
35,232	2.06	1.96	93.41	2.53	99.71
36,232	5.99	4.40	83.20	4.37	99.97
37,232	15.57	9.65	56.63	11.10	92.96
38,232	5.69	8.32	74.87	6.29	95.19
39,232	13.95	9.23	59.96	9.92	93.07
40,232	7.64	9.94	65.35	7.92	90.87
41,232	13.24	17.74	33.52	16.66	81.17
42,232	7.71	10.55	60.87	8.42	87.56
43,232	5.75	6.23	72.85	7.02	91.86
44,232	7.96	12.21	38.03	24.19	82.40
45,232	1.67	3.23	88.13	2.76	95.81
46,232	5.12	4.92	71.89	8.10	90.0

Note: The time-integrated flux in the ^{18}O + ^{16}O–^{16}O (J_1) and ^{18}O–^{16}O + ^{16}O (J_2) channels and population distribution in the ground and excited states after 250 fs are also provided.

the UV frequency of 39 069 cm^{-1}. However, the sum of the ground and excited state populations and flux flowing through both channels deviates significantly from unity at the frequency, which enables sufficient population transfer to the excited state. This may be due to the probability flow in the total dissociation ^{16}O + ^{16}O + ^{18}O channel to where flux is not being measured.

The results of the UV dynamics using the |1,0> state (Table 7.6), at the lower and higher ends of the Hartley (32,000–55,000 cm^{-1}) band, reveal that there is nil or negligible population transfer to the excited state, but at the optimal frequency of 38,562 cm^{-1}, there is maximum flux in both channels. Unlike HOD, there is no frequency corresponding to which dissociation of the ^{18}O–^{16}O bond is favored. Moreover, as seen in the case of the |0,0> state dynamics, here, the sum of the total population plus flux also deviates from 1 as soon as the flux starts leaking from both channels for the entire range of frequencies used in this investigation. The probable reason for this may again be the untracked ^{18}O + ^{16}O + ^{16}O channel.

More or less similar is the story for UV dynamics in the case of the |2,0> state of the $^{18}O^{16}O^{16}O$ molecule. It is seen from Table 7.7 that, although we have used the same field intensity as that of the |0,0> state (E_0 = 0.02 a.u.), there is minimal flux flow through both the ^{18}O + ^{16}O–^{16}O and the ^{18}O–^{16}O + ^{16}O channels. The reason for the minimal flux may be due to the small population transfer from ground to excited states seen in the entire frequency range considered here.

7.4 CONCLUDING REMARKS

The results from the selective control of simple diatomics, viz., HI and IBr photodissociation using the optimal superpositions selected by the Rayleigh–Ritz variational procedure for maximization of flux out of the desired channel for the chosen field, confirm the utility of the FOIST approach advocated here. The results reveal that the selective maximization is put in effect through the localization of the probability density at internuclear distances, which enable Frank–Condon transitions to the appropriate region of the excited states. Transfer of the wave function to the steeper repulsive part of the excited potential energy curves favors high-velocity diabatic exit into the higher channel. Localization away from the repulsive wall favors slow adiabatic exit into the lower channel. The nascent mechanistic notions linking selectivity to the appropriate modification of the initial state have been further examined by an analysis of the resulting absorption spectrum and branching ratio, and this central role for the modified spatial profile in selective control provides a new possibility for experimental exploration.

The experimental realization of the optimal initial states is, however, a completely uncharted area. In an earlier paper,[30] we have presented the formulae to obtain field parameters required to achieve these FOISTs, and the optimal control approach may also feasibly and profitably be employed to attain this FOIST, which comprises only three vibrational levels. We however believe that, while the theoretical tools are useful, the central results from our investigation[14,30–34] are that, instead of putting the entire onus of selective control on a theoretically designed laser pulse that may not be easy to realize in practice, the approach where different vibrational population mixes are experimentally obtained and subjected to readily attainable photolysis pulses, leading to an empirical experimental correlation between selectivity attained for diverse photolysis pulses and initial vibrational population mix used, represents a more promising and desirable alternative. Our results, we hope, will spur experimental tests, and a concerted partnership between field and initial state shaping is required to better realize the chemical dream[21,23] of using lasers as molecular scissors and tweezers to control chemical reactions.

It is our hope that the approach advocated here will merit experimental attention. Instead of attempting selective control by using an active field manipulating a passive molecule in the ground vibrational state, experiments that use a variety of population mixes as the initial state merit detailed attention on the basis of results adduced here. Should a pattern of the kind seen in our results is experimentally vindicated, then, the task of generating altered spatial probability density profiles of the type studied here can be reduced to finding a suitable linear combination of known vibrational eigenfunctions which reproduce these spatial probability density profiles without requiring any time-dependent quantum mechanical calculation whatsoever.

Further, the Rayleigh–Ritz variational maximization of flux by generating an optimal spatial profile for the initial wave function offers a new and flexible alternative for laser-assisted selective control of chemical reactions. It is our hope that the FOIST-based approach presented here will attract requisite experimentation, and a concerted partnership between the field and the initial state shaping advocated here will assist in keeping the dream of controlling chemical reactions by modifying the underlying quantal dynamics alive and attractive for further pursuit.[14] A detailed study of the vibrational properties of the time-integrated flux operator underlying our FOIST scheme to correlate field attributes with the resulting optimal mix of field-free vibrational eigenstates in ψ_1^{max} and ψ_2^{max} will provide additional mechanistic insights.

We have also investigated the HOD photodissociation for selective dissociation of O–H and O–D bonds using simple field profiles. A detailed analysis based on the examination of population transfer, flux, and probability density flows on the ground and excited surfaces has been employed to corroborate the microdynamical quantal picture.

With the ground vibrational state of the ground electronic state |0,0> as the initial state, the O–H dissociation is favored substantively over the O–D bond dissociation for a single-color laser, and preferential dissociation of the O–H bond may be achieved without additional excitation in the O–H mode.

Using |0,1> as the initial state with one quantum of excitation in the O–D mode, we find that favored dissociation of the O–H bond is reversed for a large range of photolysis frequencies, and H–O + D flux predominates over the H + O–D flux. Starting with |0,2> as the initial state with two quanta of excitation in the O–D mode, there is dominant dissociation of the O–D bond for a large interval of frequencies.

A limited investigation of selective control of HOD photodissociation using the optimal superpositions selected by the Rayleigh–Ritz variational procedure for the maximization of flux out of the desired channel for the chosen field indicates that further enhancement of 6%–8% in selective maximization of O–D dissociation may be possible through FOIST-based selection of the initial state. Furthermore, the mixing of vibrational states may be supplanted by mixing of colors, and the kinematic bias in favor of O–H dissociation from the ground vibrational state |0,0> can be substantially reversed with a combination of two lasers, with 54,920 and 52,303 cm^{-1} as carrier frequencies providing approximately thrice as much H–O + D as H + O–D.

We have also presented results from a quantum dynamical full two-surface calculation[60] on the HOD molecule to show that the use of a simple IR pulse tuned to the fundamental frequency of the O–H/O–D bond, followed by an UV pulse with carrier frequency that deposits the resulting linear combination of pure O–H/O–D modes below the saddle point barrier, can provide an effective pathway for selective dissociation of the desired bond. The dominant stretching mode in the linear combination resulting from the vibrational churn determines the optimal carrier frequency of the UV pulse for maximum yield with reasonable selectivity, which is found to be just enough to deposit the HOD molecule near the saddle point on the repulsive excited surface. Systematic lowering of this carrier frequency gives rise to greater selectivity with reasonable yield and further lowering and to near-total selectivity with reduced yield. The enhanced selectivity with reduction in the UV carrier frequency is linked to modification of the cross talk between the ground and the excited surfaces in a manner that removes transmodal leakage in the competing channel and keeps the population transferred to the upper surface from going over the saddle point barrier separating the two channels.

From a detailed investigation[61] of the change in average momentum and stretch in the O–H and O–D bonds using different initial states and UV pulses most suited for their transfer from the ground to the repulsive upper surface, we see that the larger stretch provided by prior vibrational excitation in the chosen bond favors selective dissociation of that bond with near-100% selectivity if the chosen bond is stretched considerably more than the other bond. Since preparation of these initial states is much easier than trying to time the UV field such that it will be concurrent with the maximum bond elongation, our results favor a prior stretched-bond–based selective photodynamic control of H + O–D ← H–O–D → H–O + D photodissociation.

The concerted use of the detailed temporal profile of expectation values of bond stretch and momentum on both the ground and the excited surfaces is seen to provide a clinching correlation between even small extra stretches in a bond facilitating the sampling of dissociative regions of the upper surface through accelerated downhill motion in the repulsive valley favoring its dissociation.[61] The detailed quantum mechanical study presented here[57–61] lends rigor to these insights into the control mechanism for HOD and, we hope, will lead to their routine use in mechanistic investigation of selective control of other triatomic/polyatomic systems.

Following the recipes developed through analyses of the diatomics and HOD mentioned above, we have attempted a fully quantal time-dependent investigation for photodynamic control of selective dissociation of the ^{18}O–^{16}O and ^{16}O–^{16}O bonds in the $^{18}O^{16}O^{16}O$ molecule using experimentally realizable Gaussian UV pulses and the ground (|0,0>), |1,0>, and |2,0> vibrational states of the ground electronic state. Results from our initial calculations have shown a ^{18}O–^{16}O + $^{16}O/^{18}O$ + ^{16}O–^{16}O flux branching ratio ~1.8, 2.0, and 0.6 from |0,0>, |1,0>, and |2,0> states, respectively, and thus demonstrate an effective route to selective photodissociation in the $^{18}O^{16}O^{16}O$ molecule as well.

An effort along these lines for simple routes to selective photodynamic control of bond dissociation is underway in our group.

ACKNOWLEDGMENTS

This research has been supported by grants from the Board of Research in Nuclear Sciences (Grant 2007/37/41-BRNS/2103) and the Department of Science and Technology (DST) (Grant SR/S1/PC-30/2006), India, given to M. K. Mishra. B. K. Shandilya acknowledges the fellowship support of CSIR, India (SRF, F. No. 09/87(0485)/2007-EMR-I). M. Sarma acknowledges the support of the Indian Institute of Technology Guwahati (Grant SG/CHM/P/MS/1). S. Adhikari is pleased to acknowledge DST, India, for the partial financial support through Project SR/S1/PC-13/2008.

REFERENCES

1. Sanz-Sanz, C., G. W. Richings, and G. A. Worth. *Faraday Discuss.* **2011**, *153*, 275–291.
2. Worth, G. A., and C. Sanz-Sanz. *Phys. Chem. Chem. Phys.* **2010**, *12*, 15570–15579.
3. Sarma, M., S. Adhikari, and M. K. Mishra. *Mol. Phys.* **2009**, *107*, 939–961.

4. Sarma, M. Ph. D. Thesis, Indian Institute of Technology Bombay, Powai, Mumbai, April 2008.
5. Lozovoy, V. V., X. Zhu, T. C. Gunaratne, D. A. Harris, J. C. Shane, and M. Dantus. *J. Phys. Chem. A* **2008**, *112*, 3789–3812.
6. Chakrabarti, R., and H. Rabitz. *Int. Rev. Phys. Chem.* **2007**, *26*, 671–735.
7. Rabitz, H. *Science* **2006**, *314*, 264–265.
8. Elles, C. G., and F. F. Crim. *Annu. Rev. Phys. Chem.* **2006**, *57*, 273–302.
9. Shapiro, M., and P. Brumer. *Phys. Rep.* **2006**, *425*, 195–264.
10. Shapiro, M., and P. Brumer. *Principles of the Quantum Control of Molecular Processes*. Wiley: New York, 2003.
11. Henriksen, N. E. *Chem. Soc. Rev.* **2002**, *31*, 37–42.
12. Rice, S. A. *Nature (London)* **2001**, *409*, 422–426.
13. Rice, S. A., and M. Zhao. *Optical Control of Molecular Dynamics*. Wiley: New York, 2000.
14. Vandana, K., and M. K. Mishra. *Adv. Quantum Chem.* **1999**, *35*, 261–281.
15. Gordon, R. J., and S. A. Rice. *Annu. Rev. Phys. Chem.* **1997**, *48*, 601–641.
16. Manz, J., and L. Wöste, Eds. *Femtosecond Chemistry*. Verlag Chemie: Weinheim, 1995.
17. Zhu, L., V. Kleiman, X. Li, S. P. Lu, K. Trentelman, and R. J. Gordon. *Science.* **1995**, *270*, 77–80.
18. Baumert, T., and G. Gerber. *Isr. J. Chem.* **1994**, *34*, 103–114.
19. Gersonde, I. H., S. Hennig, and H. Gabriel. *J. Chem. Phys.* **1994**, *101*, 9558–9564.
20. Wang, X., R. Bersohn, K. Takahashi, M. Kawasaki, and H. L. Kim. *J. Chem. Phys.* **1993**, *105*, 2992–2997.
21. Warren, W. S., H. Rabitz, and M. Dahleh. *Science.* **1993**, *259*, 1581–1589.
22. Kalyanaraman, C., and N. Sathyamurthy. *Chem. Phys. Lett.* **1993**, *209*, 52–56.
23. Crim, F. F. *Annu. Rev. Phys. Chem.* **1993**, *44*, 397–428.
24. Brumer, P., and M. Shapiro. *Annu. Rev. Phys. Chem.* **1992**, *43*, 257–282.
25. Chu, S.-I. *J. Chem. Phys.* **1991**, *94*, 7901–7909.
26. Park, S. M., S. P. Lu, and R. J. Gordon. *J. Chem. Phys.* **1991**, *94*, 8622–8624.
27. Crim, F. F. *Science.* **1990**, *249*, 1387–1392.
28. Tannor, D. J., and S. A. Rice. *Adv. Chem. Phys.* **1988**, *70*, 441–524.
29. Levy, I., and M. Shapiro. *J. Chem. Phys.* **1988**, *89*, 2900–2908.
30. Gross, P., A. K. Gupta, D. B. Bairagi, and M. K. Mishra. *J. Chem. Phys.* **1996**, *104*, 7045–7051.
31. Bairagi, D. B., P. Gross, and M. K. Mishra. *J. Phys. Chem. A.* **1997**, *101*, 759–763.
32. Vandana, K., D. B. Bairagi, P. Gross, and M. K. Mishra. *Pramana J. Phys.* **1998**, *50*, 521–534.
33. Vandana, K., and M. K. Mishra. *J. Chem. Phys.* **1999**, *110*, 5140–5148.
34. Vandana, K., and M. K. Mishra. *J. Chem. Phys.* **2000**, *113*, 2336–2342.
35. Segev, E., and M. Shapiro. *J. Chem. Phys.* **1982**, *77*, 5604–5623.
36. Engel, V., and R. Schinke. *J. Chem. Phys.* **1988**, *88*, 6831–6837.
37. Zhang, J., and D. G. Imre. *Chem. Phys. Lett.* **1988**, *149*, 233–238.
38. Zhang, J., D. G. Imre, and J. H. Frederick. *J. Phys. Chem.* **1989**, *93*, 1840–1851.
39. Shafer, N., S. Satyapal, and R. Bersohn. *J. Chem. Phys.* **1989**, *90*, 6807–6808.
40. Imre, D. G., and J. Zhang. *Chem. Phys.* **1989**, *139*, 89–121.
41. Hartke, B., J. Manz, and J. Mathis. *Chem. Phys.* **1989**, *139*, 123–146.
42. Vander Wal, R. L., J. L. Scott, and F. F. Crim. *J. Chem. Phys.* **1990**, *92*, 803–805.
43. Bar, I., Y. Cohen, D. David, S. Rosenwaks, and J. J. Valentini. *J. Chem. Phys.* **1990**, *93*, 2146–2148.
44. Vander Wal, R. L., J. L. Scott, F. F. Crim, K. Weide, and R. Schinke. *J. Chem. Phys.* **1991**, *94*, 3548–3555.

45. Bar, I., Y. Cohen, D. David, T. Arusi-Parper, S. Rosenwaks, and J. J. Valentini. *J. Chem. Phys.* **1991**, *95*, 3341–3346.
46. Amstrup, B., and N. E. Henriksen. *J. Chem. Phys.* **1992**, *97*, 8285–8295.
47. Shapiro, M., and P. Brumer. *J. Chem. Phys.* **1993**, *98*, 201–205.
48. Henriksen, N. E., and B. Amstrup. *Chem. Phys. Lett.* **1993**, *213*, 65–70.
49. Cohen, Y., I. Bar, and S. Rosenwaks. *J. Chem. Phys.* **1995**, *102*, 3612–3616.
50. Brouard, M., and S. R. Langford. *J. Chem. Phys.* **1997**, *106*, 6354–6364.
51. Campolieti, G., and P. Brumer. *J. Chem. Phys.* **1997**, *107*, 791–803.
52. Meyer, S. and V. Engel. *J. Phys. Chem. A.* **1997**, *101*, 7749–7753.
53. Elghobashi, N., P. Krause, J. Manz, and M. Oppel. *Phys. Chem. Chem. Phys.* **2003**, *5*, 4806–4813.
54. Henriksen, N. E., B. Møller, and V. Engel. *J. Chem. Phys.* **2005**, *122*, 204320/1–204320/6.
55. Akagi, H., H. Fukazawa, K. Yokoyama, and A. Yokoyama. *J. Chem. Phys.* **2005**, *123*, 184305/1–184305/7.
56. Møller, K. B., H. C. Westtoft, and N. E. Henriksen. *Chem. Phys. Lett.* **2006**, *419*, 65–69.
57. Sarma, M., S. Adhikari, and M. K. Mishra. *Chem. Phys. Lett.* **2006**, *420*, 321–329.
58. Adhikari, S., S. Deshpande, M. Sarma, V. Kurkal, and M. K. Mishra. *Radiat. Phys. Chem.* **2006**, *75*, 2106–2118.
59. Sarma, M., S. Adhikari, and M. K. Mishra. *J. Chem. Phys.* **2007**, *127*, 024305/1–024305/5.
60. Sarma, M., and M. K. Mishra. *J. Phys. Chem. A* **2008**, *112*, 4895–4905.
61. Sarma, M., S. Adhikari, and M. K. Mishra. *J. Phys. Chem. A* **2008**, *112*, 13302–13307.
62. Hay, P. J., R. T. Pack, R. B. Walker, and E. J. Heller. *J. Phys. Chem.* **1982**, *86*, 862–865.
63. Sheppard, M. G., and R. B. Walker. *J. Chem. Phys.* **1983**, *78*, 7191–7199.
64. Adler-Golden, S. M., S. R. Langhoff, C. W. Bauschlicher, and G. D. Carney. *J. Chem. Phys.* **1985**, *83*, 255–264.
65. Chasman, D., D. J. Tannor, and D. G. Imre. *J. Chem. Phys.* **1988**, *89*, 6667–6675.
66. Le Quéré, F., and C. Leforestier. *J. Chem. Phys.* **1990**, *92*, 247–253.
67. Le Quéré, F., and C. Leforestier. *J. Chem. Phys.* **1991**, *94*, 1118–1126.
68. Balakrishnan, N., and G. D. Billing. *J. Chem. Phys.* **1994**, *101*, 2968–2977.
69. Leforestier, C., F. Le Quéré, K. Yamashita, and K. Morokuma. *J. Chem. Phys.* **1994**, *101*, 3806–3818.
70. Miller, R. L., A. G. Suits, P. L. Houston, R. Toumi, J. A. Mack, and A. M. Wodtke. *Science.* **1994**, *265*, 1831–1838.
71. Svanberg, M., J. B. C. Pettersson, and D. Murtagh. *J. Chem. Phys.* **1995**, *102*, 8887–8896.
72. Amstrup, B., and N. E. Henriksen. *J. Chem. Phys.* **1996**, *105*, 9115–9120.
73. Barinovs, Ğ., N. Marković, and G. Nyman. *Chem. Phys. Lett.* **1999**, *315*, 282–286.
74. Parlant, G. *J. Chem. Phys.* **2000**, *112*, 6956–6958.
75. Lin, S. Y., K. L. Han, and G. Z. He. *J. Chem. Phys.* **2001**, *114*, 10651–10661.
76. Lin, S. Y., K. L. Han, and G. Z. He. *Chem. Phys.* **2001**, *273*, 169–174.
77. Lee, S. K., D. Townsend, O. S. Vasyutinskii, and A. G. Suits. *Phys. Chem. Chem. Phys.* **2005**, *7*, 1650–1656.
78. Qu, Z.-W., H. Zhu, S. Yu. Grebenshchikov, and R. Schinke. *J. Chem. Phys.* **2005**, *122*, 191102/1–191102/4.
79. Baloïtcha, E., and G. G. Balint-Kurti. *J. Chem. Phys.* **2005**, *123*, 014306/1–014306/11.
80. Qu, Z.-W., H. Zhu, S. Yu. Grebenshchikov, and R. Schinke. *J. Chem. Phys.* **2005**, *123*, 074305/1–074305/12.
81. Grebenshchikov, S. Yu., R. Schinke, Z.-W. Qu, and H. Zhu. *J. Chem. Phys.* **2006**, *124*, 204313/1–204313/13.
82. Garashchuk, S., V. A. Rassolov, and G. C. Schatz. *J. Chem. Phys.* **2006**, *124*, 244307/1–244307/8.

83. Brouard, M., R. Cireasa, A. P. Clark, G. C. Groenenboom, G. Hancock, S. J. Horrocks, F. Quadrini, G. A. D. Ritchie, and C. Vallance. *J. Chem. Phys.* **2006**, *125*, 133308/1–133308/16.
84. Grebenshchikov, S. Yu., Z.-W. Qu, H. Zhu, and R. Schinke. *Phys. Chem. Chem. Phys.* **2007**, *9*, 2044–2064.
85. Brouard, M., A. Goman, S. J. Horrocks, A. J. Johnsen, F. Quadrini, and W.-H. Yuen. *J. Chem. Phys.* **2007**, *127*, 144304/1–144304/14.
86. Tapavicza, E., I. Tavernelli, U. Rothlisberger, C. Filipi, and M. E. Casida. *J. Chem. Phys.* **2008**, *129*, 124108/1–124108/19.
87. Schinke, R., G. C. McBane, L. Shen, P. C. Singh, and A. G. Suits. *J. Chem. Phys.* **2009**, *131*, 011101/1–011101/4.
88. Schinke, R., and G. C. McBane. *J. Chem. Phys.* **2010**, *132*, 044305/1–044305/16.
89. Ndengué, S. A., F. Gatti, R. Schinke, H.-D. Meyer, and R. Jost. *J. Phys. Chem. A* **2010**, *114*, 9855–9863.
90. McBane, G. C., L. T. Nguyen, and R. Schinke. *J. Chem. Phys.* **2010**, *133*, 144312/1–144312/10.
91. Heller, E. J. *Acc. Chem. Res.* **1981**, *14*, 368–375.
92. Das, S., and D. J. Tannor. *J. Chem. Phys.* **1989**, *91*, 2324–2332.
93. Marston, C. C., and G. G. Balint-Kurti. *J. Chem. Phys.* **1989**, *91*, 3571–3576.
94. Kosloff, R. *J. Phys. Chem.* **1988**, *92*, 2087–2100.
95. Feit, M. D., J. A. Fleck, and A. Steiger. *J. Comput. Phys.* **1982**, *48*, 412–433.
96. Gross, P., D. Neuhauser, and H. Rabitz. *J. Chem. Phys.* **1992**, *96*, 2834–2845.
97. Mishima, K., and K. Yamashita. *J. Chem. Phys.* **1998**, *109*, 1801–1809.
98. Reimers, J. R., and R. O. Watts. *Mol. Phys.* **1984**, *52*, 357–381.
99. Staemmler, V., and A. Palma. *Chem. Phys.* **1985**, *93*, 63–69.
100. Engel, V., R. Schinke, and V. Staemmler. *J. Chem. Phys.* **1988**, *88*, 129–148.
101. Dutta, P., S. Adhikari, and S. P. Bhattacharyya. *Chem. Phys. Lett.* **1993**, *212*, 677–684.
102. Kosloff, D., and R. Kosloff. *J. Comput. Phys.* **1983**, *52*, 35–53.
103. Leforestier, C., R. H. Bisseling, C. Cerjan, M. D. Feit, R. Friesner, A. Guldberg, A. Hammerich, G. Jolicard, W. Karrlein, H.-D. Meyer, N. Lipkin, O. Roncero, and R. Kosloff. *J. Comput. Phys.* **1991**, *94*, 59–80.
104. Gross, P., D. B. Bairagi, M. K. Mishra, and H. Rabitz. *Chem. Phys. Lett.* **1994**, *223*, 263–268.
105. De Vries, M. S., N. J. A. Van Veen, and A. E. De Vries. *Chem. Phys. Lett.* **1978**, *56*, 15–17.
106. Guo, H. *J. Chem. Phys.* **1993**, *99*, 1685–1692.
107. Bony, H., M. Shapiro, and A. Yogev. *Chem. Phys. Lett.* **1984**, *107*, 603–608.
108. Vandana, K., N. Chakrabarti, N. Sathyamurthy, and M. K. Mishra. *Chem. Phys. Lett.* **1998**, *288*, 545–552.

8 Theoretical Framework for Charge Carrier Mobility in Organic Molecular Solids

S. Mohakud, Ayan Datta, and S. K. Pati

CONTENTS

8.1 INTRODUCTION TO ORGANIC MATERIALS

Organic electronics, which is a vibrant field of research spanning physics, chemistry, material science, engineering, and technology, has long been a subject of immense interest due to the realization that Si electronics would reach the physical limits very soon [1]. A few major breakthroughs, particularly the realization of molecule-based conductors, together with the miniaturization of devices from microscale to nanoscale and the discovery of electroluminescence, which opens the way for the fabrication of light-emitting diodes, have further fueled interest in this area of research. The main challenge in the field of organic electronics is to design efficient optoelectronic devices by making use of organic materials, instead of traditional inorganic materials. Organic materials are emerging as promising candidates for the fabrication of various electronic devices, such as light-emitting diodes [2–4], field-effect transistors [5,6], solar cells, and photovoltaic [7,8]. In these materials, the overlap of unhybridized p_z orbitals form extended conjugation with delocalized π-electrons. The interplay between the π-electron and the geometric structure in conjugated materials uncovers a rich variety of new concepts, giving rise to many fascinating properties [8,9]. Owing to many attractive features, such as ease of synthetic modification, fabrication, processing, and fine-tuning, these π-conjugated materials exhibit potential advantages over inorganic materials and have become active elements for many electronic devices.

These π-conjugated organic materials are mainly categorized into two groups: (1) small oligomers/molecules and (2) long polymers. Small π-conjugated oligomers or crystals are processed by vacuum sublimation techniques under controlled conditions, which results in well-defined crystal structures with limited impurities. These crystals are the ideal test bed for the investigation of fundamental parameters affecting charge transport phenomena. Over the past few decades, there have been wide investigations on various single-molecular crystals/oligomers. Among them, the oligo-acenes and its derivatives, which include anthracene, tetracene, pentacene, and rubrene [10–12], oligothiophenes (particularly sexithiophene) [13,14], triphenylamines [15,16], perylenes [17,18], tetrathiafulvalene [19,20], and fullerenes [21], have found wide applications.

Alongside their potential advantages, organic materials also pose serious challenges that prevent their industrial applications. Most of conjugated materials are p-type materials with much smaller mobilities ($10^{-2}cm^2V^{-1}s^{-1}$) than inorganic materials, except for single crystals like perylene and its related derivatives. This has been a major stumbling block for the integration of many electronic devices, such as organic field-effect transistors (OFETs). Another key challenge is to improve the lifetime, as well as charge transport, of these organic materials, which are very critical for electronic applications.

Such challenges, in recent years, have generated great scientific and technological interest for the microscopic understanding and improvement of charge transport phenomena in π-conjugated organic materials [3,4,9,17,22,23]. The key quantity that characterizes the charge transport phenomena is carrier mobility. The accurate estimation of carrier mobility has been a fundamental and challenging issue from both experimental and theoretical perspectives [3,4,5,9,14,16,17,22,23–25].

8.2 CHARGE CARRIER MOBILITY (μ)

The charge carrier mobility (μ) of a material is defined as the ratio between the drift velocity of the charge carrier (v) induced by the electric field and the amplitude of the applied electric field (F) ($\mu = v/F$). In general, carrier mobility is dictated by the diffusion coefficient (D) since the charge transport follows a diffusive mechanism. Carrier mobility is related to the diffusion coefficient via the Nernst–Einstein equation:

$$\mu = \frac{eD}{K_B T}. \tag{8.1}$$

The carrier mobility in conjugated organic materials is very low as compared to that in inorganic materials, although there have been reports on high room-temperature mobility of a few tens of $cm^2V^{-1}s^{-1}$ for single organic crystals. However, such carrier mobilities strongly depend on the chemical structure and preparation of the sample, the processing conditions, and the measurement techniques. Various experimental techniques have been developed to characterize charge carrier mobilities. The most widely referred approaches are the time of flight [26,27], field-effect transistor configuration [5,28,29], diode configuration [30], and pulse radiolysis time-resolved microwave conductivity techniques [31,32], which have elaborately been discussed in several literatures. On the other hand, theoretically, there have been two basic models, namely (1) band and (2) hopping, which describe charge transport and, hence, carrier mobility.

8.3 BAND MODEL

In case of traditional inorganic semiconductors, which are covalently bonded, the formation of bands, i.e., valence and conduction bands, is strong and distinct with a typical band gap of 1–3 eV. The interactions between the electron and lattice vibration (phonons) in these materials are generally smaller as compared to electronic interactions leading to the scattering of only delocalized charge carriers. Thus, the charge transport in inorganic materials is mainly realized due to the wavelike propagation of charge carriers in their well-constructed valence or conduction bands. Hence, the carrier mobilities in such materials achieve very large values in excess of 100 $cm^2V^{-1}s^{-1}$. However, occasional scattering with lattice vibrations and dislocations is always expected during the coherent motion of delocalized charge carriers. As the phonon populations increase with rise in temperature, the degree of scattering increases, which, in turn, reduces the carrier mobility.

In case of organic single crystals or polymers, the weak van der Waals forces and π–π interactions play the major role in holding the constituent molecules bound together and control the packing of the molecules in crystals or thin films. Because of these weak forces, the self-organization and band formation in the organic materials are quite poor. The formation of narrow band, presence of disorder, and electron–phonon interaction restrict the validity of a bandlike charge transfer mechanism in such π-conjugated organic materials. However, there have been reports from several measurements on a few single crystals that obey the band mechanism at low temperature, followed by a band-to-hopping crossover at high temperature. For example, the bandlike transport behavior has been observed in the c′ direction of a naphthalene single crystal below 100 K [33–35]. A similar behavior in the single crystals of rubrene and purified pentacene has also been demonstrated by several recent measurements within temperature ranges of 170–300 K and 225–340 K, respectively [10–12,36]. This bandlike transport behavior has successfully been described within the Holstein–Peierls model, coupled with first-principle calculations, and has widely been studied [25,36]. However, the major drawback of this approach is overestimation of the carrier mobility in case of pure and ordered single crystals, that is, the calculated mobility is two orders of magnitude larger than the experimental value. To overcome such limitations and to understand this bandlike transport behavior in detail, many new and modified approaches that take the quantum corrections and effect of thermal fluctuations into consideration have been developed in recent years [36,37].

8.4 HOPPING MODEL

At high temperature, the phonon populations become quite larger in strong electron–phonon interactions. The existence of narrow band, as explained earlier, and the strong electron–phonon interactions, along with the structural disorder of the system, cause the confinement of charge carriers in localized polaronic states [9,14,16,17,22–24]. For this reason, the mean-free path for the scattering of charge carriers becomes comparable to the order of intermolecular spacing, which enforces the charge carriers to hop between adjacent localized states, leading to a process known as the thermally activated hopping mechanism. The thermally activated hopping process is the dominant mechanism of charge transport in organic materials at room temperature, where the rate of hopping in each step is described within the semiclassical Marcus theory [38,39].

8.5 THEORETICAL FORMALISM

The charge transfer process between two spatially separated identical molecules can be understood from the following reaction, which is of the type

$$M + M^{\pm} \rightarrow M^{\pm} + M \tag{8.2}$$

for holes and electrons, respectively. M represents the molecule or oligomer undergoing charge transfer. Each hopping process in the π-conjugated organic materials can be understood as a nonadiabatic charge transfer reaction within the semiclassical Marcus theory formalism [38,39], which was originally formulated in 1956. The rate of charge transfer (W) between the initial and the final states can be derived from the Fermi's golden rule, taking the Born Oppenheimer approximation and Franck–Condon principle into consideration. The details of the derivation of the semiclassical Marcus theory are out of the scope of this chapter and have been given elsewhere [38,39].

Now, the rate (W) of charge transfer between the pair of molecules (m, n) at a fixed temperature (T) is expressed as

$$W = \frac{2H_{mn}^2}{h}\left(\frac{\pi^3}{\lambda K_B T}\right)^{\frac{1}{2}} \exp\left(-\frac{\lambda}{4K_B T}\right), \tag{8.3}$$

where H_{mn} is the coupling matrix element between the pair (m, n) of molecules, λ is the reorganization energy, and K_B is the Boltzmann constant. From the above expression, it is clear that the rate of hopping would be high if the reorganization energy is low and the intermolecular coupling is high.

8.5.1 Reorganization Energy

The reorganization energy quantifies the energy cost by locally charging a single molecule within the molecular crystals during the charge transfer process. Mainly, the total reorganization energy of the material includes modification of the molecular geometry (inner sphere) and the surrounding medium due to polarization effect (outer sphere), with the addition and removal of a charge carrier, and is expressed as $\lambda = \lambda_i + \lambda_o$, where λ_i and λ_o are the inner- and outer-sphere reorganization energy, respectively. The outer-sphere reorganization energy is difficult to quantify as it involves both electronic polarization and electron–phonon coupling of the surrounding molecules and, thus, becomes computationally expensive. Moreover, it has been predicted that the contribution of the outer-sphere reorganization energy is very small as compared to that of the inner-sphere reorganization energy. Because of its high computational cost and small magnitude, the outer-sphere contribution has generally been neglected. However, it is noteworthy that the reorganization energies of the molecules inside the crystalline environment are different from those in the gas phase. Thus, to consider the environmental effect explicitly, the embedded cluster approach, which reflects the structural modification of the desired molecule in the presence of the surrounding molecules, has to be adopted. In this approach, the geometry of the desired ionic (cationic or anionic) molecule is relaxed in the presence of neighboring molecules by freezing their positions. This method provides a very precise estimate of the reorganization energy of a molecule in its crystal and has extensively been used in the literature [20]. The inner-sphere contribution also consists of two relaxation energy terms [14,38,39] ($\lambda_i = \lambda_1 + \lambda_2$), such as the difference between the energies of the neutral molecule in its equilibrium geometry and that in the relaxed ionic geometry (λ_1) and the energy difference between the radical ion in its equilibrium geometry and that in the neutral geometry (λ_2).

Figure 8.1 presents the potential energy surfaces for both the neutral and the charged electronic states as a function of the reaction coordinates. As can be seen,

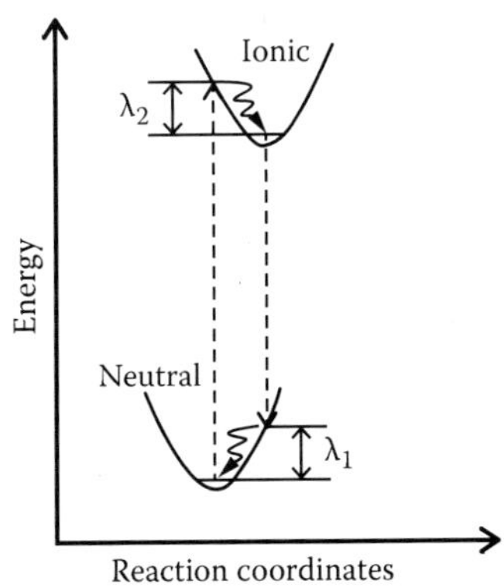

FIGURE 8.1 Potential energy surfaces of the neutral and ionic molecules represent the charge transfer.

the charge transfer occurs vertically according to the Franck–Condon principle. The reorganization energies are calculated as

$$\lambda_{\text{hole(electron)}} = \left(E^*_{\text{cation(anion)}} - E\right) + \left(E^*_{+(-)} - E_{+(-)}\right), \tag{8.4}$$

where E is the optimized ground-state energy of the neutral molecule, $E^*_{\text{cation(anion)}}$ is the energy of the neutral molecule in cationic (anionic) geometry, $E^*_{+(-)}$ is the energy of the cationic (anionic) molecule in neutral geometry, and $E_{+(-)}$ is the optimized energy of the cationic (anionic) molecule.

8.5.2 Transfer Integral

The charge transfer integral reflecting the strength of interaction between the molecular pairs plays a key role in understanding the charge transport properties. At the molecular level, with these integrals being greatly affected by nature, the size and relative orientations of the interacting monomer units establish the structure–property relationship. The charge transfer integral is defined by the matrix element $H_{mn} = \langle \psi_m | H | \psi_n \rangle$, where H is the electronic Hamiltonian of the system and ψ_m and ψ_n are the wave functions of two charge localized states. Although accurate determination of these coupling matrix elements is a very tedious and challenging issue in this area of research, there have been reports on a few simplified approaches that provide the most reliable estimation of this parameter. Here, in this chapter, a few most simplified and widely studied approaches known as the dimer-splitting method [9,17,22–24], which is based on Koopman's Theorem [40], and the fragment orbital approach for the evaluation of charge transfer integrals are presented.

8.5.2.1 Dimer-Splitting Method

The dimer-splitting method is based on the realization that the absolute value of the transfer integral for hole (electron) is half of the valence (conduction) bandwidth, that is, the energy difference between the two highest occupied (lowest unoccupied) molecular orbitals [HOMO and HOMO-1 (LUMO and LUMO+1)] in a dimer. In Figure 8.2, a schematic representation of the molecular-level splitting of a dimer is shown. This approach provides reasonable estimation of the charge transfer integrals on symmetric dimers, where spatial overlap between the molecular orbitals is negligible.

8.5.2.2 Fragment Orbital Approach

The fragment orbital approach, considering the spatial overlap between the molecular orbitals, provides accurate estimation of the charge transfer integrals [41,42]. Within this approach, the dimer molecular levels are expressed as the linear combination of individual monomer molecular levels (fragment orbitals), and the charge transfer integral $\left(H'_{mn}\right)$ can be obtained as the off-diagonal elements of the Kohn–Sham Hamiltonian matrix, which is expressed as

$$\mathrm{H}_{KS} = \mathrm{SCEC}^{-1}, \tag{8.5}$$

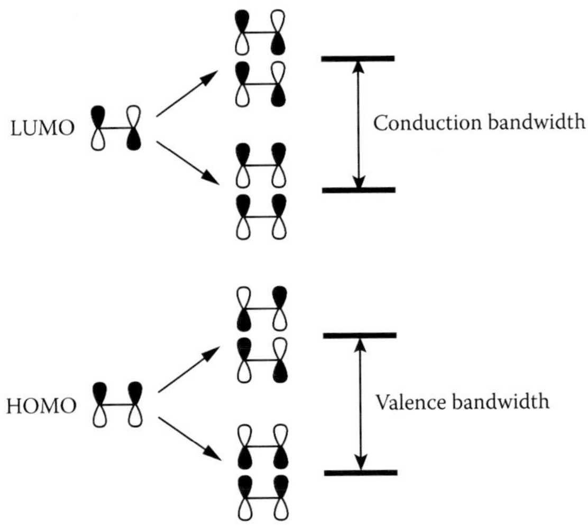

FIGURE 8.2 Schematic representation of splitting of dimer molecular levels.

where S is the intermolecular overlap matrix, C is the molecular orbital coefficient, and E is the molecular orbital energy. This procedure allows direct calculations of the charge transfer integrals, including signs, without invoking the assumption of negligible spatial overlap. The generalized charge transfer integral in the orthogonal basis can then be calculated using Lowdin transformation, which is expressed as

$$H_{mn} = H'_{mn} - \frac{1}{2} S\left(E_m + E_n\right). \tag{8.6}$$

8.5.3 Estimation of Mobility

With the estimated values of the control parameters, the rate of charge transfer can be computed. Since the charge transfer phenomena are of the diffusive type in the absence of any external potential, the diffusion coefficient, which is related to the hopping rate between pairs of the molecules, can be calculated as

$$D = \frac{1}{2d} \frac{\sum_i r_i^2 W_i^2}{\sum_i W_i}, \tag{8.7}$$

where d is the dimensionality, r is the distance between the pairs of molecules considered, and W_i is the probability for the charge carrier to a particular ith neighbor, normalized over the total hopping rate $\left(\sum_i W_i\right)$. At a given temperature, the final drift mobility due to hopping can then be evaluated from the Einstein relation, as mentioned earlier. However, this drift mobility is strongly influenced by many factors, including molecular packing, impurities, temperature, pressure, external field, carrier density, molecular weight, and size.

8.6 CALCULATIONS ON ORGANIC MOLECULAR SOLIDS

In this chapter, the control parameters and charge carrier mobilities of a few molecular crystals are estimated using the density functional theory and semiempirical methods. By performing extensive computation, the variation in electron and hole mobilities for different polymorphs of benzene, naphthalene, and octathio[8]circulene molecular crystals are studied systematically.

8.6.1 Mobility in Polymorphs of Benzene and Naphthalene

In Figure 8.3, the unit cells for the two polymorphs each of the benzene and naphthalene crystals, as retrieved from the Cambridge crystallographic database, are shown. For benzene, while the low-pressure phase (benzene I) is in Pbca group, the high-pressure phase (benzene II) crystallizes in the P21/c group [43]. Similarly, for naphthalene, the low-pressure phase (naphthalene I) has the P21/a group, and the high-pressure phase (naphthalene II) crystallizes in the P21/c group [44]. As clearly seen from the unit cells, the arrangements of the molecules are substantially different in different polymorphs.

For a detailed understanding of the nature of intermolecular interactions and important molecular contacts, 3 × 3 × 3 supercells for each polymorph of benzene and naphthalene are constructed, and the radial distribution functions $g(r)$ are calculated. A center-of-mass distance of 10 Å is maintained as a cutoff. As can be seen

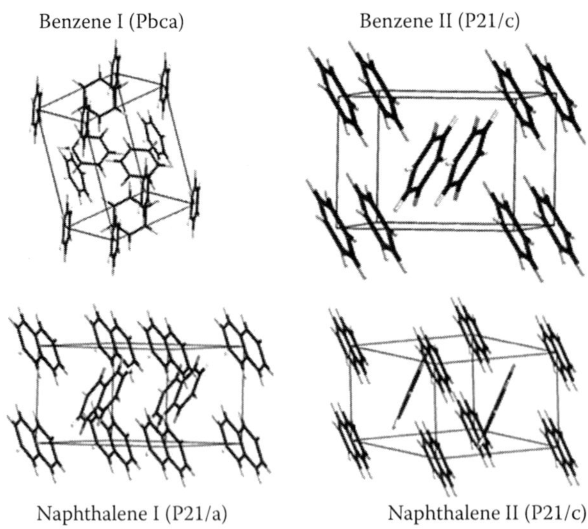

FIGURE 8.3 Unit cells and groups for the two polymorphs of benzene and naphthalene.

from the radial distribution functions for both the polymorphs in Figure 8.4, the most significant contributions appear at short intermolecular contacts. For benzene I, the most significant contribution arises from the contact at 5.08 Å, followed by the contact at 5.89 Å. For the benzene-II phase, the important contacts are at 4.55 and 5.42 Å, respectively. Similarly, while naphthalene I has three nearest-neighbor contacts [at r (center-of-mass distances) = 4.96, 5.9, and 8.66 Å], the three nearest-neighbor contacts in naphthalene II are substantially different [at r = 5.94, 5.03, and 7.78 Å]. It is very important to realize that contact pairs at larger distances do not contribute to conductance as the molecular orbitals of the monomers practically do not interact. As a consequence of this, as shown below, the essential parameter, namely H_{mn}, almost vanishes beyond a certain intermolecular distance for all dimer conformations. Thus, the computation of mobility for both the polymorphs should involve the short intermolecular contact pairs only.

However, the nature of intermolecular interactions not only depends on the center-of-mass distances between the relevant pairs but also on the orientation of the molecules with respect to each other. In Figure 8.5, the unique intermolecular contacts and their molecular structures, as derived from $g(r)$ for benzene in both the polymorphs, are shown. Benzene I has two contacts consisting of L-shaped (A) and V-shaped (B) dimers. In benzene II, the two contacts are T-shaped (A) and slipped-parallel (B) dimers. Similarly, naphthalene I consists of one T-shaped (A) dimer and two slipped-parallel (B) and (C) dimers, whereas naphthalene II consists of one slipped-parallel (A) dimer, one V-shaped (B) dimer, and one twisted slipped-parallel (C) dimer, respectively. Such different molecular orientations with respect to each other at a given distance would govern the diffusion of holes and electrons, which requires detailed modeling.

For each of the orientations of the dimers in the polymorphs of benzene and naphthalene, the center-of-mass distances between the rings are varied to understand the

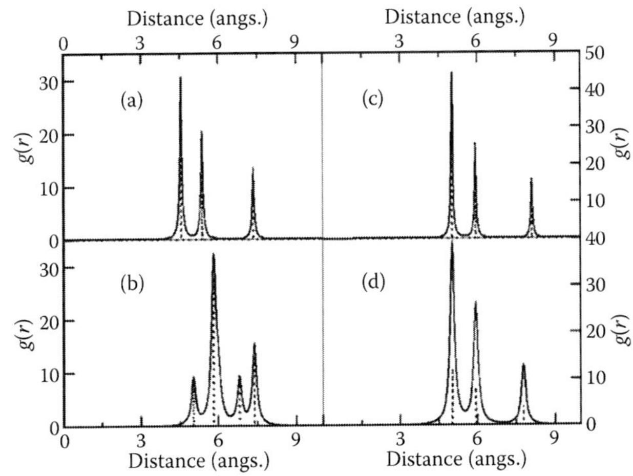

FIGURE 8.4 Radial distribution function for (a) benzene I, (b) benzene II, (c) naphthalene I, and (d) naphthalene II in 3 × 3 × 3 supercells. Note that a Lorentzian broadening (Γ) of 0.1 is used for smoothening the crystalline δ functions at the peak positions (dotted histograms).

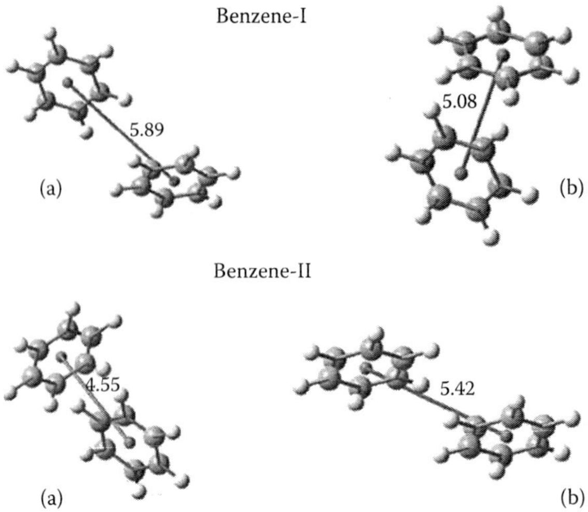

FIGURE 8.5 Unique stacking arrangements for benzene in the two phases. The balls at the center represent the centers of mass in each ring. The intermolecular distances are reported in angstroms. (a) and (b) represent the local picture of the packing configurations in the crystals.

variations of H_{mn}^{hole} and H_{mn}^{electron}, with increase or decrease in the molecular distances. However, since the dimers for both benzene and naphthalene are held together through weak dispersion forces, the suitability of nonlocal correlations within the density functional theory is known to be poor [45]. Although weak dispersion forces are accurately captured through perturbative improvement of the HF wave functions at the MP2 level, the suitability of the MP2 method needs to be verified for the excited state, particularly since H_{mn}^{electron} depends on the splitting of the monomer LUMO level into dimer LUMO and LUMO+1 levels, and the perturbation is practically valid only for the ground state. However, the semiempirical ZINDO/S Hamiltonian is well parameterized to capture the low-energy excited state properties of organic π-conjugated systems since ZINDO considers the delocalization of charges over the entire length of the system [46]. To demonstrate the selection of methods, in Figure 8.6, the H_{mn}^{hole} and H_{mn}^{electron} are plotted for a benzene dimer of type A in the benzene-I phase, as calculated from the HOMO and HOMO-1 gaps and LUMO and LUMO+1 gaps at the ZINDO/S and MP2/6-31++G (d, p) level.

As can be seen, although the MP2/6-31++G(d,p) level of calculations underestimates the H_{mn}^{hole} by 50% compared to the ZINDO/S level, it does provide a qualitatively similar picture of the monotonically decreasing behaviors of H_{mn}^{hole} with increase in the center-of-mass distances between the benzene rings. However, the MP2/6-31++G(d,p) level of calculations fails miserably for calculations of H_{mn}^{electron}. While the ZINDO/S calculations predict monotonically decreasing behaviors of H_{mn}^{electron} with increase in the center-of-mass distances, the MP2/6-31++G (d, p) level of calculations shows a flat profile. Of course, a simple electrostatic theory predicts that, as one increases the distance between monomers, the dipolar splitting becomes negligible, leading to the degeneracy of the HOMO and HOMO-1 levels and the

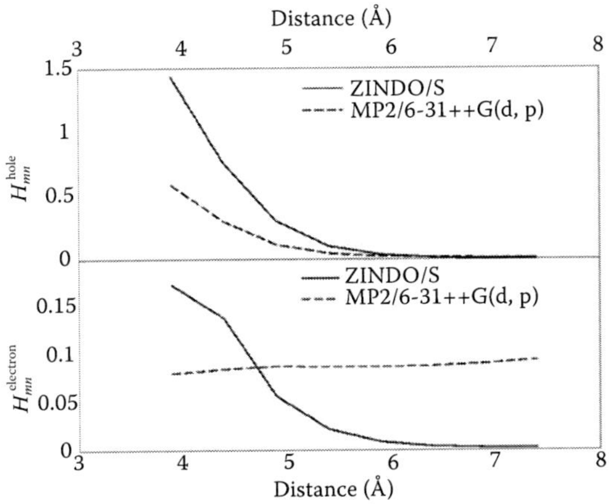

FIGURE 8.6 Comparison of the transfer matrix element for hole conductance $\left(H_{mn}^{\text{hole}}\right)$ and electron conductance $\left(H_{mn}^{\text{electron}}\right)$ (both in electron volts) at the ZINDO/S and MP2/6-31++G (d, p) levels for type A stacking for the benzene-I phase with increase in the center-of-mass distances (in angstroms).

LUMO and LUMO+1 levels, essentially to monomer levels [47]. This is indeed well reproduced through the ZINDO/S calculations. Thus, all the calculations for H_{mn}^{hole} and H_{mn}^{electron} are performed at the ZINDO/S level. It is also verified that the time-dependent density functional theory calculations at the B3LYP/6-31++G (d, p) level fail to account for the correct LUMO and LUMO+1 gaps at large separation of the monomers. Another inference from the comparison of H_{mn}^{hole} and H_{mn}^{electron} is that H_{mn}^{hole} is approximately ten times larger than H_{mn}^{electron}. This can be easily understood by comparing the number of nodes in the frontier orbitals for the dimers. The LUMO and LUMO+1 levels in the dimer have a total of four nodes, i.e., two from each monomer, whereas the HOMO-1 and HOMO levels in the dimer have two nodes, i.e., one node from each molecule. This leads to larger stabilization/destabilization of the HOMO-1/HOMO levels compared to the stabilization/destabilization in the LUMO/LUMO+1 levels, respectively. Thus, the HOMO-1 and HOMO gap is larger than the LUMO and LUMO+1 gap in the dimer. This effectively translates into larger (smaller) HOMO (LUMO) bandwidths in such stacking environments. However, this may not be the case in all the different stacks considered for benzene and naphthalene, and essentially requires computations of H_{mn}^{hole} and H_{mn}^{electron} in all the unique dimer pairs. The increase in the intermolecular distances in a dimer leads to the decrease in the transfer matrix element for both H_{mn}^{hole} and H_{mn}^{electron}.

Closed-shell calculations for the singlet neutral structures of benzene and naphthalene and open-shell calculations for the doublet cationic and anionic geometries at the 6-31G++(d,p) basis set level are carried out. Electron correlation has been included according to the Density Functional Theory (DFT) method using Becke's three-parameter hybrid formalism and the Lee–Yang–Parr functional, B3LYP, available in the Gaussian electronic structure set of codes [48–50]. The geometries of benzene and naphthalene are optimized with the removal of vibrational instabilities at the

B3LYP/6-31G++(d,p) level [49]. The aromatic rings for both benzene and naphthalene remain planar in both the cationic and anionic geometries. However, an important inference obtained by comparing the geometries of the cationic and anionic analogs of benzene and naphthalene with their neutral geometries is that, whereas the anionic analog of benzene shows bond lengths and bond angles that are quite close to the neutral geometry, the cationic geometries show a substantial bond length alteration. On the contrary, in the case of naphthalene, the cation retains the geometry closer to the neutral form, whereas the naphthalene anion shows substantial distortions. This suggests that a smaller nuclearity aromatic ring, such as benzene, has a preference to form anions, and for a larger nuclearity molecule with fused aromatic rings, such as naphthalene, formation of a cation is more favorable. Calculations by Deng and Goddard for larger rings such as anthracene, tetracene, and pentacene also suggest that larger rings have a propensity to form cations, instead of anions [51].

The reorganization energies for the cationic and anionic conductors are calculated as explained earlier. For benzene, $\lambda_{hole} = 0.285$ eV and $\lambda_{electron} = 0.0039$ eV. For naphthalene, $\lambda_{hole} = 0.178$ eV and $\lambda_{electron} = 0.2344$ eV. For a more reliable estimate of the reorganization energies of the molecules within the crystal, the embedded cluster approach was used, wherein the molecules are relaxed within the shell of nearest neighbors. Within the embedded cluster approach, we find that $\lambda_{hole} = 0.22$ eV and $\lambda_{electron} = 0.0024$ eV for benzene and $\lambda_{hole} = 0.09$ eV and $\lambda_{electron} = 0.15$ eV for naphthalene. Thus, consideration of the nearest-neighbor shell around a molecule reduces the reorganization energies, but qualitatively, the picture remains the same. The calculated reorganization energies clearly suggest that attachment of an electron is much favorable for benzene than naphthalene and that the latter would rather prefer to form a cation instead.

In Table 8.1, the overall calculated magnitudes of μ_{hole} and $\mu_{electron}$ for the two polymorphs of benzene and naphthalene are reported. This is computed by considering contributions from all pairs of unique interactions and their populations, which were determined from $g(r)$. For both benzene I and benzene II, $\mu_{electron}$ is larger than μ_{hole} by five to eight times. This is a direct consequence of the fact that $\lambda_{electron}$ is much smaller than λ_{hole} and overwhelms the smaller $H_{mn}^{electron}$. Both benzene I and benzene II have similar magnitudes for μ_{hole} and $\mu_{electron}$, suggesting that the crystalline environments for both phases of benzene are similar in the context of mobility. However, for both the polymorphs of naphthalene, the hole mobilities are an

TABLE 8.1
Hole and Electron Mobilities for the Two Polymorphs of Benzene and Naphthalene Crystals

Molecule	μ_{hole}	$\mu_{electron}$
Benzene I	0.51	2.84
Benzene II	0.49	3.27
Naphthalene I	1.76	0.11
Naphthalene II	0.88	0.15

Note: Mobilities are reported in units of $cm^2V^{-1}S^{-1}$.

order of magnitude larger than the electron mobilities, arising from the fact that λ_{hole} is smaller than $\lambda_{electron}$ and H_{mn}^{hole} is larger than $H_{mn}^{electron}$. Thus, for naphthalene, both transfer-integral terms and reorganization energies favor hole conductance. Naphthalene I has a μ_{hole} twice in magnitude than that of naphthalene II. The larger μ_{hole} in naphthalene I is due to the presence of the stacking pairs of type A (intermolecular separation = 4.96 Å), which leads to larger H_{mn}^{hole} [52].

Similarly, calculations were also performed on the α and β phases of perylene crystals [53]. The μ_{hole} and $\mu_{electron}$ values for the α-phase of perylene are calculated as 23.4 and 67.2 $cm^2V^{-1}s^{-1}$, respectively. For the β-phase, μ_{hole} and $\mu_{electron}$ are found to be 8.3 $cm^2V^{-1}s^{-1}$ and 20.4 $cm^2V^{-1}s^{-1}$, respectively. Thus, for both the polymorphs, the electron mobilities are at least two orders of magnitude larger than the hole mobilities. The α-phase has $\mu_{electron}$ three times that for the β-phase. Such a large electron mobility in the α-phase is due to the slipped-parallel arrangement of molecules at short intermolecular distances. Note that, although the β-phase has a slipped-parallel orientation of the monomers, the intermolecular distances are larger (d = 5.84 Å), which results in smaller $H_{mn}^{electron}$ values. In fact, the presence of such slipped-parallel π-stacked arrangements of molecules (d = 3.4 Å) for perylene derivatives like dicyanoperylene-3,4,9,10-bis(dicarboximide) (PDI-CN2) has been found to give rise to large preferential electron mobilities [18,54].

8.6.2 Mobility in Octathio[8]circulene: Sulflower

Octathio[8]circulene, a member of the family of oligothiophene systems, has been synthesized in 2007 [55]. The octathio[8]circulene, comprising eight fused π-conjugated annulated thiophene rings, is highly symmetric and is popularly known as sulflower. As can be seen in Figure 8.7a, this is a heterocyclic compound with a highly planar structure of large surface area, which has been shown theoretically for a large weight percentage of hydrogen adsorption [56]. Figure 8.7b presents the molecular packing of the crystal. The molecules are situated on the top of each other at a distance of about 3.9 Å in a slipped-parallel manner in one columnar layer and between the columnar layers; they make a tilting angle of 131° with a distance of 10.08 Å.

For a detailed characterization and understanding of different dimeric arrangements in the crystal, the radial distribution function $g(r)$ is calculated, considering a super cell of dimension 3 × 3 × 3 $Å^3$ and maintaining a cutoff distance of 15 Å.

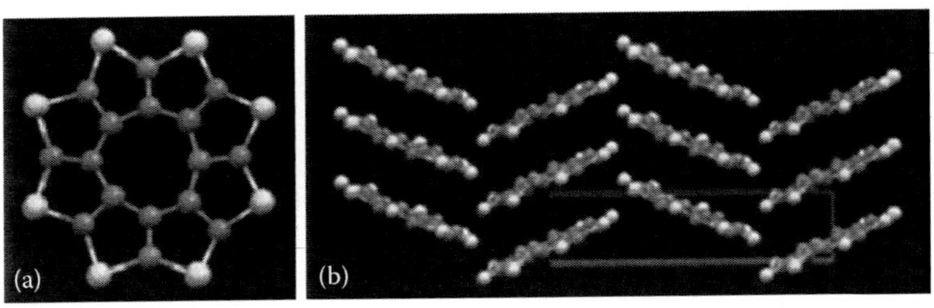

FIGURE 8.7 (a) Structure of a single octathio[8]circulene molecule. (b) 3-D view of the molecular packing.

As can be seen in Figure 8.8, the most significant distribution of molecules arises at molecular separations of 3.9, 7.81, 10.08, and 11.15 Å. Interestingly, the probability at 7.81 Å corresponds to the next nearest molecular distance, arranged in slipped-parallel manner within the columnar layer. Depending on the intermolecular separations and molecular orientations, three unique molecular pairs are selected, namely slipped-parallel (3.9 Å), tilted (10.08 Å), and axial (11.15 Å) dimers for the calculations. The hole and electron transfer matrix elements for each pair of molecules are calculated within the framework of the dimer-splitting method by varying the center-of-mass distances between them and keeping their relative orientations unchanged.

The well-established semiempirical ZINDO/S method has been used to calculate the hole and electron transfer matrix elements for all the selected dimers. Both H_{mn}^{hole} and H_{mn}^{electron} show a monotonic decreasing behavior due to the fact that the interaction between the molecular orbitals of the two molecules reduces with the increase in intermolecular distance, resulting in a negligible exciton splitting with degeneracy in the individual monomeric orbitals [57].

In case of slipped-parallel dimers, since the molecular distance is small and two molecules are situated in a stacked geometric configuration, the molecular orbitals interact very strongly and give larger excitonic splitting at the dimer level. In such a dimer, the LUMO and LUMO+1 splitting (0.108 eV) is more compared to HOMO and HOMO-1 splitting (0.062 eV). However, in case of tilted dimers, the monomers are oriented at an angle of 131° with a large separation between the centers of masses to minimize the electrostatic repulsions between the sulfur atoms. Because of these spatial orientations, the interactions become very small, leading to a diminished orbital splitting. The dimer orbital splitting of HOMO and HOMO-1 is found to be 0.022 eV, whereas the splitting of LUMO and LUMO+1 is 0.036 eV for this case. The dimer with intermolecular separation of 7.81 Å, being the next nearest neighbors, interacts very weakly and gives negligibly small splitting of molecular orbitals ($H_{mn}^{\text{electron}} = 0.00008$ eV and $H_{mn}^{\text{hole}} = 0.0001$ eV). For the largest distance molecular

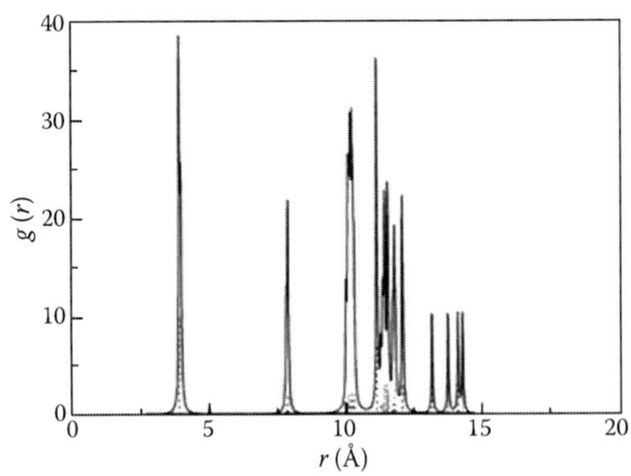

FIGURE 8.8 Radial distribution function $g(r)$ for the octathio[8]circulene crystal in supercells of dimension 3 × 3 × 3 $Å^3$. Note that a Lorentzian broadening (Γ) of 0.06 is used for smoothening the crystalline δ-functions at the peak positions (dotted histogram).

arrangement (at a distance 11.15 Å), the dimer molecular-level splitting is also quite small: $H_{mn}^{electron} = 0.0139$ eV and $H_{mn}^{hole} = 0.015$ eV. Due to the negligible contribution of the dimer with 7.81-Å distance, the contributions from only three types of molecular dimers are considered for the calculations. It is found that the LUMO splitting is larger than the HOMO splitting for each of the dimer orientation. To verify this model in the bulk, periodic DFT-based calculations were performed, and the densities of states are computed using generalized gradient approximation for the Perdew, Burke, and Ernzerho exchange–correlation energy functional within a SIESTA set of codes [58]. A double ξ-basis set with the polarized orbitals has been included for all the atoms. Figure 8.9 represents the density of states (DOS) for the octathio[8] circulene crystal. It is clearly seen that the conduction (LUMO) bandwidth (0.074 eV) in this crystal is larger than the valence (HOMO) bandwidth (0.036 eV), which supports the qualitative estimations of results obtained through the ZINDO/S level of calculations.

The reorganization energy for the electron ($\lambda_{electron}$) and hole (λ_{hole}) carriers are calculated by optimizing all the anionic and cationic structures of individual molecular geometry with B3LYP exchange and correlations using three different basis sets 3-21G, 6-31G, and 6-31G++(d, p), as implemented in a *Gaussian03* set of codes [48–50]. The calculations indicate adequate consistency in the results obtained through different levels of calculations, which has been shown in Table 8.2.

The reorganization energy for the electron carriers is found to be larger than that for the hole carriers. This could be explained in terms of the geometrical changes that occur in different optimized structures, as well as the details of the HOMO. The neutral and cationic molecules possess highly planar geometries, whereas the structure of the anionic species is puckered. Thus, putting an extra electron to the neutral molecule, leading to the anionic one, results in the puckering of the geometry. Because of the puckered geometry, the anionic species show more electron density on the convex side than that on the concave side. The drastic change in geometry, as well as the asymmetric electron distribution in the anionic structure, enhances the electron reorganization energy than that for the hole.

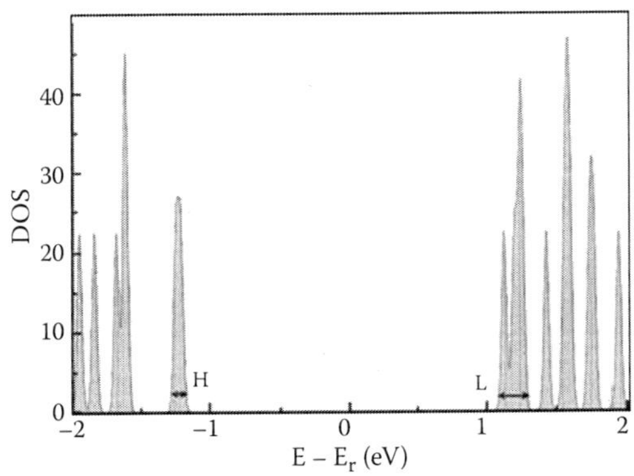

FIGURE 8.9 DOS as a function of energy, scaled with respect to the Fermi energy (EF). L and H are the LUMO and HOMO bandwidths of the sulflower crystal, respectively.

TABLE 8.2
Electron and Hole Reorganization Energies (in Units of Electron Volts) of a Single Octathio[8]circulene Molecule Obtained Using Different Basis Set Calculations

Basis Sets	$\lambda_{electron}^{molecule}$	$\lambda_{hole}^{molecule}$
3-21G	0.34	0.14
6-31G	0.34	0.13
6-31G++(d,p)	0.40	0.13

The embedded cluster approach was utilized for the calculations to include the environmental effect, as it is computationally less expensive. Within the embedded cluster method, the magnitude of reorganization energies for both the charge carriers are further reduced in comparison to individual molecules due to the presence of intermolecular interactions: $\lambda_{electron} = 0.118\text{eV}$ and $\lambda_{hole} = 0.04\text{eV}$.

With the estimated values of the transfer matrix elements and reorganization energy for both the charge carriers, the electron and hole mobilities are calculated, considering the weighted average over all the dimeric configurations with their probability density within the crystal (as shown in Figure 8.8). The electron and hole mobilities for the octathio[8]circulene crystal are found to be 2.6 and 3.5 $cm^2V^{-1}S^{-1}$, respectively. The larger hole conductance in the crystal is primarily due to the smaller hole reorganization energy. However, as can be seen, even the electron carrier mobility is also quite substantial compared to most of the organic molecular crystals, except for perylene. On the other hand, the hole mobility is an order of magnitude larger than that in perylene, benzene, naphthalene, and pentacene molecular crystals. Furthermore, the important inference is that, in this crystal, both the charge carrier mobilities are quite substantial, suggesting a high degree of ambipolar charge transport. Such π-conjugated organic semiconductors with large ambipolar carrier mobilities would be effective in photoluminescence in the radioactive decay process of the singlet excitons, formed through the recombination of electrons and holes. Organic light-emitting diodes are fabricated based on this principle. Hence, the results suggest that organic crystals like sulflower with large and similar magnitudes of charge carrier mobilities would provide better efficiency to electronic devices like OFETs [59].

8.7 CONCLUSIONS AND FUTURE DIRECTIONS

Charge transport in organic conjugated molecules has emerged as a technologically important phenomenon. However, the processes involved in this mechanism are complicated due to intermolecular forces, molecular packing, disorder, and strong coupling. In this chapter, we have tried to provide a general strategy to estimate the mobility of such aggregates. Figure 8.10 shows a flowchart where the important steps for such a computation are depicted. Such a method can be readily implemented in a molecular simulation.

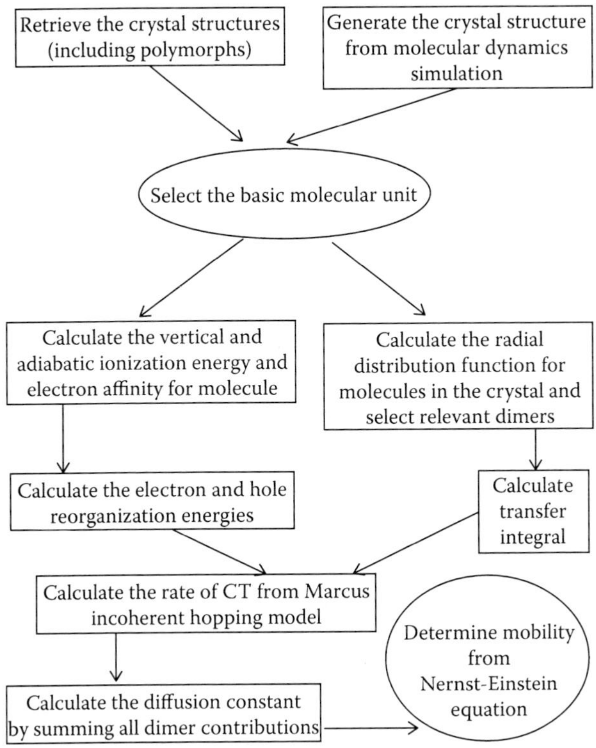

FIGURE 8.10 Flowchart showing the technique to implement the computation of mobility in a molecular solid.

In the next few years, one expects a rapid progress in both simulations in materials and better experimental control over parameters that control processes like molecular crystallization and seeding. Hence, the strategy that we are proposing here will act as a review for calculating and benchmarking new molecular crystals for their charge transfer efficiency.

ACKNOWLEDGMENTS

S. K. Pati and S. Mohakud would like to thank the Department of Science and Technology (DST), Government of India, for the research support. A. Datta would like to thank the Council of Scientific and Industrial Research and DST for the partial funding.

REFERENCES

1. Aviram, A., and M. A. Ratner. *Chem. Phys. Lett.* **1974**, 29, 277–283; Lakshmi, S., S. Dutta, and S. K. Pati. *J. Phys. Chem. C* **2008**, 112, 14718–14730, and references therein.
2. Tang, C. W., and S. A. V. Slyke. *Appl. Phys. Lett.* **1987**, 51, 913–915.
3. Burroughes, J. H., D. D. Bradley, C. A. R. Brown, R. N. Marks, K. Mackey, R. H. Friend, P. L. Burns, and A. B. Holmes. *Nature* **1990**, 347, 539–541.

4. Friend, R. H. et al. *Nature* **1999**, 397, 121–128.
5. Horowitz, G. *Adv. Mater.* **1998**, 10, 365–377.
6. Nelson, S. F., Y. Y. Lin, D. J. GunYdlach, and T. N. Jackson. *Appl. Phys. Lett.* **1998,** 72, 1854–1856.
7. Yu, G., J. Wang, J. McElvain, and A. Heeger. *J. Adv. Mater.* **1998**, 17, 1431–1434.
8. Brabec, C. J., N. S. Sariciftci, and C. J. Hummelen. *Adv. Funct. Mater.* **2001**, 11, 15–26.
9. Bredas, J. L., D. Beljonne, V. Coropceanu, and J. Cornil. *Chem. Rev.* **2004**, 104, 4971–5004.
10. Sundar, V. C., J. Zaumseil, V. Podzorov, E. Menard, R. L. Willett, T. Someya, M. E. Gershenson, and J. A. Rogers. *Science* **2004**, 303, 1644–1646.
11. Ostroverkhova, O., D. G. Kooke, F. A. Hegmann, J. E. Anthony, V. Podzorov, M. E. Gershenson, O. D. Jurchescu, and T. T. M. Palstra. *Appl. Phys. Lett.* **2006**, 88, 162101–162103.
12. Podzorov, V., E. Menard, J. A. Rogers, and M. E. Gershenson. *Phys. Rev. Lett.* **2005**, 95, 226601–226604.
13. Siegrist, T., C. Kloc, R. A. Laudise, and H. E. Katz. *Adv. Mater.* **1998**, 10, 379–382.
14. Yang, X., L. Wang, C. Wang, W. Long, and Z. Shuai. *Chem. Mater.* **2008**, 20, 3205–3211.
15. Shirota, Y. *J. Mater. Chem.* **2005**, 15, 75–93.
16. Yang, X., Q. Li, and Z. Shuai. *Nanotechnology* **2007**, 18, 424029(1–6).
17. Bredas, J. L., J. P. Calbert, D. A. daSilvaFilho, and J. Cornil. *PNAS* **2002**, 99, 5804–5809.
18. Jones, B. A., M. J. Ahrens, M. H. Yoon, A. Facchetti, T. J. Marks, and M. R. Wasielewski. *Angew. Chem.* Int. Ed. **2004**, 43, 6363–6366.
19. Rovira, C. *Chem. Rev.* **2004**, 104, 5289–5318.
20. Mas-Torrent, M., M. Durkut, P. Hadley, X. Ribas, and C. Rovira. *J. Am. Chem. Soc.* **2004**, 126, 984–985.
21. Haddon, R. C., A. S. Perel, R.C. Morris, T. T. M. Palstra, A. F. Hebard, and R. M. Fleming. *Appl. Phys. Lett.* **1995**, 67, 121–123.
22. Coropceanu, V., J. Cornil, D. A. da Silvo Filho, Y. Olivier, R. Silbey, J. L. Bredas. *Chem. Rev.* **2007**, 107, 926–952.
23. Hutchinson, G. R., M. A. Ratner, T. J. Marks, and Hopping. *J. Am. Chem. Soc.* **2005**, 127, 2339–2350.
24. Hutchinson, G. R., M. A. Ratner, and T. J. Marks. *J. Am. Chem. Soc.* **2005**, 127, 16866–16881.
25. Holstein, T. *Ann. Phys. (N. Y.)* **1995**, 8, 325–342.
26. Kepler, R. G. *Phys. Rev.* **1960**, 119, 1226–1229.
27. Leblanc, O. H. *J. Chem. Phys.* **1960**, 33, 626.
28. Horowitz, G. *J. Mater. Res.* **2004**, 19, 946.
29. Dodabalapur, A., A. Torsi, and H. E. Katz. *Science* **1995**, 268, 270–271.
30. Blom, P. W. M., M. J. M. de Jong, and J. J. M. Vleggaar. *Appl. Phys. Lett.* **1996**, 68, 3308–3310.
31. Prins, P. et al. *Adv. Mater.* **2005**, 17, 718–723.
32. Hoofman, R. J. O. M., M. P. de Haas, L. D. A. Siebbeles, and J. M. Warman. *Nature* **1998**, 392, 54–56.
33. Schein, L. B. *Phys. Rev. B* **1977**, 15, 1024–1034.
34. Schein, L. B., C. B. Duke, and A. R. McGhie. *Phys. Rev. B* **1978,** 40, 197–200.
35. Schein, L. B., and A. R. Mc Ghie. *Phys. Rev. B* **1979**, 20, 1631–1639.
36. Nan, G., X. Yang, L. Wang, Z. Shuai, and Y. Zhao. *Phys. Rev. B* **2009**, 79, 115203–115211.
37. Troisi, A., and G. Orlandi. *Phys. Rev. Lett.* **2006,** 96, 086601–086604.
38. Marcus, R. A. *Rev. Mod. Phys.* **1993**, 65, 599–610.
39. Marcus, R. A. *J. Chem. Phys.* **1956**, 24, 966–978.
40. Lin, B. C., C. P. Cheng, Z. Q. You, and C. P. Hsu. *J. Am. Chem. Soc.* **2005**, 127, 66–67.

41. Senthilkumar, K., F. C. Grozema, F. M. Bickelhaupt, and L. D. A. Siebbels. *J. Chem. Phys.* **2003**, 119, 9809–9817.
42. Senthilkumar, K., F. C. Grozema, C. F. Guerra, F. M. Bickelhaupt, F. D. Lewis, M. A. Ratner, and L. D. A. Siebbeles. *J. Am. Chem. Soc.* **2005**, 127, 14894–14903.
43. Piermarini, G. J., A. D. Michell, C. E. Weir, and S. Block. *Science* **1969**, 165, 1250–1255.
44. Block, S., C. E. Weir, and G. Piermarini. *J. Science* **1970**, 169, 586–587.
45. Grimme, S. *J. Comput. Chem.* **2006**, 27, 1787–1799.
46. Ridley, J., and M. C. Zerner. *Theor. Chim. Acta* **1973**, 32, 111–134.
47. Datta, A., and S. K. Pati. *J. Chem. Phys.* **2003**, 118, 8420–8427.
48. Becke, A. D. *Phys. Rev. A* **1988**, 38, 3098–3100.
49. Lee, C., W. Yang, and R. G. Parr. *Phys. Rev. B* **1988**, 37, 785–789.
50. Frisch, M. J. et al. Gaussian03, revisionC.02; Gaussian, Inc., Walling-ford 2004.
51. Deng, W.-Q., and W. A. Goddard III. *J. Phys. Chem. B* **2004**, 108, 8614–8621.
52. Datta, A., S. Mohakud, and S. K. Pati. *J. Chem. Phys.* **2007**, 126, 144710–144716.
53. Datta, A., S. Mohakud, and S. K. Pati. *J. Mater. Chem.* **2007**, 17, 1933–1938.
54. Ahrens, M. J., and M. J. Fuller, and M. R. Wasielewski. *Chem. Mater.* **2003**, 15, 2684–2686.
55. Chernichenko, K. Y., V. V. Sumerin, R. V. Shpanchenko, E. S. Balenkova, and V. Nenajdenko. *Angew. Chem.* Int. Ed. **2006**, 45, 7367–7370.
56. Datta, A., and S. K. Pati. *J. Phys. Chem.* **2007**, C 111, 4487–4490.
57. Datta, A., and S. K. Pati. *Chem. Soc. Rev.* **2006**, 35, 1305–1323.
58. Perdew, J. P., K. Burke, and M. Ernzerhof. *Phys. Rev. Lett.* **1996**, 77, 3865–3868.
59. Mohakud, S., and S. K. Pati. *J. Mater. Chem.* **2009**, 19, 4356–4361.

9 Quantum Brownian Motion in a Spin-Bath

Sudarson Sekhar Sinha, Arnab Ghosh, Deb Shankar Ray, and Bidhan Chandra Bag

CONTENTS

9.1 INTRODUCTION

Quantum Brownian motion is a frontier area of research in quantum optics (Louisell 1973; Sargent, Scully, and Lamb 1974; Meystre and Sargent 1991), condensed matter (Grabert, Schramm, and Ingold 1988; Weiss 1999), and chemical physics (Hänggi, Talkner, and Borkovec 1990; Zwanzig 1973) over the last several decades. The fundamental paradigm is the irregular movement of a particle in a thermal bath kept at finite temperature. The bath constitutes many, in principle, infinite degrees of freedom interacting with the particle. In general, the bath is considered to be a set of harmonic oscillators of varied frequency ranges and is therefore bosonic in character. The particle follows the reduced stochastic dynamics obtained after appropriate elimination of these reservoir degrees of freedom under suitable approximation schemes. The dynamical evolution is characterized by quantum dissipation and bosonic noise related to each other by fluctuation–dissipation (FD) relation, ensuring a detailed balance in the system. This chapter concerns a system–reservoir model where the reservoir is constituted by a set of two-level atoms or spin-1/2 systems.

This is basically a generalization of a model of a harmonic oscillator coupled to a set of two-level atoms used a couple of decades ago by Sargent et al. (Sargent, Scully, and Lamb 1974) for treatment of quantum dissipation of a cavity mode in the presence of an atomic beam reservoir. The primary focus of this chapter is to understand the quantum dissipation induced by a spin-bath within a theoretical framework applicable to problems of condensed matter and chemical physics.

We begin by noting that quantum noise due to spin-bath is distinctly different from that of a harmonic bath. First, a spin-1/2 system has no classical analog. The FD relation for the spin-bath therefore does not reduce to the classical FD relation in high temperature limit just as that for a bosonic bath. Second, since the two-level bath atoms are described by Pauli spin matrices and anticommutation relations, the statistical properties of a spin-bath differ from those of a bosonic bath. These differences originate from an effective temperature-dependent spectral density function for the bath, relating its noise correlation to the quantum dissipation noted earlier by Caldeira et al. (Caldeira and Leggett 1981; Caldeira, Castro Neto, and Oliveira de Carvalho 1993; Caldeira and Leggett 1983a, 1983b) in the functional integrals in their spin-spin-bath model. While, in the limit $T \to 0$, the behavior of the spin-bath merges to that of a harmonic bath, their behaviors began to differ at finite temperatures. As the temperature rises, the bath atoms get thermally saturated, so that the effective spectral density function or system–bath coupling tends to decrease. As a result, quantum dissipation is dominated by coherence. Shao and Hänggi (Shao and Hänggi 1998) have shown in this context that temperature assists in the suppression of decoherence. Spin-bath has been the subject of study in several earlier occasions, e.g., for computation of optical conductivity and direct current resistivity of charge carriers in the presence of an applied electric field (Ferrer, Caldeira, and Smith 2006), for explaining dynamic localization (Ferrer and Smith 2007) of a particle at a low temperature and magnetic relaxation of molecular crystals (Prokof'ev and Stamp 1996, 1998), etc. Recently, in a series of papers (Sinha, Ghosh, and Ray 2011a, 2011b; Sinha et al. 2010; Ghosh, Sinha, and Ray 2012), we have proposed a scheme for the quantum Brownian motion of a particle in a spin-bath of two-level atoms. This is based on the description of the bath by spin-coherent state representation of the noise operators and a canonical thermal distribution of the associated c-numbers. The reduced dynamics is governed by a generalized Langevin equation of the quantum mechanical mean position of the particle in an external field. The treatment has been successfully applied (Sinha, Ghosh, and Ray 2011a; Ghosh, Sinha, and Ray 2011b; Ghosh, Sinha, and Ray 2011a) to understand several anomalous features of relaxation in optically excited systems, e.g., narrowing of line width and decrease in integrated absorption coefficient with temperature, temperature-assisted coherence in thermal activation, and tunneling in the decay of a metastable state in an anharmonic potential. Our aim in this chapter is twofold: (1) to present a brief overview of this development and (2) to introduce a scheme for generation of c-number noise for numerical simulation of the quantum Langevin equation for a given potential. The underlying idea lies on expressing the c-number noise of the bath as a superposition of several Ornstein–Uhlenbeck noise processes. Using two illustrative examples, we show how temperature assists coherence in the reduction of the distribution width of the position of a particle in a harmonic well and suppresses the rate of barrier crossing in a thermally activated process due to thermal saturation of the bath. The theoretical analysis has been corroborated by numerical simulation.

The outlay of this chapter is as follows: In Section 9.2, we introduce the system/spin-bath model and derive the operator Langevin equation for the particle. This is followed by a discussion on stochastic dynamics in the presence of c-number noise, highlighting the role of the spectral density function in the high- and low-temperature regimes. A scheme for the generation of spin-bath noise as a superposition of several Ornstein–Uhlenbeck noise processes and its implementation in numerical simulation of the quantum Langevin equation are described in Section 9.3. Two examples have been worked out in Section 9.4 to illustrate the basic theoretical issues. This chapter is concluded in Section 9.5.

9.2 QUANTUM BROWNIAN MOTION IN A TWO-LEVEL SPIN-BATH

9.2.1 The Model and the Operator Langevin Equation

We consider a particle of unit mass coupled to a set of two-level atoms with characteristic frequencies $\{\omega_k\}$. This is represented by the following Hamiltonian:

$$\hat{H} = \frac{\hat{p}^2}{2} + V(\hat{q}) + \hbar\sum_k \omega_k \hat{\sigma}_k^\dagger \hat{\sigma}_k + \frac{1}{2}\sum_k g_k\,\hat{q}\left(\frac{g_k}{\omega_k^2}\hat{q} - \sqrt{\frac{2\hbar}{\omega_k}}\left(\hat{\sigma}_k^\dagger + \hat{\sigma}_k\right)\right), \tag{9.1}$$

where $\hat{q}$ and $\hat{p}$ are the coordinate and momentum operators of the particle. $\hat{\sigma}_k^\dagger\left(\hat{\sigma}_k\right)$ is the creation (annihilation) operator for the kth two-level atom. The external force field acting on the particle is defined by the potential energy operator $V\left(\hat{q}\right)$. g_k is the coupling constant between the kth atom and the particle. $\hat{q}$ and $\hat{p}$ follow the usual commutation relation $\left[\hat{q},\hat{p}\right] = i\hbar$, and the two-level atoms obey the anticommutation rule $\left\{\hat{\sigma}_k, \hat{\sigma}_k^\dagger\right\} = 1$ and the following algebra:

$$\left[\hat{\sigma}_k^\dagger, \hat{n}_k\right] = -\hat{\sigma}_k^\dagger; \left[\hat{\sigma}_k, \hat{n}_k\right] = \hat{\sigma}_k; \hat{\sigma}_k^2 = \hat{\sigma}_k^{\dagger 2} = 0 \text{ and } \left[\hat{\sigma}_k^\dagger, \hat{\sigma}_k\right] = \hat{\sigma}_{zk}, \tag{9.2}$$

where the number operator for the atom is defined as $\hat{n}_k = \hat{\sigma}_k^\dagger \hat{\sigma}_k$. Furthermore, it follows from the above relations that $\hat{\sigma}_{zk} = 2\hat{n}_k - 1$. The counter term in Equation 9.1 $\left(\frac{g_k^2}{2\omega_k^2}\hat{q}^2\right)$ is for ensuring that the particle feels the potential $V(\hat{q})$, which remains unaffected by the interaction during the dynamic evolution of the particle.

The Heisenberg equations of motion for the particle and the spin-bath variables may be written as follows:

$$\dot{\hat{q}} = \hat{p} \tag{9.3}$$

$$\dot{\hat{p}} = -V'(\hat{q}) + \sum_k g_k \sqrt{\frac{\hbar}{2\omega_k}} \left(\hat{\sigma}_k^\dagger + \hat{\sigma}_k\right) - \sum_k \frac{g_k^2}{\omega_k^2}\hat{q} \tag{9.4}$$

$$\dot{\hat{\sigma}}_k^\dagger = i\omega_k \hat{\sigma}_k^\dagger + i g_k \hat{q}\sqrt{\frac{1}{2\hbar\omega_k}}\hat{\sigma}_{zk} \tag{9.5}$$

$$\dot{\hat{\sigma}}_k = -i\omega_k \hat{\sigma}_k - i g_k \hat{q}\sqrt{\frac{1}{2\hbar\omega_k}}\hat{\sigma}_{zk} \tag{9.6}$$

$$\dot{\hat{\sigma}}_{zk} = i g_k \hat{q}\sqrt{\frac{1}{2\hbar\omega_k}}\left(\hat{\sigma}_k^\dagger - \hat{\sigma}_k\right). \tag{9.7}$$

After formally integrating Equations 9.5 and 9.6 and using the results in Equation 9.4, we arrive at the operator equation for the particle:

$$\begin{aligned}\ddot{\hat{q}} = {} & -V'(\hat{q}) - \sum_k \frac{g_k^2}{\omega_k^2}\int_0^t dt'\, \dot{\hat{q}}(t')\cos\omega_k(t-t') \\ & + \sum_k g_k \sqrt{\frac{\hbar}{2\omega_k}} \left\{ \left(\hat{\sigma}_k^\dagger(0) - \frac{g_k}{\omega_k}\sqrt{\frac{1}{2\hbar\omega_k}}\hat{q}(0) \right) e^{i\omega_k t} \right. \\ & \left. + \left(\hat{\sigma}_k(0) - \frac{g_k}{\omega_k}\sqrt{\frac{1}{2\hbar\omega_k}}\hat{q}(0) \right) e^{-i\omega_k t} \right\}. \end{aligned} \tag{9.8}$$

In deriving the above equation, we approximate $\hat{\sigma}_{zk}(t') \sim \hat{\sigma}_{zk}(0)$ since Equations 9.5 through 9.7 suggest that the polarization operators $\hat{\sigma}_k^\dagger, \hat{\sigma}_k$ are governed by the time scale of free evolution (ω_k), whereas the population difference operators $\hat{\sigma}_{zk}$ follow a much slower time scale determined by the coupling constant g_k. We further note that $\hat{\sigma}_{zk}(0) = 2\hat{n}_k(0) - 1$ and assume that the system–reservoir interaction sets in at $t = 0$ through the ground levels of most of the bath atoms, i.e., $\hat{\sigma}_{zk}(0) \approx -1$. This leads to an effective linearization of the damping term in Equation 9.8. Based on these considerations, we rewrite Equation 9.8 as

$$\ddot{\hat{q}} = -V'(\hat{q}) - \sum_k \frac{g_k^2}{\omega_k^2}\int_0^t dt' \dot{\hat{q}}(t')\cos\omega_k(t-t') + \sum_k g_k \sqrt{\frac{\hbar}{2\omega_k}} \left\{ \hat{S}_k^\dagger(0) e^{i\omega_k t} + \hat{S}_k(0) e^{-i\omega_k t} \right\}, \tag{9.9}$$

where

$$\hat{S}_k(0) = \hat{\sigma}_k(0) - \frac{g_k}{\omega_k}\sqrt{\frac{1}{2\hbar\omega_k}}\hat{q}(0) \tag{9.10}$$

$$\hat{S}_k^\dagger(0) = \hat{\sigma}_k^\dagger(0) - \frac{g_k}{\omega_k}\sqrt{\frac{1}{2\hbar\omega_k}}\hat{q}(0). \tag{9.11}$$

$\hat{S}_k^\dagger(0)$ and $\hat{S}_k(0)$ are the shifted bath operators. The operator equation for the particle is then given by the following form:

$$\ddot{\hat{q}} = -V'(\hat{q}) - \int_0^t dt' \kappa(t-t')\dot{\hat{q}}(t') + \hat{f}(t). \tag{9.12}$$

Here, the memory kernel and the noise operator are expressed as

$$\kappa(t-t') = \sum_k \frac{g_k^2}{\omega_k^2} \cos\omega_k(t-t') \tag{9.13}$$

and

$$\hat{f}(t) = \sum_k g_k \sqrt{\frac{\hbar}{2\omega_k}}\left(\hat{S}_k(0)\,e^{-i\omega_k t} + \hat{S}_k^\dagger(0)\,e^{i\omega_k t}\right) = \hat{F}(t) + \hat{F}^\dagger(t), \tag{9.14}$$

respectively. The noise properties of the operator $\hat{f}(t)$ can be derived with the help of the canonical thermal distribution of bath operators at $t = 0$ as follows:

$$\left\langle \hat{f}(t)\right\rangle_{qs} = 0 \tag{9.15}$$

$$Re\left\{\left\langle \hat{F}(t)\hat{F}^\dagger(t') - \hat{F}^\dagger(t)\hat{F}(t')\right\rangle_{qs}\right\} = \frac{\hbar}{2}\sum_k \frac{g_k^2}{\omega_k}\tanh\left(\frac{\hbar\omega_k}{2KT}\right)\cos\omega_k(t-t'). \tag{9.16}$$

Here $\langle\ldots\rangle_{qs}$ refers to the quantum statistical average and is defined as

$$\left\langle \hat{A}\right\rangle_{qs} = \frac{Tr\,\hat{A}\,e^{-\hat{H}_{\text{bath}}/KT}}{Tr\,e^{-\hat{H}_{\text{bath}}/KT}}. \tag{9.17}$$

Equation 9.17 is true for any bath operator $\hat{A}$, where $\hat{H}_{\text{bath}} = \hbar \sum_k \omega_k \hat{\sigma}_k^\dagger \hat{\sigma}_k$ at $t = 0$.

In defining Equation 9.15, it is necessary to set the quantum mechanical mean position $\langle \hat{q}(0) \rangle = 0$ without any loss of generality. This allows further simplification:

$$\left\langle \hat{S}_k(0) \right\rangle_{qs} = \left\langle \hat{\sigma}_k(0) \right\rangle_{qs} \tag{9.18}$$

$$\left\langle \hat{S}_k^\dagger(0) \right\rangle_{qs} = \left\langle \hat{\sigma}_k^\dagger(0) \right\rangle_{qs}. \tag{9.19}$$

Equation 9.16 is the FD relation expressed in terms of noise operators ordered appropriately. The origin of the temperature-dependent contribution $\tanh\left(\frac{\hbar\omega_k}{2KT}\right)$ can be traced to the following averages:

$$\left\langle \hat{n}_k \right\rangle_{qs} = \frac{\sum_{n_k=0,1} n_k e^{-n_k \hbar\omega_k / KT}}{\sum_{n_k=0,1} e^{-n_k \hbar\omega_k / KT}} = \frac{1}{e^{\hbar\omega_k / KT} + 1} = \bar{n}_F(\omega_k) \tag{9.20}$$

$$\left\langle \hat{\sigma}_{zk}(0) \right\rangle_{qs} = 2\left\langle \hat{n}_k \right\rangle - 1 = -\tanh\frac{\hbar\omega_k}{2KT}. \tag{9.21}$$

In the above equations, $\bar{n}_F$ can be identified as the Fermi–Dirac distribution function denoting the average thermal excitation number of the spin-bath. This distribution does not contain any contribution due to the chemical potential, which implies that our stating Hamiltonian Equation 9.1 does not conserve the spin –1/2 particle number. Ideally, the typical system/spin-bath model for dissipation may be considered as an ion in an environment of two-level quantum dots (Bras et al. 2002; Favero et al. 2007; Santori and Yamamoto 2009; Xu and Teichert 2005) with characteristic frequencies governed by the size distributions of the dots. To preclude the possibility of any recurrence and to ensure irreversibility associated with the notion of dissipation, a large number is an essential requirement.

9.2.2 Quantum Langevin Dynamics in c-Number Spin-Bath

In this section, we will discuss a c-number formalism to describe the spin-bath. To this end, we proceed as follows: We return to Equation 9.12 and carry out the quantum mechanical average $\langle \ldots \rangle$ to obtain

$$\left\langle \ddot{\hat{q}} \right\rangle + \int_0^t dt' \left\langle \dot{\hat{q}}(t') \right\rangle \kappa(t - t') + \left\langle V'(\hat{q}) \right\rangle = \left\langle \hat{f}(t) \right\rangle. \tag{9.22}$$

In Equation 9.22, the quantum mechanical average is taken over the initial product separable quantum states of the particle and the spin-bath at $t = 0$, $|\vartheta\rangle|\xi_1\rangle|\xi_2\rangle\ldots|\xi_N\rangle$. Here, $|\vartheta\rangle$ denotes any arbitrary initial state of the particle, and $|\xi_k\rangle$ corresponds to the initial coherent state of the kth two-level atom of the bath. Existence of such coherent states had already been proven a couple of decades ago by Radcliffe (Radcliffe 1971). These states are analogous to harmonic oscillator coherent states. Typically, for a spin-1/2 system, such a state is generated by the action of the annihilation operator on vacuum, which is defined as the spin state with minimal projection. Applications of spin-coherent states are well known in the context of ferromagnetic spin wave, phase transition in the Dicke model of superradiance, equilibrium statistical mechanics of radiation–matter interaction, etc. For details, we refer to the work of Klauder and Skagerstam (Klauder and Skagerstam 1985).

The main purpose of using these spin-coherent states for the quantum mechanical averaging of the bath operators is to formulate Equation 9.22 as a classical-looking Langevin equation for the quantum mechanical mean position of the particle. We now denote the quantum mechanical averages as

$$\langle \hat{q}(t) \rangle = q(t) \tag{9.23}$$

$$\langle \hat{f}(t) \rangle = \eta(t) \tag{9.24}$$

where

$$\begin{aligned}\eta(t) &= \sum_k g_k \sqrt{\frac{\hbar}{2\omega_k}} \left\{ \langle \hat{S}_k(0) \rangle e^{-i\omega_k t} + \langle \hat{S}_k^\dagger(0) \rangle e^{i\omega_k t} \right\} \\ &= \sum_k g_k \sqrt{\frac{\hbar}{2\omega_k}} \left\{ \xi_k(0) e^{-i\omega_k t} + \xi_k^*(0) e^{i\omega_k t} \right\}.\end{aligned} \tag{9.25}$$

The last equality follows from Equations 9.18 and 9.19. $\langle \hat{S}_k(0) \rangle = \langle \hat{\sigma}_k(0) \rangle = \xi_k(0)$, and $\langle \hat{S}_k^\dagger(0) \rangle = \langle \hat{\sigma}_k^\dagger(0) \rangle = \xi_k^*(0)$. Here, $\xi_k(0)$ and $\xi_k^*(0)$ are the associated c-numbers (note that we have set $\langle \hat{q}(0) \rangle = (0)$) for the two-level bath atoms. Equation 9.22 may then be rewritten as

$$\ddot{q} + \int_0^t \dot{q}(t')\kappa(t-t')\,dt' + \langle V'(\hat{q}) \rangle = \eta(t). \tag{9.26}$$

Now, to realize $\eta(t)$ as an effective c-number noise, we introduce the ansatz that $\xi_k(0)$ and $\xi_k^*(0)$ are distributed according to a thermal canonical distribution of Gaussian form as follows:

$$P_k\left(\xi_k(0), \xi_k^*(0)\right) = N \exp\left\{ -\frac{|\xi_k(0)|^2}{2\tanh\left(\frac{\hbar\omega_k}{2KT}\right)} \right\}, \tag{9.27}$$

where N is a normalization constant. The width of the distribution is defined by $\tanh\left(\frac{\hbar\omega_k}{2KT}\right)$. For any arbitrary quantum mechanical mean value of a bath operator $\langle \hat{A}_k \rangle$, which is a function of $\xi_k(0), \xi_k^*(0)$, its statistical average can then be written as

$$\langle\langle \hat{A}_k \rangle\rangle_s = \int \langle \hat{A}_k \rangle P_k\left(\xi_k^*(0), \xi_k(0)\right) d\xi_k^*(0) d\xi_k(0). \tag{9.28}$$

The ansatz Equation 9.27 and the definition of statistical average Equation 9.28 can be used to show that c-number noise $\eta(t)$ satisfies the following relations:

$$\langle \eta(t) \rangle_s = 0 \tag{9.29}$$

$$\langle \eta(t)\eta(t') \rangle_s = \frac{\hbar}{2} \sum_k \frac{g_k^2}{\omega_k} \cos \omega_k (t - t') \tanh\left(\frac{\hbar\omega_k}{2KT}\right). \tag{9.30}$$

Equations 9.29 and 9.30 imply that c-number noise $\eta(t)$ is such that it is zero-centered and follows the FD relation as expressed in Equation 9.16. Therefore, Equations 9.16 and 9.30 are equivalent. However, the distinct advantage of the formulation of c-number noise is that it allows us to bypass the operator-ordering prescription for deriving the noise properties of the bath. We also point out that the temperature dependence as derived here for the noise correlation function matches exactly that obtained in earlier work (Sinha, Ghosh, and Ray 2011a, 2011b). The quantum Langevin equation with c-number noise $\eta(t)$, as defined by Equations 9.26, 9.29, and 9.30, is classical looking in form but entails a quantum character. Therefore, the scheme can be used to implement the techniques of classical nonequilibrium statistical mechanics for various purposes. It is also pertinent to note that, for a traditional harmonic oscillator heat bath, i.e., the bosonic reservoir, one can proceed exactly in a similar fashion (Banerjee, Bag et al. 2002; Banerjee et al. 2004; Banerjee, Banik et al. 2002; Barik, Banerjee, and Ray 2009) and use harmonic oscillator coherent states. The canonical thermal distribution function for bosonic c-number noise corresponding to its spin-bath counterpart is the well-known Wigner thermal distribution function (Hillery et al. 1984). This has been extensively used in several earlier occasions.

In order to quantify the properties of the thermal bath, it is convenient to introduce a spectral density function (Leggett et al. 1987) $J(\omega)$ associated with the system–bath interaction

$$J(\omega) = \frac{\pi}{2} \sum_k \frac{g_k^2}{\omega_k} \delta(\omega - \omega_k). \tag{9.31}$$

With the help of $J(\omega)$, we may rewrite the expressions for memory kernel (Equation 9.13)

$$\kappa(t) = \frac{2}{\pi} \int_0^\infty d\omega \frac{J(\omega)}{\omega} \cos \omega t \tag{9.32}$$

and for the FD relation (Equation 9.30)

$$\begin{aligned}\langle \eta(t)\eta(t')\rangle_s &= Re\left\{\left\langle \hat{F}(t)\hat{F}^{\dagger}(t') - \hat{F}^{\dagger}(t)\hat{F}(t')\right\rangle_{qs}\right\} \\ &= \frac{2}{\pi}\int_0^{\infty} d\omega \frac{J(\omega)}{\omega}\left[\frac{\hbar\omega}{2}\tanh\left(\frac{\hbar\omega}{2KT}\right)\right]\cos\omega(t-t').\end{aligned} \quad (9.33)$$

The relaxation induced by the bath is determined by the properties of the spectral density function, which connects the noise correlation function to dissipative kernel. The thermal behavior of the spin-bath can be studied under two distinct situations: (1) the high-temperature regime and (2) the low-but-finite-temperature regime.

High-temperature regime: From Equation 9.21, it follows that $\left\langle \hat{\sigma}_{zk}(0)\right\rangle_{qs} = -\tanh\left(\frac{\hbar\omega_k}{2KT}\right) = \frac{1}{e^{\frac{\hbar\omega_k}{KT}}+1} - \frac{1}{2}$; $\left\langle \hat{\sigma}_{zk}(0)\right\rangle_{qs}$ is a measure of the population difference between the two levels of the kth bath atom. At high temperature, $\left(1+e^{\frac{\hbar\omega}{KT}}\right) \to 2$, which implies $\left\langle \hat{\sigma}_{zk}(0)\right\rangle_{qs} \to 0$, that is, the bath atoms get thermally saturated and the effective system–bath coupling is largely suppressed since the noise correlation tends to vanish. This gives rise to the emergence of the coherent behavior of the system dynamics, which corroborates the earlier observation of Shao and Hänggi et al. (Shao and Hänggi 1998).

Low-but-finite-temperature regime: On the other hand, in case of low but finite temperature, i.e., much below the saturation temperature, we have $\left\langle \hat{\sigma}_{zk}(0)\right\rangle_{qs} \approx -1$. This implies that the effective spin-bath/system coupling is large, and the noise correlation gets enhanced. Thus, quantum dissipation is more facilitated as the temperature is lowered. As $T \to 0$, the noise correlation of the spin-bath (Equation 9.33) tends to that of the bosonic bath.

9.2.3 Quantum Correction Equations and Quantum Statistical Averages

The quantum Langevin Equation 9.26 may be expressed in a more convenient form for interpretation. We add the force term $V'(q)$ on both sides. The resulting equation is given by

$$\ddot{q} + \int_0^t \dot{q}(t')\kappa(t-t')\,dt' + V'(q) = \eta(t) + Q, \quad (9.34)$$

where

$$Q = V'(q) - \left\langle V'(\hat{q})\right\rangle, \quad (9.35)$$

which represents the quantum correction due to the system potential. Equation 9.34 thus describes the motion of a particle (quantum mechanical mean position) in a force field simultaneously driven by the c-number quantum noise of the spin-bath and quantum dispersion Q, characteristic of the nonlinearity of the potential. Q may be expressed in a more explicit and useful form by recognizing the operator nature of the system variables $\hat{q}$ and $\hat{p}$ as

$$\hat{q}(t) = q(t) + \delta\hat{q}(t) \tag{9.36}$$

$$\hat{p}(t) = p(t) + \delta\hat{p}(t), \tag{9.37}$$

where $\delta\hat{q}$ and $\delta\hat{p}$ are the quantum correction operators around the corresponding quantum mechanical mean values of $q\left(\equiv\langle\hat{q}\rangle\right)$ and $p\left(\equiv\langle\hat{p}\rangle\right)$. By construction, $\langle\delta\hat{q}\rangle = \langle\delta\hat{p}\rangle = 0$ and $\left[\delta\hat{q}, \delta\hat{p}\right] = i\hbar$. Using Equation 9.36 in $V'(\hat{q})$ and a Taylor series expansion around q, we may express Q as

$$Q = -\sum_{n\geq 2} \frac{1}{n!} V^{(n+1)}(q)\langle\delta\hat{q}^n(t)\rangle. \tag{9.38}$$

Here, $V^{(m)}(q)$ is the mth derivative of the potential $V(q)$ with respect to q. For example, the lowest order correction (n = 2) is given by $Q = -\frac{1}{2}V'''(q)\langle\delta\hat{q}^2(t)\rangle$. The determination of Q in quantum Langevin Equation 9.34 therefore depends on $\langle\delta\hat{q}^2(t)\rangle$, which may be estimated by solving quantum correction equations. To achieve this, we consider the quantum operator Equation 9.12 and use Equations 9.36 and 9.37. Furthermore, using Equation 9.34 in the resulting equation, we obtain

$$\delta\ddot{\hat{q}} + \int_0^\infty dt'\, \kappa(t-t')\, \delta\dot{\hat{q}}(t') + V''(q)\delta\hat{q} + \sum_{n\geq 2} \frac{1}{n!} V^{(n+1)}(q)\left(\delta\hat{q}^n(t) - \langle\delta\hat{q}^n(t)\rangle\right) = \delta\hat{\eta}(t) \tag{9.39}$$

where $\delta\hat{\eta}(t) = \hat{f}(t) - \eta(t)$.

Equation 9.39 forms the basis for the calculation of quantum mechanical correction $\langle\delta\hat{q}^n(t)\rangle$. However, it is analytically untractable for an exact solution. Depending on the nonlinearity of the potential and memory kernel, systematic approximation schemes may be employed. The typical familiar case of the exponential memory kernel is illustrated in the next section.

It is thus evident that the quantum Brownian motion of a particle in a spin-bath may be calculated, in principle, as a stochastic process by solving the Langevin Equation 9.34 for quantum mechanical mean values simultaneously with quantum correction equations, which describe the quantum mechanical fluctuation or dispersion around them. Before closing this section, we mention that an essential element of this approach is to express the quantum statistical average as a sum of statistical averages over a set of functions of the quantum mechanical mean values and dispersion. To illustrate, we calculate, for example, the quantum statistical averages $\langle \hat{q} \rangle_{qs}$ and $\langle \hat{q}^2 \rangle_{qs}$. Making use of Equation 9.36, we write

$$\begin{aligned}\langle \hat{q} \rangle_{qs} &= \langle q + \delta\hat{q} \rangle_{qs} \\ &= \langle q \rangle_s + \langle \langle \delta\hat{q} \rangle \rangle_s \\ &= \langle q \rangle_s\end{aligned} \tag{9.40}$$

and

$$\begin{aligned}\langle \hat{q}^2 \rangle_{qs} &= \left\langle (q + \delta\hat{q})^2 \right\rangle_{qs} \\ &= \langle q^2 \rangle_s + \langle \delta\hat{q}^2 \rangle_{qs} \\ &= \langle q^2 \rangle_s + \langle \langle \delta\hat{q}^2 \rangle \rangle_s .\end{aligned} \tag{9.41}$$

It is necessary to distinguish between the quantum mechanical mean of an operator $\hat{A}$ expressed as $\langle \hat{A} \rangle (= A)$ from the statistical average of the quantum mechanical mean A expressed as $\langle A \rangle_s$ and that from the quantum statistical average of $\hat{A}$ denoted as $\langle \hat{A} \rangle_{qs}$.

The quantum nature of the dynamics manifests in two different ways: First, the spin-bath reservoir is quantum mechanical in character, whose noise properties are expressed through the quantum FD relation. Second, the nonlinearity of the system potential gives rise to the quantum correction terms.

9.3 SCHEME FOR NUMERICAL SIMULATION OF QUANTUM LANGEVIN EQUATION

9.3.1 Generation of Spin-Bath Noise

Equation 9.33 expresses the FD relation, which is the key element for the generation of c-number noise. $\langle \eta(t)\eta(t') \rangle_s$ is the correlation function, which is classical in form but quantum mechanical in its content. We now show that c-number noise $\eta(t)$ can be generated as a superposition of several Ornstein–Uhlenbeck noise processes (Banerjee et al. 2004). To this end, we begin by expressing $\langle \eta(t)\eta(t') \rangle_s = C(t - t')$ in the continuum limit as follows:

$$C(t-t') = \frac{2}{\pi} \int_0^\infty d\omega \frac{J(\omega)}{\omega} \frac{\hbar\omega}{2} \tanh\left(\frac{\hbar\omega}{2KT} \right) \cos\omega(t-t'). \tag{9.42}$$

It is essential to know a priori the spectral density function $J(\omega)$. We assume the spectral density function of Lorentzian form, i.e.,

$$J(\omega) = \frac{1}{\pi}\frac{\Gamma\omega}{1+(\omega-\omega_0)^2\tau_c^2}, \tag{9.43}$$

where Γ and τ_c are the dissipation constants in the Markovian limit and correlation time, respectively. ω_0 is the linearized system frequency. When Equation 9.43 is used, Equation 9.32 yields an exponential memory kernel $\kappa(t) = (\Gamma/\tau_c)e^{-|t|/\tau_c}$. For a given set of parameters Γ and τ_c and temperature T, we first numerically evaluate the integral in Equation 9.42 as a function of time. The results are shown in Figure 9.1 by dots. The calculated correlation function is then numerically fitted using a superposition of several exponentials with parameter set $\{D_i,\tau_i\}$ as

$$C(t-t') = \sum_i \frac{D_i}{\tau_i}\exp\left(-\frac{|t-t'|}{\tau_i}\right), \quad i = 1,2,3.... \tag{9.44}$$

The fitted curves are shown by continuous lines in Figure 9.1. In this numerical fitting, the Levenberg–Marquardt (LM) algorithm (Marquardt 1963; Press et al.

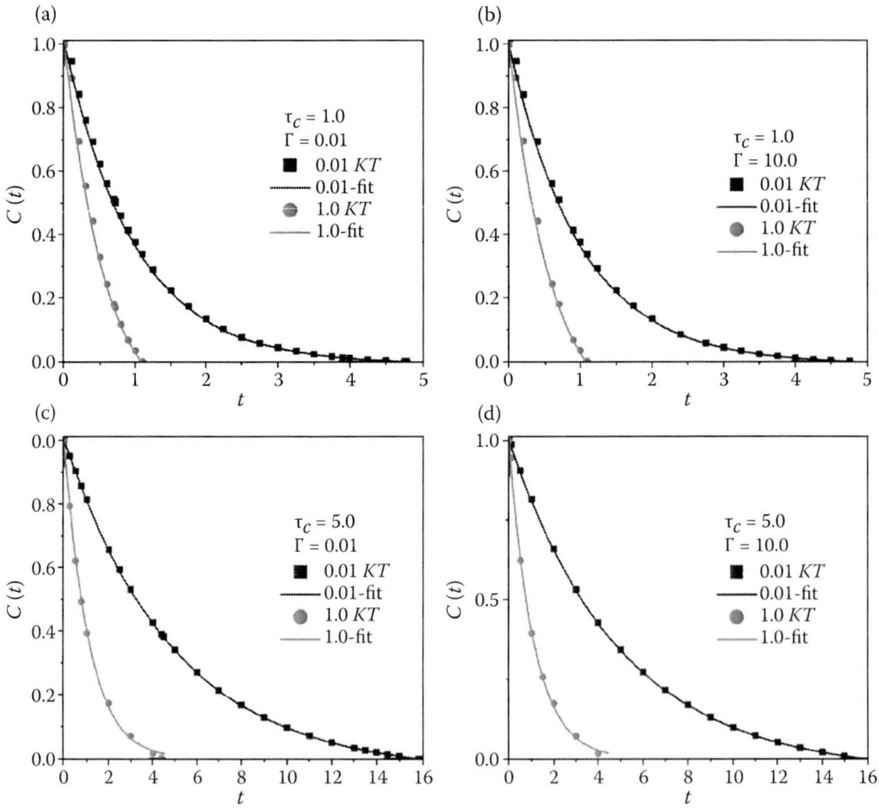

FIGURE 9.1 Variation of normalized c-number noise correlation function $C(t)$ with time for different parameter sets of Γ and τ_c and T. The dots represent the analytical variation based on Equation 9.42, and the lines represent numerical fitting using Equation 9.44 (scale: arbitrary).

2007), which is very efficient for minimizing a function (generally nonlinear) over a set of parameters, is used.

The numerically obtained parameter set D_i and τ_i are temperature dependent. Now, with the help of these parameters, a set of exponentially correlated color noise variables ϕ_i is generated by the following set of equations:

$$\dot{\phi}_i = -\frac{\phi_i}{\tau_i} + \frac{1}{\tau_i}\epsilon_i(t) \tag{9.45}$$

where

$$\langle \epsilon_i(t) \rangle = 0$$

$$\langle \epsilon_i(0)\epsilon_j(\tau) \rangle = 2D_i\delta_{ij}\delta(\tau),\ i = 1, 2, 3\ldots, \tag{9.46}$$

i.e., $\epsilon_i(t)$ is a Gaussian white noise obeying Equation 9.46. Each noise ϕ_i is thus an Ornstein–Uhlenbeck process with properties

$$\langle \phi(t) \rangle = 0$$

$$\left\langle \phi_i(t)\phi_j(t') \right\rangle = \delta_{ij}\frac{D_i}{\tau_i}\exp\left(-\frac{|t-t'|}{\tau_i}\right), \quad i = 1,2,3\ldots. \tag{9.47}$$

The c-number noise $\eta(t)$ due to spin-bath is therefore given by a superposition of several Ornstein–Uhlenbeck noise processes as follows:

$$\eta(t) = \sum_i^n \phi_i \tag{9.48}$$

where τ_i and D_i can now be interpreted as the correlation time and strength of the color noise variable ϕ_i. Having derived the scheme for the generation of c-number noise $\eta(t)$, we now proceed to solve the quantum Langevin equation with quantum correlations.

9.3.2 Solution of Quantum Langevin Equation

In order to solve the c-number Langevin Equation 9.34, we first rewrite it in the following equivalent form:

$$\dot{q} = p$$

$$\dot{p} = -V'(q) + Q\left(q, \left\langle \delta\hat{q}^n \right\rangle\right) + \sum_i \phi_i(t) + z, \quad i = 1,2,3\ldots$$

$$\dot{z} = -\Gamma\frac{p}{\tau_c} - \frac{z}{\tau_c} \tag{9.49}$$

$$\dot{\phi}_i = -\frac{\phi_i}{\tau_i} + \frac{1}{\tau_i}\epsilon_i(t),$$

where the auxiliary variable z has been introduced to bypass the convolution integral in Equation 9.34. To estimate the $Q\left(q,\langle \delta\hat{q}^n\rangle\right)$ required for solving the above set of equations, we now return to the operator Equation 9.39 and rewrite it after introducing the exponential memory kernel $\kappa(t) = \frac{\Gamma}{\tau_c}\exp(-|t|/\tau_c)$ and an auxiliary operator $\delta\hat{z}$ to avoid the convolution integral in Equation 9.39. The resulting equation after quantum mechanical averaging over the coherent bath states are given by

$$\begin{aligned}
\delta\dot{\hat{q}} &= \delta\hat{p} \\
\delta\dot{\hat{p}} &= -V''(q)\delta\hat{q} - \sum_{n\geq 2}\frac{1}{n!}V^{n+1}(q)\left(\delta\hat{q}^n - \langle\delta\hat{q}^n\rangle\right) + \delta\hat{z} \\
\delta\dot{\hat{z}} &= -\frac{\Gamma}{\tau_c}\delta\hat{p} - \frac{1}{\tau_c}\delta\hat{z}.
\end{aligned} \tag{9.50}$$

Making use of the above set of operator equations, we derive the following equations for quantum corrections up to the second order:

$$\begin{aligned}
\frac{d}{dt}\langle\delta\hat{q}^2\rangle &= \langle\delta\hat{q}\delta\hat{p} + \delta\hat{p}\delta\hat{q}\rangle \\
\frac{d}{dt}\langle\delta\hat{q}\delta\hat{p} + \delta\hat{p}\delta\hat{q}\rangle &= 2\langle\delta\hat{p}^2\rangle - 2V''(q)\langle\delta\hat{q}^2\rangle + \langle\delta\hat{q}\delta\hat{z} + \delta\hat{z}\delta\hat{q}\rangle \\
\frac{d}{dt}\langle\delta\hat{q}\delta\hat{z} + \delta\hat{z}\delta\hat{q}\rangle &= -\frac{\Gamma}{\tau_c}\langle\delta\hat{q}\delta\hat{p} + \delta\hat{p}\delta\hat{q}\rangle - \frac{1}{\tau_c}\langle\delta\hat{q}\delta\hat{z} + \delta\hat{z}\delta\hat{q}\rangle + \langle\delta\hat{p}\delta\hat{z} + \delta\hat{z}\delta\hat{p}\rangle \\
\frac{d}{dt}\langle\delta\hat{p}^2\rangle &= -V''(q)\langle\delta\hat{q}\delta\hat{p} + \delta\hat{p}\delta\hat{q}\rangle + \langle\delta\hat{p}\delta\hat{z} + \delta\hat{z}\delta\hat{p}\rangle \\
\frac{d}{dt}\langle\delta\hat{p}\delta\hat{z} + \delta\hat{z}\delta\hat{p}\rangle &= -2\frac{\Gamma}{\tau_c}\langle\delta\hat{p}^2\rangle - \frac{1}{\tau_c}\langle\delta\hat{p}\delta\hat{z} + \delta\hat{z}\delta\hat{p}\rangle - V''(q)\langle\delta\hat{q}\delta\hat{z} + \delta\hat{z}\delta\hat{q}\rangle + 2\langle\delta\hat{z}^2\rangle \\
\frac{d}{dt}\langle\delta\hat{z}^2\rangle &= -\frac{\Gamma}{\tau_c}\langle\delta\hat{p}\delta\hat{z} + \delta\hat{z}\delta\hat{p}\rangle - \frac{2}{\tau_c}\langle\delta\hat{z}^2\rangle \\
\frac{d}{dt}\langle\delta\hat{z}\rangle &= -\frac{1}{\tau_c}\langle\delta\hat{z}\rangle.
\end{aligned} \tag{9.51}$$

The integration of Equations 9.49 and 9.51 is carried out using the second-order Heun's algorithm, with a very small time step of 0.001. These equations differ from the corresponding classical equations in two ways: First, the noise correlation of c-number spin-bath variables $\eta(t)$ are quantum mechanical in nature, as evident from the correlation function in Equation 9.42, which is numerically fitted by the superposition of exponential functions with D_i and τ_i. Second, the knowledge of Q requires the quantum correction equations that yield quantum dispersion around the quantum mechanical mean values q and p for the system. Statistical averaging over noise is performed for 5000 trajectories for the specified initial values of q and p. The initial values of dispersion are set as $\langle \delta\hat{q}^2(0)\rangle = \langle \delta\hat{p}^2(0)\rangle = 0.5$ and $\langle \delta\hat{q}\delta\hat{p} + \delta\hat{p}\delta\hat{q}\rangle = 1.0$, with other averages set as zero. We have estimated the time average contribution of the quantum corrections up to $1/\Gamma$ time for each trajectory in solving the Langevin equation.

9.4 ILLUSTRATIVE EXAMPLES: RESULTS AND DISCUSSIONS

A major focus of the theory of quantum dissipation in a spin-bath is the conspicuous thermal behavior of the reservoir. Our analysis clearly shows that, at temperatures close to zero, a spin-bath behaves almost in the same way as a bosonic bath, implying a universality in the nature of bath as $T\rightarrow 0$. At higher temperatures (below saturation temperature), the system–bath coupling tends to diminish, which is reflected in the emergence of coherence in the dynamics, and the behavior of a spin-bath differs significantly from that of a bosonic bath. In what follows, we consider two specific examples to illustrate these aspects.

9.4.1 Harmonic Oscillator

We begin with the motion of the particle in a harmonic potential $V(q) = \frac{\omega_0^2}{2} q^2$. Figure 9.2 exhibits the results on the spatial distribution of probability density $P(x)$

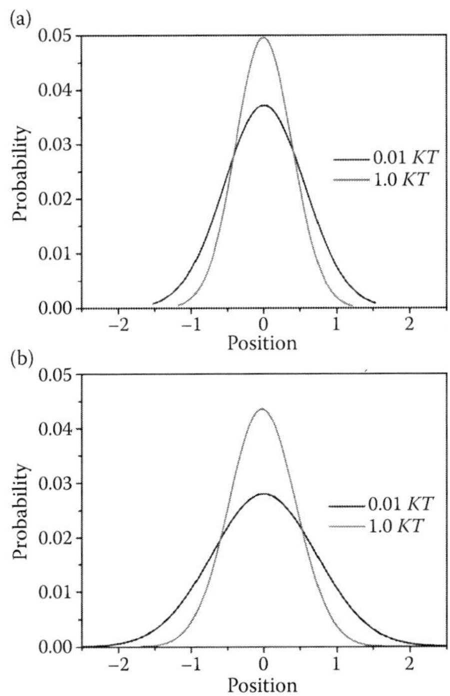

FIGURE 9.2 Probability density functions for position for $\tau_c = 1.0$ at (a) $\Gamma = 0.01$ and (b) $\Gamma = 10.0$ (scale: arbitrary).

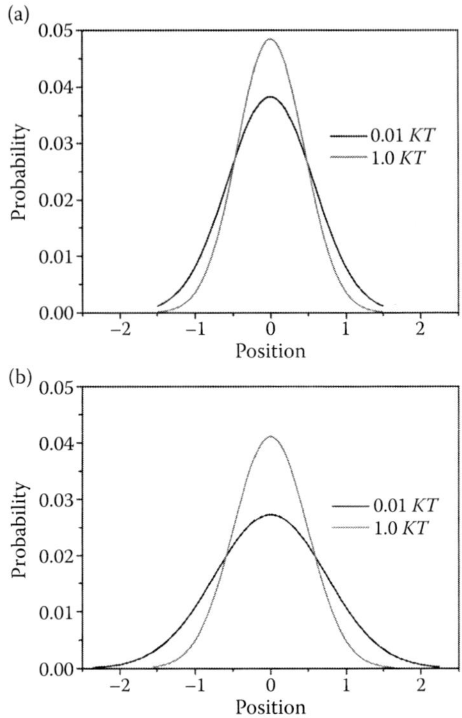

FIGURE 9.3 Probability density function for position for $\tau_c = 5.0$ at (a) $\Gamma = 0.01$ and (b) $\Gamma = 10.0$ (scale: arbitrary).

of the position at two temperatures in the strong ($\Gamma = 1$) and weak ($\Gamma = 0.01$) damping regime for short τ_c. The width decreases with increase in temperature, which is an immediate consequence of the reduction of system–bath coupling.

In Figure 9.3, we plot the probability density function for long correlation time τ_c. A comparison between Figure 9.2a and Figure 9.3a and between Figure 9.2b and Figure 9.3b indicates that the nature of the distribution or the width practically remains insensitive to τ_c or the memory of the dynamics, at least for the range of the damping constant relevant for this study. The reduction in line width with the increase in temperature in optical systems had been studied earlier by several groups in the context of photoabsorption and emission experiments (Bras et al. 2002; Favero et al. 2007) and others (Ghosh, Sinha, and Ray 2011a; Ghosh, Sinha, and Ray 2011b).

9.4.2 Anharmonic Oscillator

Next, we consider the case of the cubic potential of the form $V(q) = \frac{a}{2}q^2 - \frac{b}{3}q^3$, where a and b are constants that determine barrier height $V_0 = a/b$. For the present purpose, we choose the parameter set as $a = 2.0$ and $b = 0.45$. The leading order

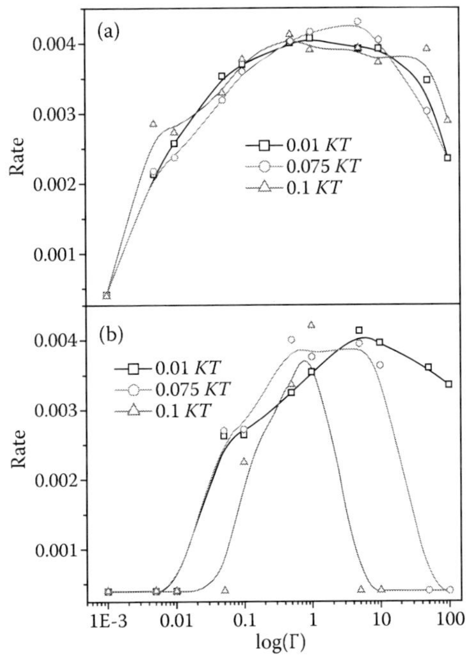

FIGURE 9.4 Kramers' turnover in a spin-bath. Variation of barrier-crossing rates with Γ for three different temperatures for (a) $\tau_c = 1.0$ and (b) $\tau_c = 5.0$ (scale: arbitrary).

quantum correction Q contains only one term, i.e., $Q = b\langle \delta\hat{q}^2 \rangle$. For the given set of parameter values, we have numerically computed the average time $\langle\tau\rangle$ taken by the particle, starting from the bottom of the metastable well at $q = 0$ to reach the barrier top for the first time (mean first passage time) for 5000 realizations of the stochastic trajectories. The simulated results of the variation of the barrier-crossing rate $(1/\langle\tau\rangle)$ against Γ for three different temperatures are illustrated in Figure 9.4a and b for two cases (small and large τ_c). It is evident that the rate initially increases with increase in Γ to reach a maximum, followed by a decrease. This qualitative feature corresponds to Kramers' turnover phenomenon, which is well known in the classical dynamic theory of reaction rates. To allow for fair comparison with the analytical results based on the Smoluchowski equation for spin-bath (Sinha, Ghosh, and Ray 2011b), we illustrate in Arrhenius plots (Figure 9.5) the numerical variation of log(rate) versus $1/T$ under overdamped condition ($\Gamma \gg \omega_0$). The agreement is fair enough to vindicate the validity of the numerical scheme. The interplay of Γ and T in the numerical simulation of the barrier-crossing rate for short τ_c and long τ_c over a wide parameter range is shown in the contour plots in Figures 9.6 and 9.7, respectively.

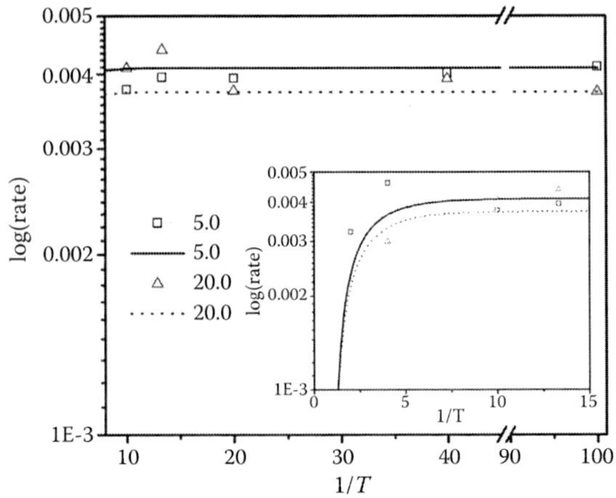

FIGURE 9.5 Comparison between (dot) numerical and (line) analytical Arrhenius plots for τ_c= 1.0 and several values of Γ (scale: arbitrary). The inset highlights the high-temperature regime.

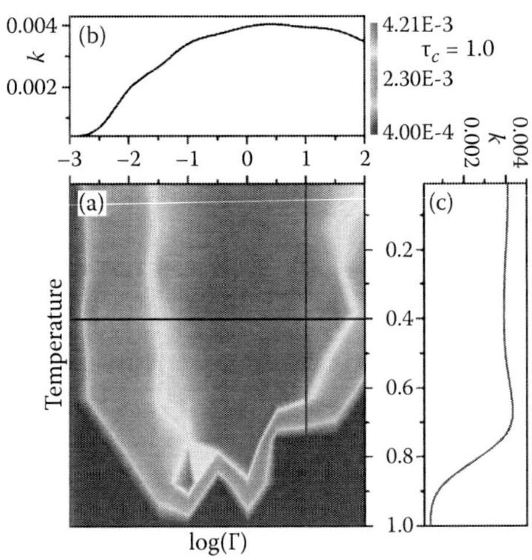

FIGURE 9.6 (a) Contour plot of the barrier-crossing rate for τ_c = 1.0. (b) Variation of the barrier-crossing rate with log(Γ). (c) Variation of the barrier-crossing rate with temperature (scale: arbitrary).

9.5 CONCLUSION

We have considered the stochastic dynamics of a particle interacting with its environment of two-level systems in the presence of an external potential field. The treatment is based on the canonical quantization procedure. This approach directly yields the dissipative term and the noise operators. It may be pertinent to mention that, although the calculation of dissipative effects is straightforward, the treatment of noise is not simple as far as the path integral techniques are concerned.

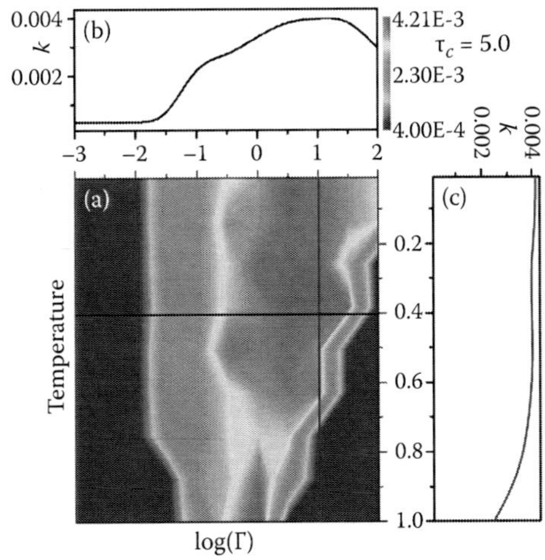

FIGURE 9.7 (a) Contour plot of the barrier-crossing rate for $\tau_c = 5.0$. (b) Variation of the barrier-crossing rate with log(Γ). (c) Variation of the barrier-crossing rate with temperature (scale: arbitrary).

Implementation of the spin-coherent state representation for the bath degrees of freedom and the thermal canonical distribution of the c-number variables allows us to realize the noise correlation function as a classical-looking FD relation with a quantum mechanical content and a generalized quantum Langevin equation for mean position of the particle. An important offshoot of this formulation is a scheme for the generation of spin-bath noise from a priori knowledge of the spectral density function for the bath. This is based on visualizing the c-number noise of the spin-bath as a superposition of several Ornstein–Uhlenbeck noise processes. The resulting Langevin equation is computationally tractable with the help of standard numerical techniques. We now summarize the main aspects of this study.

1. The FD relation is characterized by a spectral density function and a temperature-dependent hyperbolic tangent factor that decreases with temperature. At 0 K, this factor merges into that for a harmonic bath, implying the universality of the nature of the bath as it approaches absolute zero.
2. With rise in temperature, the hyperbolic tangent factor begins to decrease until the bath reaches thermal saturation. This results in suppression of system–bath coupling or quantum dissipation and signals the emergence of the thermally induced coherent behavior of the dynamics. Application of the scheme on the stochastic dynamics of a particle in a harmonic well captures the coherent behavior in the narrowing of the distribution width of position. This coherence is also reflected in the suppressions of thermal activation and Kramers' turnover in barrier-crossing dynamics in an anharmonic potential.

Spin-bath is an interesting object for study of quantum dissipation and noise for which classical limit is not realizable, in principle. However, atoms and molecules embedded in a sea of two-level quantum dots of varying sizes may offer themselves as worthwhile candidates for investigation in this direction. In view of recent advances in nanotechnology with quantum dots, these observations may be the subject of concrete experimental tests.

ACKNOWLEDGMENTS

S. S. Sinha would like to thank A. Baura for the discussion. A. Ghosh would like to thank the Council of Scientific and Industrial Research, Government of India, for the partial financial support.

REFERENCES

Banerjee, D., B. C. Bag, S. K. Banik, and D. S. Ray. 2002. Approach to quantum Kramers' equation and barrier crossing dynamics. *Physical Review E* 65 (2):021109.

Banerjee, D., B. C. Bag, S. K. Banik, and D. S. Ray. 2004. Solution of quantum Langevin equation: Approximations, theoretical and numerical aspects. *Journal of Chemical Physics* 120 (19):8960–8972.

Banerjee, D., S. K. Banik, B. C. Bag, and D. S. Ray. 2002. Quantum Kramers equation for energy diffusion and barrier crossing dynamics in the low-friction regime. *Physical Review E* 66 (5):051105.

Barik, D., D. Banerjee, and D. S. Ray. 2009. *Quantum Brownian Motion in c-Numbers: Theory and Application*. NY: Nova-Science Publishers.

Bras, F., P. Boucaud, S. Sauvage, G. Fishman, and J. M. Gérard. 2002. Temperature dependence of intersublevel absorption in InAs/GaAs self-assembled quantum dots. *Applied Physics Letters* 80:4620.

Caldeira, A. O., A. H. Castro Neto, and T. Oliveira de Carvalho. 1993. Dissipative quantum systems modeled by a two-level-reservoir coupling. *Physical Review B* 48 (18):13974–13976.

Caldeira, A. O., and A. J. Leggett. 1981. Influence of dissipation on quantum tunneling in macroscopic systems. *Physical Review Letters* 46 (4):211–214.

Caldeira, A. O., and A. J. Leggett. 1983a. Path integral approach to quantum Brownian motion. *Physica A: Statistical Mechanics and its Applications* 121 (3):587–616.

Caldeira, A. O., and A. J. Leggett. 1983b. Quantum tunnelling in a dissipative system. *Annals of Physics* 149 (2):374–456.

Favero, I., A. Berthelot, G. Cassabois, C. Voisin, C. Delalande, Ph Roussignol, R. Ferreira, and J. M. Gérard. 2007. Temperature dependence of the zero-phonon linewidth in quantum dots: An effect of the fluctuating environment. *Physical Review B* 75 (7):073308.

Ferrer, A. V., A. O. Caldeira, and C. M. Smith. 2006. Optical conductivity of charge carriers interacting with a two-level systems reservoir. *Physical Review B* 74 (18):184304.

Ferrer, A. V., and C. M. Smith. 2007. Dynamical localization of a particle coupled to a two-level system thermal reservoir. *Physical Review B* 76 (21):214303.

Ghosh, A., S. S. Sinha, and D. S. Ray. 2011a. Dissipation in a spin bath: Thermally induced coherent intensity and spectral splitting. *Physical Review E* 83 (6):061154.

Ghosh, A., S. S. Sinha, and D. S. Ray. 2011b. Langevin–Bloch equations for a spin bath. *Journal of Chemical Physics* 134:094114.

Ghosh, A., S. S. Sinha, and D. S. Ray. 2012. Canonical formulation of quantum dissipation and noise in a generalized spin bath. *Physical Review E* 86 (1):011122.

Grabert, H., P. Schramm, and G.-L. Ingold. 1988. Quantum Brownian motion: The functional integral approach. *Physics Reports* 168 (3):115–207.

Hänggi, P., P. Talkner, and M. Borkovec. 1990. Reaction-rate theory: fifty years after Kramers. *Reviews of Modern Physics* 62 (2):251–341.

Hillery, M., R. F. O'Connell, M. O. Scully, and E. P. Wigner. 1984. Distribution functions in physics: Fundamentals. *Physics Reports* 106 (3):121–167.

Klauder, J. R., and B. S. Skagerstam. 1985. *Coherent States: Applications in Physics and Mathematical Physics*. Singapore: World Scientific.

Leggett, A. J., S. Chakravarty, A. T. Dorsey, M. P. A. Fisher, A. Garg, and W. Zwerger. 1987. Dynamics of the dissipative two-state system. *Reviews of Modern Physics* 59 (1):1–85.

Louisell, W. H. 1973. *Quantum Statistical Properties of Radiation*. New York: John Wiley.

Marquardt, D. W. 1963. An algorithm for least-squares estimation of nonlinear parameters. *SIAM Journal on Applied Mathematics* 11:431–441.

Meystre, P., and M. Sargent. 1991. *Elements of Quantum Optics*. New York: Springer-Verlag.

Press, W. H., S. A. Teukolsky, W. T. Vetterling, and B. P. Flannery. 2007. *Numerical Recipes: The Art of Scientific Computing*. 3rd ed. New York: Cambridge University Press.

Prokof'ev, N. V., and P. C. E. Stamp. 1996. Quantum relaxation of magnetisation in magnetic particles. *Journal of Low Temperature Physics* 104 (3):143–209.

Prokof'ev, N. V., and P. C. E. Stamp. 1998. Low-temperature quantum relaxation in a system of magnetic nanomolecules. *Physical Review Letters* 80 (26):5794–5797.

Radcliffe, J. M. 1971. Some properties of coherent spin states. *Journal of Physics A: General Physics* 4 (3):313.

Santori, C., and Y. Yamamoto. 2009. Quantum dots: Driven to perfection. *Nature Physics* 5 (3):173–174.

Sargent, M., M. O. Scully, and W. E. Lamb. 1974. *Laser Physics*. Massachusetts: Addison-Wesley Publishing Company.

Shao, J., and P. Hänggi. 1998. Decoherent dynamics of a two-level system coupled to a sea of spins. *Physical Review Letters* 81 (26):5710–5713.

Sinha, S. S., A. Ghosh, and D. S. Ray. 2011a. Decay of a metastable state induced by a spin bath. *Physical Review E* 84 (4):041113.

Sinha, S. S., A. Ghosh, and D. S. Ray. 2011b. Quantum Smoluchowski equation for a spin bath. *Physical Review E* 84 (3):031118.

Sinha, S. S., D. Mondal, B. C. Bag, and D. S. Ray. 2010. Quantum diffusion in a fermionic bath. *Physical Review E* 82 (5):051125.

Weiss, U. 1999. *Quantum Dissipative Systems*. Singapore: World Scientific.

Xu, M., and C. Teichert. 2005. Size distribution and dot shape of self-assembled quantum dots induced by ion sputtering. *Physica E: Low-Dimensional Systems and Nanostructures* 25 (4):425–430.

Zwanzig, R. 1973. Nonlinear generalized Langevin equations. *Journal of Statistical Physics* 9 (3):215–220.

10 Excitation Energy Transfer from Fluorophores to Graphene

R. S. Swathi and K. L. Sebastian

CONTENTS

10.1 INTRODUCTION

A fluorophore in an electronically excited state (which we denote by F^*) can return to the ground state by a radiative pathway like fluorescence or by nonradiative pathways like internal conversion and intersystem crossing. In addition to these intrinsic pathways of de-excitation of F^*, the interaction of F^* with another molecule, e.g., Q, can lead to the return of F^* to the ground state. Due to the presence of such an additional deactivation pathway by means of the interaction of F^* with Q, the excited state population and, hence, the fluorescence intensity decrease more rapidly than in the absence of such interaction. This loss of fluorescence intensity is called fluorescence quenching [1,2]. Excitation energy transfer (EET) is one of the major intermolecular photophysical processes that affect the excited state properties of a fluorophore, leading to fluorescence quenching. EET involves nonradiative transfer of excitation energy from an excited fluorophore, which is the donor, to the acceptor molecule and occurs due to the electrostatic coulombic interaction between the transition charge densities of the donor and of the acceptor. It requires that the emission spectrum of the donor overlaps with the absorption spectrum of the acceptor, so that (several) vibronic transitions of the donor are in resonance with those of the acceptor. Therefore, the process is also often referred to as fluorescence resonance energy transfer (FRET). In the absence of EET, the excited state lifetime of the donor is given by $\tau_0 = \frac{1}{k_0} = \frac{1}{k_r + k_{nr}}$, where k_r is the radiative rate constant and k_{nr} is the sum of the rate constants of nonradiative processes like internal conversion and intersystem crossing. Due to the interaction of the donor with the acceptor, leading to EET, the excited state lifetime of the donor decreases and is modified to $\tau = \frac{1}{k} = \frac{1}{k_r + k_{nr} + k_{ET}}$. Thus, the efficiency of energy transfer can be measured by time-resolved fluorescence spectroscopy experiments using

$$\Phi_T = \frac{k_{ET}}{k} = 1 - \frac{k_0}{k} = 1 - \frac{\tau}{\tau_0}. \tag{10.1}$$

It is also possible to perform steady-state fluorescence spectroscopy experiments and to monitor the decrease in fluorescence intensity of the donor, in both the absence and the presence of the acceptor to estimate the transfer efficiency using

$$\Phi_T = \frac{k_{ET}}{k} = 1 - \frac{k_0}{k} = 1 - \frac{\Phi}{\Phi_0}, \tag{10.2}$$

where Φ and Φ_0 are the fluorescence quantum yields of the donor in the presence and absence of the acceptor, respectively.

It is possible to evaluate the rate of energy transfer k_{ET} using the Fermi golden rule [3] of quantum mechanics. At the lowest level of calculation, the matrix element for the process is just the electrostatic interaction between the transition densities of the donor and of the acceptor. At large distances, this may be approximated by the interaction between the corresponding transition dipoles, which is proportional to R^{-3}, where R is the distance between the donor and the acceptor [3]. The rate of energy transfer is proportional to the square of it, and it is usual to write the rate as

$$k_{ET} = \frac{1}{\tau_0}\left(\frac{R_F}{R}\right)^6. \tag{10.3}$$

R_F is known as the Förster distance and is defined as the distance at which the rate of transfer is equal to the decay rate of the donor in the absence of the energy transfer process. Thus, the transfer efficiency is 50% when $R = R_F$. This R^{-6} dependence of the rate has first been suggested theoretically by Förster in 1947 [4]. Subsequently, Stryer and Haugland [5] used polyproline chains of various lengths to separate an energy donor and an energy acceptor and studied the distance dependence of the rate of transfer experimentally. Their results were found to be in agreement with the R^{-6} dependence of the Förster expression for the rate. They suggested that the process can be used as a spectroscopic ruler for measuring distances and has since been used extensively (see Figure 10.1) for understanding the conformational dynamics of biological molecules like proteins, RNA, etc. [6,7]. If the donor and the acceptor are two dye molecules, FRET is found to be effective when the separation distances are in the range of 10–100 Å. For separation distances greater than 100 Å, in FRET, the excited donor decays according to its natural lifetime, rather than by energy transfer.

The R^{-6} dependence of the rate in the Förster expression was derived within the dipolar approximation, in the regime when the sizes of the donor and the acceptor are smaller in comparison with the distances separating the donor and the acceptor. However, when the distances between the donor and the acceptor are not much larger in comparison with their sizes, the dipolar approximation to the interaction is not a very good approximation, thereby leading to deviations from the traditional R^{-6} dependence. For instance, if the energy transfer process involves extended electronic systems wherein the electronic excitations are delocalized, one expects the breakdown of the dipolar approximation. In recent times, a variety of materials like polymers [8], nanoparticles [9], quantum wells [10], quantum wires [11], metal surfaces [12], etc., have been used as energy donors/acceptors in the EET process, and deviations from the R^{-6} dependence have been studied. A non-R^{-6} dependence is of great interest due to the need to develop nanoscopic rulers that can measure distances well beyond 100 Å.

The EET from the molecular excited states to the electronic states of metal surfaces has been well studied in the past few decades. Drexhage et al. [13] measured

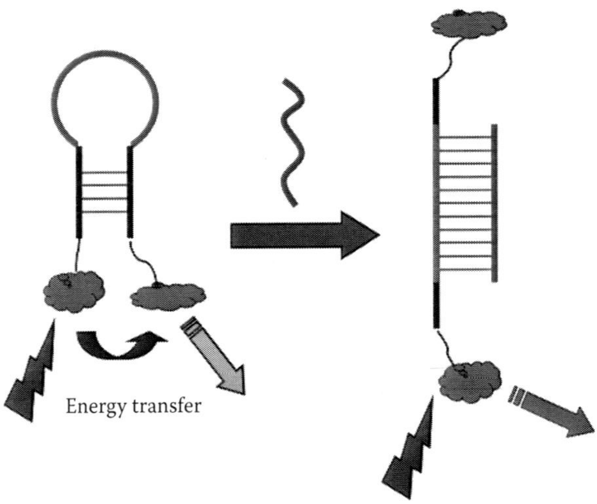

FIGURE 10.1 Schematic representation of a molecular beacon, a hair pin–shaped molecule containing a fluorophore, and a quencher at its two ends. On excitation of the fluorophore, it is possible to observe emissions from the quencher because of EET. However, when a target nucleotide sequence is added, the fluorophore and the quencher are separated, and the fluorescence of the fluorophore is recovered. The process can be used to probe the hybridization of oligonucleotides.

the fluorescence lifetime of an excited molecule as a function of the distance from a metal surface. Oscillations were found in the fluorescence lifetime at large distances, whereas a monotonic decrease was observed at shorter distances. The oscillations were due to the interference of the incident electric field wave of the emitting fluorophore and the wave reflected from the metal surface. The decrease at short distances was due to the nonradiative transfer of energy from the molecule to the metal. Chance et al. [14] theoretically analyzed this process of energy transfer from an excited molecule fluorescing near a metallic film. They consider the fluorophore to be an oscillating dipole and use the classical theory of an oscillating charge distribution near a dielectric interface to study the distance dependence of the rate. They found a d^{-3} dependence of the rate of energy transfer on the distance between the molecule and the metal for very thick films. However, for the case of thin metallic films, they found a d^{-4} dependence. The above-mentioned distance dependencies can be easily understood: In FRET, with the donor and acceptor transition densities being point dipoles, transfer occurs from a zero-dimensional donor to a zero-dimensional acceptor, leading to an R^{-6} dependence. For the case of transfer to the bulk modes of a metal, the acceptor excitations are in three dimensions. Therefore, one has a d^{-3} dependence. When the transfer occurs to the surface modes, one has a d^{-4} dependence because of the confinement of the acceptor excitations to two dimensions [12,15–17].

There are several reports in the literature of carbon nanotubes (CNTs), which are found to be efficient quenchers of the electronic excited states of dyes [18–20]. In studies on the energy transfer from pyrene molecules to the CNTs [18,19], the molecules were attached to CNTs using flexible spacers. Fluorescence lifetime measurements indicated a decrease in the lifetime of the fluorophore after tethering to the CNTs due to the EET. A single-walled CNT can be thought of as a rolled-up sheet of a single layer of graphite known as graphene [21]. After the preparation of graphene by Novoselov et al. [22] using the micromechanical cleavage of graphite, it has been attracting a great deal of attention from both the experimental and the theoretical communities [23,24]. Since CNTs (which are rolled-up sheets of graphene) were found to quench the fluorescence of dyes, we asked the following questions:

1. Can graphene quench the fluorescence of fluorophores?
2. If so, what would be the distance dependence of the rate of transfer from fluorophores to graphene?
3. How does the distance dependence compare with that obtained for the case of transfer to a metal surface?

The motivation for us to look at the EET from fluorophores to graphene is threefold. 1) Graphene has very interesting electronic properties [25], which makes it an ideal candidate for the theoretical investigation of the EET from the excited states of molecules to its electronic energy levels. 2) Although fluorescence quenching of dyes near nanotubes is known, the distance dependence of the rate has not been measured. The starting point for the theoretical calculations on CNTs is graphene, and hence, it is worthwhile to study the rate of energy transfer to graphene. 3) Metal surfaces have been shown to be effective quenchers of both electronic and vibrational excitations [12,15].

Like in a metal, continuum of electronic excitations is possible for graphene. However, graphene is a zero-gap semiconductor, and the Fermi surface for graphene is a set of points that are referred to as the K-points. A metal has a finite density of states at the Fermi level, whereas graphene has a zero density of states at the Fermi level. Therefore, it would be interesting to see if energy transfer to graphene follows the same dependence as that for the surface excitations of a metal surface. Interestingly, although fluorescence quenching of dyes near CNTs was known, there were no reports of fluorescence quenching by graphene at the time when we thought of this possibility.

In view of the above, we have studied the EET from a dye molecule that is kept at distance z above a layer of graphene to the electronic energy levels of graphene [26,27]. The objective of our work is to consider the extended electronic charge density of graphene and understand the distance dependence of the rate [28,29]. We use the Fermi golden rule as the starting point and use the tight-binding model for graphene [21,30] to obtain analytic expressions for the rate of energy transfer within the "Dirac cone" approximation. We find that the long-range behavior of the rate of transfer has a z^{-4} dependence on the distance between the dye and the sheet. Numerical results are obtained for the decay of the π–π^* excitation of the pyrene molecule, which has been demonstrated to be efficiently quenched by CNTs. Our analysis suggests that, for the case of pyrene, energy transfer is possible up to a distance of ~300 Å, which is well beyond the FRET limit (100 Å). Recent experiments that have been performed after our theory was reported have in fact observed the fluorescence quenching of dyes near graphene. Further, the process has been found to be very useful in fabricating devices based on graphene [31], in eliminating fluorescence signals in resonance Raman spectroscopy [32], and in visualizing graphene-based sheets using fluorescence quenching microscopy [33]. The process has also been found to be useful in quantitative deoxyribonucleic acid (DNA) analysis [34,35].

The Fermi surface of undoped graphene is a set of six points known as the K-points. As a result of this, the density of states at the Fermi level is zero. It is possible to shift the Fermi level of graphene away from the K-point experimentally, either by electrical or chemical doping [36,37]. This will make the density of states at the new Fermi level nonzero. With nonzero density of states at the Fermi level, it becomes easier to transfer excitation energy to doped graphene. Thus, we study the effect of shifting the Fermi level on the distance dependence of the rate of energy transfer from fluorophores to graphene [28,38]. We imagine that the Fermi level is shifted into the conduction band to a level with the magnitude of wave vector k_F as a result of doping. We use the Dirac cone approximation, which allows us to get analytical expressions for the rate at large distances. We find a crossover of the distance dependence of the rate from z^{-4} to exponential as the Fermi level is increasingly shifted into the conduction band, with the crossover occurring at a shift of the Fermi level by an amount of $\hbar\Omega/2$. The analysis presented here for the EET to graphene has also been extended by us to understand the process of energy transfer to CNTs [39] but shall not be discussed in this chapter. We believe that the EET involving carbon-based materials is very interesting due to the fact that, using such materials, it is possible to measure distances well beyond the traditional FRET limit.

10.2 MODEL FOR THE RATE OF ENERGY TRANSFER

We consider the process of energy transfer from an energy donor D to an energy acceptor A. The donor is initially in an excited state, and it transfers its energy to the acceptor, which means that an electron that was in the orbital ψ_e^D makes a transition to ψ_g^D. In the acceptor, an electron that was sitting in the orbital ψ_g^A gets excited to ψ_e^A. The matrix element for the process is given by

$$U = \frac{e^2}{4\pi\varepsilon}\int d\mathbf{r}_1 \int d\mathbf{r}_2 \frac{\psi_e^D(\mathbf{r}_1)\psi_g^{D*}(\mathbf{r}_1)\psi_g^A(\mathbf{r}_2)\psi_e^{A*}(\mathbf{r}_2)}{|\mathbf{r}_1 - \mathbf{r}_2|}. \tag{10.4}$$

Note that this is actually an electrostatic interaction between two transition densities $\psi_e^D\psi_g^{D*}$ and $\psi_g^A\psi_e^{A*}$. In the case of the EET from dyes to graphene, the orbitals $\psi_g^A(\mathbf{r}_2)$ and $\psi_e^A(\mathbf{r}_2)$ on the acceptor are extended in space with delocalized charge densities (in comparison with the separation between D and A) [29]. In such a situation, we can approximate the interaction matrix element as that between the transition dipole of the dye, with $\boldsymbol{\mu}_{eg}^D$ located at the center of charge and given by $\boldsymbol{\mu}_{eg}^D = -e\int d\mathbf{r}_1\psi_e^D(\mathbf{r}_1)\mathbf{r}_1\psi_g^{D*}(\mathbf{r}_1)$, and the transition charge density $-e\psi_g^A(\mathbf{r}_2)\psi_e^{A*}(\mathbf{r}_2)$ of graphene. Therefore, the matrix element for interaction becomes

$$U = \boldsymbol{\mu}_{eg}^D \cdot \nabla\Phi, \tag{10.5}$$

where Φ is the electrostatic potential at point $\mathbf{r}$ (the position of the donor) due to the charge density $-e\psi_g^A(\mathbf{r}_2)\psi_e^{A*}(\mathbf{r}_2)$. The electronic energy states of graphene are characterized by wave vector $\mathbf{k}$. As a result of the EET from the dye, an electron in an energy level with wave vector $\mathbf{k}_i$ is excited to a level with wave vector $\mathbf{k}_f$. The Fermi golden rule expression for the rate of energy transfer is given by

$$k(\hbar\Omega) = \frac{2\pi}{\hbar}\sum_{\mathbf{k}_i}\sum_{\mathbf{k}_f} |U_{\mathbf{k}_i,\mathbf{k}_f}|^2\,\delta(E_{\mathbf{k}_f} - E_{\mathbf{k}_i} - \hbar\Omega), \tag{10.6}$$

where $\hbar\Omega$ is the emission energy of the fluorophore. We define $\mathbf{k}_f = \mathbf{k}_i + \mathbf{q}$, where $\mathbf{q}\hbar$ is the momentum transferred to graphene. The rate can therefore be written as

$$k(\hbar\Omega) = \frac{2\pi}{\hbar}\sum_{\mathbf{k}_i}\sum_{\mathbf{q}} |U_{\mathbf{k}_i,\mathbf{q}}|^2\,\delta(E_{\mathbf{k}_i+\mathbf{q}} - E_{\mathbf{k}_i} - \hbar\Omega). \tag{10.7}$$

The transition density given by $-e\psi_g^A(\mathbf{r}_2)\psi_e^{A*}(\mathbf{r}_2) = -e\psi_{\mathbf{k}_i+\mathbf{q}}^{A*}(\mathbf{r}_2)\psi_{\mathbf{k}_i}^A(\mathbf{r}_2)$ then has a periodicity with wave vector $\mathbf{q}$.

Here, we consider the transfer of energy to the π system of graphene. We first use a nearest neighbor tight-binding model to obtain the wave functions and energies for the graphene lattice [21,30]. The unit cell of the hexagonal lattice of graphene is a parallelogram and contains two atoms. Therefore, it is convenient to imagine the lattice to be made up of two sets of carbon atoms, which we label as *A* and *B*. The nearest neighbors of an A-type atom are three B-type atoms, and the next nearest neighbors are six A-type atoms (see Figure 10.2). Vectors $\mathbf{a}_1$ and $\mathbf{a}_2$ are the real lattice vectors of the lattice. $|\mathbf{a}_1| = |\mathbf{a}_2| = a$, where a is the lattice constant and is given by $a = l_{C-C}\sqrt{3}$, with $l_{C-C} =$ 1.42 Å, which is the C–C bond length in graphene. The carbon atoms of the lattice are sp^2 hybridized. Each carbon atom uses its three sp^2 hybrid orbitals to form bonds with the three nearest neighbor carbon atoms. This leaves an extra electron sitting in the $2p_z$ orbital of each carbon atom. These electrons on each atom combine to form π bonding and π antibonding molecular orbitals. We first form two tight-binding Bloch functions generated from the $2p_z$ orbital wave functions of the carbon atoms of the lattice as

$$\psi_{\mathbf{k}_A} = \frac{1}{\sqrt{N}} \sum_A e^{i\mathbf{k}.\mathbf{s}_A} \chi_A(\mathbf{r} - \mathbf{s}_A) \tag{10.8}$$

and

$$\psi_{\mathbf{k}_B} = \frac{1}{\sqrt{N}} \sum_B e^{i\mathbf{k}.\mathbf{s}_B} \chi_B(\mathbf{r} - \mathbf{s}_B). \tag{10.9}$$

N is the number of atoms of each type in the lattice, and χ is the $2p_z$ wave function on the carbon atoms. $\mathbf{s}_A(\mathbf{s}_B)$ are two-dimensional vectors lying in the plane of graphene and specifying the positions of the A(B)-type atoms. The total wave function

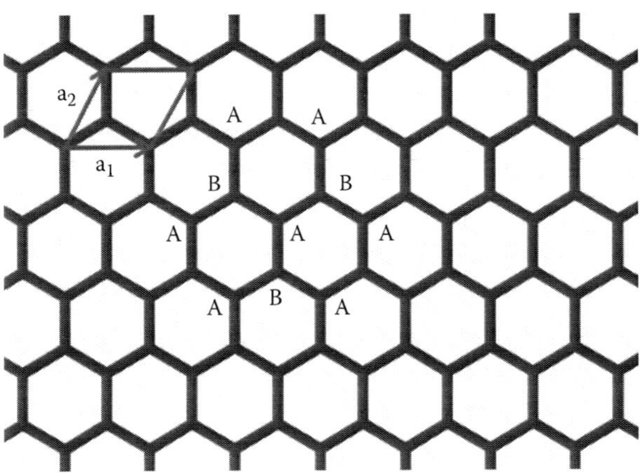

FIGURE 10.2 Schematic of the graphene lattice showing the unit cell and the real lattice vectors. The nearest neighbors and the next nearest neighbors of one of the carbon atoms are also shown.

for the π system of the lattice for a particular two-dimensional wave vector $\mathbf{k} \equiv (k_x, k_y)$ may be written as a linear combination of the two Bloch functions written above, i.e.,

$$\psi_{\mathbf{k}}(\mathbf{r}) = c_A \psi_{\mathbf{k}_A} + c_B \psi_{\mathbf{k}_B}. \tag{10.10}$$

We substitute the above form for the wave function into the Schrödinger equation $\hat{H}\psi = E\psi$ and determine coefficients c_A and c_B variationally [30]. The energy bands are given by $E_{\mathbf{k}}^{\pm} = H_{AA} \pm | H_{AB} |$, where within the tight-binding model,

$$H_{AA} = \left\langle \psi_{\mathbf{k}_A} \middle| \hat{H} \middle| \psi_{\mathbf{k}_A} \right\rangle \simeq \left\langle \chi_A(\mathbf{r} - \mathbf{s}_A) \middle| \hat{H} \middle| \chi_A(\mathbf{r} - \mathbf{s}_A) \right\rangle = \alpha \tag{10.11}$$

and

$$H_{AB}(\mathbf{k}) = \left\langle \psi_{\mathbf{k}_A} \middle| \hat{H} \middle| \psi_{\mathbf{k}_B} \right\rangle \simeq t \left[e^{-\frac{ik_x a}{\sqrt{3}}} + 2e^{\frac{ik_x a}{2\sqrt{3}}} \cos(k_y a/2) \right]. \tag{10.12}$$

In the above, t is the nearest neighbor matrix element, which is known as the hopping integral, and is given by $t = \left\langle \chi_A(\mathbf{r} - \mathbf{s}_A) \middle| \hat{H} \middle| \chi_B(\mathbf{r} - \mathbf{s}_B) \right\rangle$. We choose $\alpha = 0$. With this, the π energy bands are given by

$$E_{\mathbf{k}}^{\pm} = \pm t \left[1 + 4\cos(k_y a/2)\cos\left(k_x \sqrt{3}a/2\right) + 4\cos^2(k_y a/2) \right]^{\frac{1}{2}}, \tag{10.13}$$

where the + sign is for the conduction band and the − sign is for the valence band. A plot of the energy bands as a function of k_x and k_y is shown in Figure 10.3. The

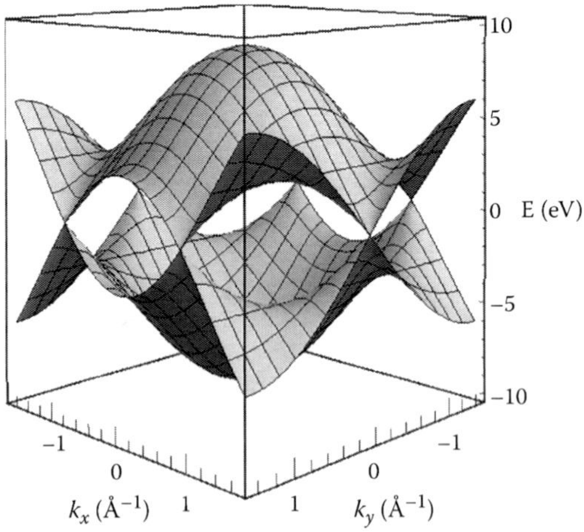

FIGURE 10.3 Electronic energy bands of the π system of graphene as a function of the wave vector components k_x and k_y. The valence and the conduction bands meet at the K-points.

valence and the conduction bands meet at the K-points. As well known, the energy follows a linear dispersion relation around the K-points. Denoting the deviation of the wave vector from a K-point as **k** and carrying out an expansion around the K-point, we get

$$E_{\mathbf{k}}^{\pm} = \pm t \frac{\sqrt{3}a}{2} k = \pm v_f k, \tag{10.14}$$

with $k = \sqrt{k_x^2 + k_y^2}$. It is convenient to work with this linear dispersion relation as the required integrals can be evaluated analytically. The wave function for the **k**th orbital of graphene then is

$$\psi_{\mathbf{k}}^{\pm}(\mathbf{r}) = \frac{1}{\sqrt{2N}} \left[\sum_A e^{i\mathbf{k}.\mathbf{s}_A} \chi_A(\mathbf{r} - \mathbf{s}_A) \mp e^{-i\delta_{\mathbf{k}}} \sum_B e^{i\mathbf{k}.\mathbf{s}_B} \chi_B(\mathbf{r} - \mathbf{s}_B) \right], \tag{10.15}$$

where the + sign in $\psi_{\mathbf{k}}^{\pm}(\mathbf{r})$ holds for the conduction band (π^* band) and the – sign holds for the valence band (π band). The phase factor $\delta_{\mathbf{k}}$ is defined by the relation

$$e^{i\delta_{\mathbf{k}}} = -\frac{H_{AB}(\mathbf{k})}{|H_{AB}(\mathbf{k})|}. \tag{10.16}$$

Again, expanding the wave vector near the K-point, we have

$$\delta_{\mathbf{k}} = \tan^{-1}\left(\frac{k_y}{k_x}\right) = \varphi_{\mathbf{k}}, \tag{10.17}$$

where $\varphi_{\mathbf{k}}$ is the angle that vector **k** makes with the x-axis. As clear from Equation 10.5, we consider the transition dipole moment of the dye $\boldsymbol{\mu}_{eg}^{D} = \mu_x \hat{\mathbf{i}} + \mu_y \hat{\mathbf{j}} + \mu_z \hat{\mathbf{k}}$ to be interacting electrostatically with the electrons of graphene [26,27,29]. As a result of such an interaction, the dye returns back to its ground state, and an electron in graphene is excited. The excitation of an electron from $\psi_{\mathbf{k}_i}^{-}(\mathbf{r})$ to $\psi_{\mathbf{k}_i+\mathbf{q}}^{+}(\mathbf{r})$ leads to a transition charge density given by

$$\rho(\mathbf{r}) = -e\psi_{\mathbf{k}_i+\mathbf{q}}^{+*}(\mathbf{r})\psi_{\mathbf{k}_i}^{-}(\mathbf{r}) \simeq \frac{e}{2N}\left[e^{i(\delta_{\mathbf{k}_i+\mathbf{q}} - \delta_{\mathbf{k}_i})} - 1\right] \sum_C e^{-i\mathbf{q}\cdot\mathbf{s}_C} \chi_C(\mathbf{r} - \mathbf{s}_C)\chi_C^{*}(\mathbf{r} - \mathbf{s}_C), \tag{10.18}$$

where the summation is over carbon atoms of either type in the lattice. Further, we have neglected the product of χs, which belong to different atoms of the lattice, as they are negligible. The electrostatic potential due to such a transition density is given by

$$\Phi(\mathbf{r}) = \frac{1}{4\pi\varepsilon}\int d\mathbf{r}_2 \frac{\rho(\mathbf{r}_2)}{|\mathbf{r}-\mathbf{r}_2|}. \tag{10.19}$$

Since the density $\chi_C(\mathbf{r}_2 - \mathbf{s}_C)\chi_C^*(\mathbf{r}_2 - \mathbf{s}_C)$ is localized near the Cth atom, we can use the multipole expansion to calculate its electrostatic potential at point $\mathbf{r}$. The lowest order term is the monopole term of Equation 10.18, leading to

$$\Phi(\mathbf{r}) = \frac{e}{8\pi\varepsilon N}\left[e^{i(\delta_{\mathbf{k}_i+\mathbf{q}}-\delta_{\mathbf{k}_i})} - 1\right]\sum_C \frac{e^{-i\mathbf{q}\cdot\mathbf{s}_C}}{|\mathbf{r}-\mathbf{s}_C|}. \tag{10.20}$$

For small values of $q(=|\mathbf{q}|)$, the sum in the above equation may be replaced by an integral so that

$$\Phi(\mathbf{r}) = \frac{e}{8\pi\varepsilon N A_u}\left[e^{i(\delta_{\mathbf{k}_i+\mathbf{q}}-\delta_{\mathbf{k}_i})} - 1\right]\int d\mathbf{s}_C \frac{e^{-i\mathbf{q}\cdot\mathbf{s}_C}}{|\mathbf{r}-\mathbf{s}_C|}, \tag{10.21}$$

where A_u is the area of the unit cell of graphene. The two-dimensional integral written above can be evaluated to get

$$\Phi(\mathbf{r}) = \frac{e}{4\varepsilon A}\left[1 - e^{i(\delta_{\mathbf{k}_i+\mathbf{q}}-\delta_{\mathbf{k}_i})}\right]\frac{e^{-qz}e^{-i\mathbf{q}\cdot\mathbf{X}}}{q}, \tag{10.22}$$

where we have used $\mathbf{r} = (\mathbf{X}, z)$, with $\mathbf{X}$ being parallel to the plane of graphene. A is the area of the graphene lattice. We now evaluate the matrix element of Equation 10.5 and find it to be

$$U = \frac{e}{4\varepsilon A}\left[e^{i(\delta_{\mathbf{k}_i+\mathbf{q}}-\delta_{\mathbf{k}_i})} - 1\right]\boldsymbol{\mu}_{eg}^{D}\cdot\left(i\hat{\mathbf{q}} + \hat{\mathbf{k}}\right)e^{-qz}e^{-i\mathbf{q}\cdot\mathbf{X}}, \tag{10.23}$$

where $\hat{\mathbf{q}} = \dfrac{\mathbf{q}}{q}$ is the unit vector in the direction of $\mathbf{q}$, and $\hat{\mathbf{k}}$ is the unit vector in the z-direction. Thus,

$$|U|^2 = \frac{e^2}{8\varepsilon^2 A^2}\left[1 - \cos\left(\varphi_{\mathbf{k}_i+\mathbf{q}} - \varphi_{\mathbf{k}_i}\right)\right]\left|\boldsymbol{\mu}_{eg}^{D}\cdot\left(i\hat{\mathbf{q}} + \hat{\mathbf{k}}\right)\right|^2 e^{-2qz}, \tag{10.24}$$

where $\varphi_{\mathbf{k}_i}$ and $\varphi_{\mathbf{k}_i+\mathbf{q}}$ are the polar angles of the two-dimensional wave vectors $\mathbf{k}_i$ and $\mathbf{k}_i + \mathbf{q}$, respectively. We now use Equation 10.7 to obtain the following expression for the rate of transfer:

$$k(\hbar\Omega) = \frac{\pi e^2}{4\hbar\varepsilon^2 A^2}\sum_{\mathbf{q}}\left|\boldsymbol{\mu}_{eg}^{D}\cdot\left(i\hat{\mathbf{q}}+\hat{\mathbf{k}}\right)\right|^2 e^{-2qz}G(\mathbf{q}), \tag{10.25}$$

where

$$G(\mathbf{q}) = \sum_{\mathbf{k_i}}[1-\cos(\varphi_{\mathbf{k}_i+\mathbf{q}}-\varphi_{\mathbf{k}_i})]\delta\left(E^{+}_{\mathbf{k}_i+\mathbf{q}}-E^{-}_{\mathbf{k}_i}-\hbar\Omega\right). \tag{10.26}$$

The sum over $\mathbf{k}_i$ in the above equation can be converted into an integral to get

$$G(\mathbf{q}) = \frac{A}{4\pi^2}\int d\mathbf{k}_i\left[1-\frac{\mathbf{k}_i\cdot(\mathbf{k}_i+\mathbf{q})}{|\mathbf{k}_i||(\mathbf{k}_i+\mathbf{q})|}\right]\delta\left(E^{+}_{\mathbf{k}_i+\mathbf{q}}-E^{-}_{\mathbf{k}_i}-\hbar\Omega\right). \tag{10.27}$$

We first evaluate $G(\mathbf{q})$. We introduce a new variable $\mathbf{k}_i'$ defined by $\mathbf{k}_i' = \mathbf{k}_i + \frac{\mathbf{q}}{2}$. Then, we use Equation 10.14 for the energy levels of graphene to get

$$G(\mathbf{q}) = \frac{A}{4\pi^2 v_f}\int d\mathbf{k}_i'\left[1-\frac{\left(\mathbf{k}_i'-\frac{\mathbf{q}}{2}\right)\cdot\left(\mathbf{k}_i'+\frac{\mathbf{q}}{2}\right)}{\left|\mathbf{k}_i'-\frac{\mathbf{q}}{2}\right|\left|\mathbf{k}_i'+\frac{\mathbf{q}}{2}\right|}\right]\delta\left(\left|\mathbf{k}_i'-\frac{\mathbf{q}}{2}\right|+\left|\mathbf{k}_i'+\frac{\mathbf{q}}{2}\right|-\frac{\hbar\Omega}{v_f}\right). \tag{10.28}$$

We choose the direction of $\mathbf{q}$ as the x-axis and then make another change of variable to $\mathbf{r}$ ($\equiv (x,y)$) given by $\mathbf{r} = \frac{\mathbf{k}_i'}{q/2}$ to get

$$G(\mathbf{q}) = \frac{Aq^2}{16\pi^2 v_f}\int d\mathbf{r}\left[1-\frac{(\mathbf{r}-\hat{\mathbf{i}}).(\mathbf{r}+\hat{\mathbf{i}})}{|\mathbf{r}-\hat{\mathbf{i}}||\mathbf{r}+\hat{\mathbf{i}}|}\right]\delta\left[\frac{q}{2}\left(|\mathbf{r}-\hat{\mathbf{i}}|+|\mathbf{r}+\hat{\mathbf{i}}|\right)-\frac{\hbar\Omega}{v_f}\right]. \tag{10.29}$$

The above equation can be rewritten as

$$G(\mathbf{q}) = \frac{Aq^2}{8\pi^2 v_f}\int_{-\infty}^{\infty}dx\int_{0}^{\infty}dy\left[1-\frac{x^2+y^2-1}{\sqrt{(x-1)^2+y^2}\sqrt{(x+1)^2+y^2}}\right]\times \delta\left[\frac{q}{2}\left(\sqrt{(x-1)^2+y^2}+\sqrt{(x+1)^2+y^2}\right)-\frac{\hbar\Omega}{v_f}\right]. \tag{10.30}$$

We now change over to elliptic coordinates defined by $x = \mu\upsilon$ and $y = \sqrt{(\mu^2-1)(1-\upsilon^2)}$. The transformation gives $dxdy = \frac{\mu^2-\upsilon^2}{\sqrt{(\mu^2-1)(1-\upsilon^2)}} d\mu d\upsilon$. Substituting the above transformation into Equation 10.30 gives

$$G(\mathbf{q}) = \frac{Aq^2}{4\pi^2 v_f} \int_1^\infty d\mu \int_{-1}^1 d\upsilon \sqrt{\frac{1-\upsilon^2}{\mu^2-1}} \delta\left(q\mu - \frac{\hbar\Omega}{v_f}\right). \tag{10.31}$$

The above integrals can now be easily performed to get

$$G(\mathbf{q}) = \frac{A}{8\pi} \frac{q^2 \Theta(\hbar\Omega - qv_f)}{\sqrt{(\hbar\Omega)^2 - q^2 v_f^2}}. \tag{10.32}$$

We now substitute the above result into Equation 10.25, replace the sum over $\mathbf{q}$ by an integral, and use $\boldsymbol{\mu}_{eg}^D = \mu_x \hat{\mathbf{i}} + \mu_y \hat{\mathbf{j}} + \mu_z \hat{\mathbf{k}}$ to get

$$k(\hbar\Omega) = \frac{e^2}{64\pi h \varepsilon^2} \int_0^\infty dqq \int_0^{2\pi} d\theta \left[\mu_z^2 + (\mu_x \cos\theta + \mu_y \sin\theta)^2\right] e^{-2qz} \frac{\Theta(\hbar\Omega - qv_f) q^2}{\sqrt{(\hbar\Omega)^2 - q^2 v_f^2}}, \tag{10.33}$$

where (q,θ) are the polar coordinates of $\mathbf{q}$. The integral over θ can be easily performed to get

$$k(\hbar\Omega) = \frac{e^2}{64 h \varepsilon^2} \left(\mu_x^2 + \mu_y^2 + 2\mu_z^2\right) \int_0^{\hbar\Omega/v_f} dq \frac{e^{-2qz} q^3}{\sqrt{(\hbar\Omega)^2 - q^2 v_f^2}}. \tag{10.34}$$

We now show the explicit orientation dependence of the rate of energy transfer. Consider that the transition dipole of the dye $\boldsymbol{\mu}_{eg}^D$ makes angles of α and β with respect to the directions $\hat{\boldsymbol{i}}$ and $\hat{\mathbf{k}}$, respectively (see Figure 10.4). We therefore have $\mu_x = \mu_{eg} \cos\alpha \sin\beta$, $\mu_y = \mu_{eg} \sin\alpha \sin\beta$, and $\mu_z = \mu_{eg} \cos\beta$, where we have defined $\mu_{eg} = \left|\boldsymbol{\mu}_{eg}^D\right|$. Using the above definitions, the rate can be written as

$$k(\hbar\Omega) = \frac{e^2}{64 h \varepsilon^2} \left(\mu_{eg}^2 \sin^2\beta + 2\mu_{eg}^2 \cos^2\beta\right) \int_0^{\hbar\Omega/v_f} dq \frac{e^{-2qz} q^3}{\sqrt{(\hbar\Omega)^2 - q^2 v_f^2}}. \tag{10.35}$$

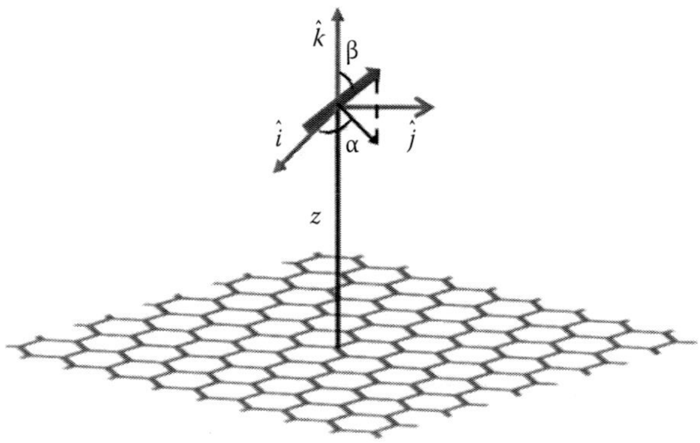

FIGURE 10.4 Schematic of the donor–acceptor system showing the transition dipole of the fluorophore and the graphene lattice. In the figure, we also show the angles that the transition dipole moment vector makes with the coordinate axes.

We now perform an averaging over all possible orientations of the donor transition dipole moment to get

$$\langle k(\hbar\Omega)\rangle = \frac{e^2\mu_{eg}^2}{48h\varepsilon^2}\int_0^{\hbar\Omega/v_f} dq \frac{e^{-2qz}q^3}{\sqrt{(\hbar\Omega)^2 - q^2v_f^2}}. \tag{10.36}$$

For large values of z, only values of $q \simeq 1/(2z)$ contribute to the above integral. In such a case, we can neglect $q^2v_f^2$ in comparison with $(\hbar\Omega)^2$ and evaluate the integral in Equation 10.36 by extending the upper limit of the integral to infinity to get

$$\langle k(\hbar\Omega)\rangle = \frac{\pi e^2}{64\Omega h^2\varepsilon^2}\frac{\mu_{eg}^2}{z^4}. \tag{10.37}$$

Thus, we show that the long-range behavior of the rate of energy transfer to graphene has a power law $[(distance)^{-4}]$ dependence. The integral in Equation 10.36 can be evaluated exactly for small values of z by making the transformation $q = \hbar\Omega/v_f \sin(t)$ and then expanding the exponential term to get

$$\langle k(\hbar\Omega)\rangle = \frac{e^2\sqrt{\pi}}{96h\varepsilon^2}\frac{(\hbar\Omega)^3\mu_{eg}^2}{v_f^4}\sum_{n=0}^{\infty}\frac{(-1)^n}{n!}\left(\frac{2z\hbar\Omega}{v_f}\right)^n\frac{\Gamma((4+n)/2)}{\Gamma((5+n)/2)}. \tag{10.38}$$

The sum in the above expression can be evaluated numerically. However, it is possible to evaluate the integral in Equation 10.36 numerically for all values of z while the asymptotic expression given by Equation 10.37 is valid for large z.

10.3 RESULTS AND DISCUSSION

We use the equations of the above section to evaluate the rate of energy transfer from a fluorophore to graphene. We choose pyrene as the fluorophore for our numerical calculations. The fluorescence emission maximum of pyrene occurs at 390 nm (3.2 eV) [18]. We have performed *ab initio* calculations for pyrene to get the transition dipole moment and obtained $\mu_{eg} = 4.5\ D$ (B3LYP/6-311++G** level of calculation) [40]. The lifetime of the excited state of pyrene is 410 ns [1] (in ethanol). Therefore, the total rate of de-excitation in the absence of graphene is $2.4 \times 10^6\ s^{-1}$. When pyrene is kept at distance z above a layer of graphene, it is possible that the excitation energy may be transferred to graphene, provided that the rate of transfer is greater than or comparable to the rates of other de-excitation processes. To explore this, we have calculated the rate of nonradiative energy transfer from the excited state of pyrene to graphene using the equations from the model that we have developed, for z varying in the range of 5–400 Å. Note that the numerical values of the rates obtained using the equations derived earlier have been multiplied by a factor of 2 to account for the two inequivalent K-points of graphene. We consider the transfer of a fixed amount of energy $\hbar\Omega$ to graphene and use $\hbar\Omega = 3.2$ eV, which is the emission maximum of pyrene. The numerical value of the hopping integral for graphene was taken to be $t = 3.0$ eV [21].

Using the above-mentioned parameters, we use Equations 10.36 and 10.37 to evaluate the exact and asymptotic long-range rates of energy transfer from pyrene to graphene. Figure 10.5 shows a comparison of the two as a function of the distance.

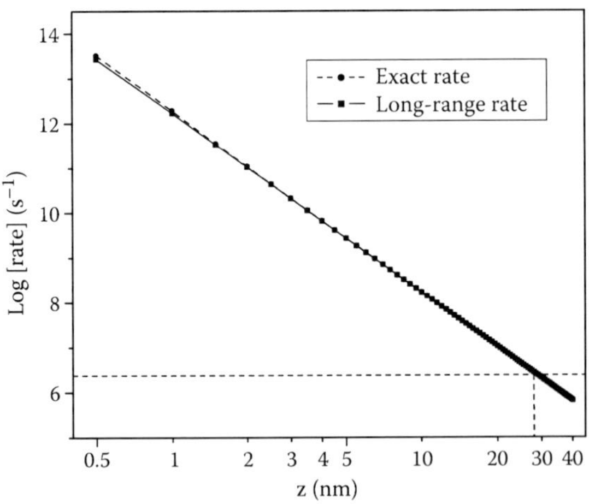

FIGURE 10.5 Comparison of the exact and long-range rates of energy transfer from pyrene to graphene. The dashed horizontal line shows the natural decay rate of pyrene in the absence of graphene. The rate of energy transfer is greater than the natural decay rate up to a distance of 290 Å. (Reprinted with permission from Swathi, R. S. and K. L. Sebastian, *J. Chem. Phys.*, 130, 086101, 2009. Copyright 2009, American Institute of Physics.)

It is clear that the rate has a z^{-4} dependence for distances greater than 20 Å, even though, for shorter distances, the deviation is not large. In addition, in Figure 10.5, we show the natural rate of decay of the excited state of pyrene (2.4×10^6 s^{-1}) as the dashed horizontal line. The rate of transfer is greater than the natural decay rate up to a distance of 290 Å. Thus, when pyrene is kept at distances in the range of 5–290 Å from graphene, the rate of energy transfer is higher than the other de-excitation pathways, and pyrene predominantly decays by the energy transfer to graphene. At 290 Å, both rates become equal, and beyond 290 Å, energy transfer is not the dominant pathway for the de-excitation of the excited state. Therefore, it is possible to think of an experiment wherein pyrene is kept at various distances from graphene, and fluorescence quenching by graphene is measured by fluorescence spectroscopy experiments. Measurements of the fluorescence intensity and the excited state lifetime of pyrene in both the presence and the absence of graphene may be used to verify this prediction.

10.4 RECENT EXPERIMENTS ON FLUORESCENCE QUENCHING BY GRAPHENE

Our theoretical study on the EET from fluorophores to graphene was the first report, theory, or experiment on fluorescence quenching by graphene. Since then, there have been experimental studies by several groups with interesting applications for the quenching process. Resonance Raman spectroscopy is a powerful spectroscopic technique for characterizing the molecular structures. Raman scattering is a very weak process. The Raman signals are usually masked by the much stronger fluorescence emission. It is possible to suppress fluorescence signals in resonance Raman measurements using currently available techniques like femtosecond stimulated Raman spectroscopy [41], coherent anti-Stokes Raman spectroscopy [42], optical Kerr gate [43], and ultrafast Raman loss spectroscopy [44]. However, these techniques involve a complex experimental setup. Xie et al. [32] have shown that the fluorescence quenching ability of graphene can be used to obtain good resonance Raman signals from fluorescent dye molecules. When the dyes were adsorbed on graphene, the fluorescence of the dyes was found to be quenched. Using graphene as the substrate, they were able to suppress the fluorescence background by ~10^3 times and obtain good resonance Raman signals from the dyes.

Visualizing graphene-based sheets is a very challenging problem because regular methods like atomic force microscopy require very special substrates and rigorous operating conditions. Therefore, a general visualization technique that allows for quick sample examination over large areas is highly desirable. Kim et al. [33] reported that graphene-based sheets can be easily visualized under a fluorescence microscope when coated with dyes because of the fluorescence quenching of the dyes by graphene. They have used a nonfluorescent polymer spacer to fix the distance between the graphene-based sheet and the dye layer. For a polymer spacer of thickness 20 nm, they have observed 23% quenching. However, for a spacer of thickness 200 nm, graphene-based sheets became invisible, meaning that graphene is no longer able to quench the fluorescence. In these experiments, the dye coating could be easily removed later on without disrupting the sheets. They also demonstrated for the first time that the method could be used for direct observation of suspended graphene sheets in solution.

Device fabrication using carbon-based materials like graphene and CNTs has been a very active area of research in the last few years. Recently, Sagar et al. [31] have reported a method for on-the-fly graphene-based device fabrication. The method is based on identifying graphene flakes using a confocal microscope making use of the property of fluorescence quenching of dyes by graphene. After locating the graphene flakes, they were able to fabricate the devices on the fly by performing the necessary lithography steps using a confocal laser scanning microscope.

There has been a recent report of the application of graphene oxide–based fluorescence quenching for quantitative DNA analysis [34]. They appended single-stranded DNA (ss-DNA) molecules to fluorophores. In the presence of graphene oxide, the fluorescence of these probes was found to be very small. This is due to the strong hydrophobic adsorption of the dyes at the graphene oxide surface and the subsequent highly efficient energy transfer from the dyes to graphene oxide. However, when a complementary target of the probe ss-DNA was added, they hybridize to form a duplex, and the fluorescence is recovered (see also [35]). This happens because of the different affinities of the ss-DNA and the duplex to graphene oxide. The interaction between the duplex and graphene oxide was found to be rather weak. This is due to the fact, that in the duplex, the nucleobases were effectively shielded within the phosphate backbone of the duplex, whereas, in case of ss-DNA, the nucleobases were found to be nearly flat at the graphene oxide surface, leading to favorable π stacking interactions. The fluorescence intensities with the complementary target and the target with single-base mismatch were found to be quite different. Therefore, the method was found to exhibit very high sequence specificity. A recent study has shown that photoexcited semiconductor nanocrystals can also transfer the excitation energy to graphene very efficiently [45]. However, in spite of a lot of effort in studying energy transfer to graphene, to the best of our knowledge, there is no detailed experimental study of the distance dependence of the rate.

10.5 EET FROM FLUOROPHORES TO DOPED GRAPHENE

We now consider the EET from fluorophores to doped graphene. We imagine that the Fermi level is shifted into the conduction band to a level with magnitude of wave vector k_F. Thus, the rate of energy transfer has contributions from two different sets of transitions in graphene. In the first set, $\mathbf{k}_i$ lies in the valence band, with $0 \leq \mathbf{k}_i \leq \infty$, and $\mathbf{k}_f$ lies in the conduction band, with $k_F < \mathbf{k}_f < \infty$. In the second set, both $\mathbf{k}_i$ and $\mathbf{k}_f$ lie in the conduction band, with $0 \leq \mathbf{k}_i \leq k_F$ and $k_F < \mathbf{k}_f < \infty$ (see Figure 10.6). The total rate can thus be written as the sum total of both contributions, i.e., $k = k_1 + k_2$. k_1 and k_2 are both given by

$$k_j = \frac{\pi e^2}{4\hbar \epsilon^2 A^2} \sum_{\mathbf{q}} \left| \boldsymbol{\mu}_{eg}^{D} . \left(i\hat{\mathbf{q}} + \hat{\mathbf{k}} \right) \right|^2 \exp(-2qz) G_j(\mathbf{q}) \tag{10.39}$$

but with differing expressions for $G_j(\mathbf{q})$.

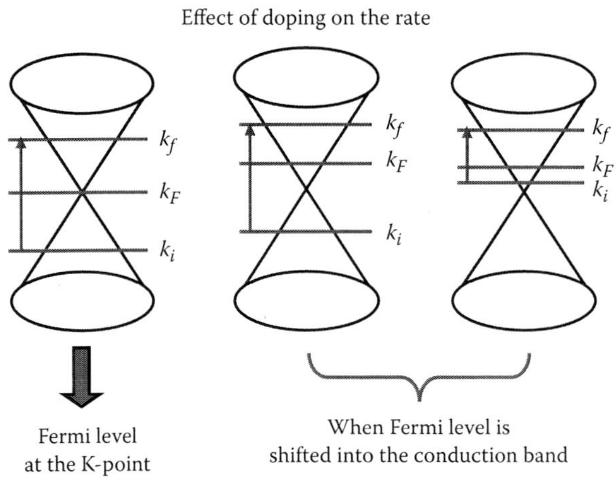

FIGURE 10.6 Schematic of the energy bands, showing the Fermi wave vectors, the initial and final wave vectors corresponding to EET. $E_F = v_f k_F$ is the location of the new Fermi level on doping graphene. (Reprinted with permission from Swathi, R. S. and K. L. Sebastian, *J. Chem. Sci.*, 124, 233, 2012.)

$$G_1(\mathbf{q}) = \sum_{\mathbf{k_i} \in \text{valence band}} [1 - \cos(\varphi_{\mathbf{k}_i+\mathbf{q}} - \varphi_{\mathbf{k}_i})]\delta\left(E^+_{\mathbf{k}_i+\mathbf{q}} - E^-_{\mathbf{k}_i} - \hbar\Omega\right)\Theta\left(|\mathbf{k}_i + \mathbf{q}| - k_F\right). \tag{10.40}$$

The theta function is introduced to satisfy the condition, $k_F < \mathbf{k}_f < \infty$. In a similar fashion,

$$G_2(\mathbf{q}) = \sum_{\mathbf{k_i} \in \text{conduction band}} [1 + \cos(\varphi_{\mathbf{k}_i+\mathbf{q}} - \varphi_{\mathbf{k}_i})]\delta\left(E^+_{\mathbf{k}_i+\mathbf{q}} - E^+_{\mathbf{k}_i} - \hbar\Omega\right)$$
$$\times\Theta\left(|\mathbf{k}_i + \mathbf{q}| - k_F\right)\Theta\left(k_F - |\mathbf{k}_i|\right). \tag{10.41}$$

The two theta functions satisfy the conditions $0 \le \mathbf{k}_i \le k_F$ and $k_F < \mathbf{k}_f < \infty$. We now evaluate $G_1(\mathbf{q})$ and $G_2(\mathbf{q})$ separately. We replace the sum over $\mathbf{k}_i$ in the expression for $G_1(\mathbf{q})$ by an integral and use the linear dispersion relation for the energy levels of graphene $\left(E^{\pm}_{\mathbf{k}} = \pm v_f k\right)$ to get

$$G_1(\mathbf{q}) = \frac{A}{4\pi^2 v_f}\int d\mathbf{k}_i\left[1 - \frac{\mathbf{k}_i\cdot(\mathbf{k}_i+\mathbf{q})}{|\mathbf{k}_i||(\mathbf{k}_i+\mathbf{q})|}\right]\delta\left(|\mathbf{k}_i| + |\mathbf{k}_i+\mathbf{q}| - \frac{\hbar\Omega}{v_f}\right)\Theta\left(|\mathbf{k}_i+\mathbf{q}| - k_F\right). \tag{10.42}$$

$G_1(\mathbf{q})$ can be evaluated using the procedure outlined in the previous section for the evaluation of $G(\mathbf{q})$. On performing the transformation to elliptical coordinates, we get

$$G_1(\mathbf{q}) = \frac{Aq^2}{4\pi^2 v_f}\int_1^\infty d\mu \int_{-1}^{1} d\upsilon \sqrt{\frac{1-\upsilon^2}{\mu^2-1}}\,\delta\left(q\mu - \frac{\hbar\Omega}{v_f}\right)\Theta\left[\frac{q}{2}(\mu+\upsilon) - k_F\right]. \quad (10.43)$$

The integral over μ can be easily performed to get

$$G_1(\mathbf{q}) = \frac{Aq^2}{4\pi^2}\frac{\Theta(\hbar\Omega - qv_f)}{\sqrt{(\hbar\Omega)^2 - q^2 v_f^2}}\int_{-1}^{1} d\upsilon\sqrt{1-\upsilon^2}\,\Theta\left[\frac{q}{2}\left(\frac{\hbar\Omega}{qv_f}+\upsilon\right) - k_F\right]. \quad (10.44)$$

The above equation can be rewritten as

$$G_1(\mathbf{q}) = \frac{Aq^2}{4\pi^2}\frac{\Theta(\hbar\Omega - qv_f)}{\sqrt{(\hbar\Omega)^2 - q^2 v_f^2}}\Theta\left[1 - \frac{2E_F - \hbar\Omega}{qv_f}\right]\int_{\mathrm{Max}\left[-1,\frac{2E_F-\hbar\Omega}{qv_f}\right]}^{1} d\upsilon\sqrt{1-\upsilon^2}. \quad (10.45)$$

We now substitute back the above expression into the rate expression of Equation 10.39 and convert the sum over $\mathbf{q}$ to an integral to get

$$k_1 = \frac{e^2}{64\hbar\varepsilon^2\pi^3}\int_0^\infty dq q^3 e^{-2qz}\frac{\Theta(\hbar\Omega - qv_f)}{\sqrt{(\hbar\Omega)^2 - q^2 v_f^2}}\Theta\left[1 - \frac{2E_F - \hbar\Omega}{qv_f}\right]\int_{\mathrm{Max}\left[-1,\frac{2E_F-\hbar\Omega}{qv_f}\right]}^{1}$$

$$\times d\upsilon\sqrt{1-\upsilon^2}\int_0^{2\pi} d\theta\left|\boldsymbol{\mu}_{eg}^{D}\cdot\left(i\hat{\mathbf{q}} + \hat{\mathbf{k}}\right)\right|^2, \quad (10.46)$$

where (q,θ) are the polar coordinates of $\mathbf{q}$. After performing the integral over θ, we average over all possible orientations of the donor transition dipole as earlier to get

$$k_1 = \frac{e^2\mu_{eg}^2}{48\hbar\varepsilon^2\pi^2}\int_0^\infty dq q^3 e^{-2qz}\frac{\Theta(\hbar\Omega - qv_f)}{\sqrt{(\hbar\Omega)^2 - q^2 v_f^2}}\Theta\left[1 - \frac{2E_F - \hbar\Omega}{qv_f}\right]\int_{\mathrm{Max}\left[-1,\frac{2E_F-\hbar\Omega}{qv_f}\right]}^{1} d\upsilon\sqrt{1-\upsilon^2}. \quad (10.47)$$

Evaluation of the integral over υ leads to

$$k_1 = \frac{e^2\mu_{eg}^2}{96\hbar\varepsilon^2\pi^2}\int_0^\infty dq q^3 e^{-2qz}\frac{\Theta(\hbar\Omega - qv_f)}{\sqrt{(\hbar\Omega)^2 - q^2v_f^2}}\Theta\left[1-\frac{2E_F-\hbar\Omega}{qv_f}\right]$$
$$\times\left[\frac{\pi}{2}-\left(u\sqrt{1-u^2}+\sin^{-1}u\right)\right], \qquad (10.48)$$

where u is defined by $u = \mathrm{Max}\left[-1,\frac{2E_F-\hbar\Omega}{qv_f}\right]$.

We now evaluate $G_2(\mathbf{q})$ defined by Equation 10.41. Using the same procedure as before, we find that

$$G_2(\mathbf{q}) = \frac{Aq^2}{4\pi^2 v_f}\int_1^\infty d\mu\int_{-1}^{1} d\upsilon\sqrt{\frac{\mu^2-1}{1-\upsilon^2}}\delta\left(q\upsilon-\frac{\hbar\Omega}{v_f}\right)\Theta\left[\frac{q}{2}(\mu+\upsilon)-k_F\right]\Theta\left[k_F-\frac{q}{2}(\mu-\upsilon)\right]. \qquad (10.49)$$

The integral over υ can be easily performed to get

$$G_2(\mathbf{q}) = \frac{Aq^2}{4\pi^2}\frac{\Theta(qv_f-\hbar\Omega)}{\sqrt{q^2v_f^2-(\hbar\Omega)^2}}\int_1^\infty d\mu\sqrt{\mu^2-1}\Theta\left[\frac{q}{2}\left(\frac{\hbar\Omega}{qv_f}+\mu\right)-k_F\right]$$
$$\times\Theta\left[k_F-\frac{q}{2}\left(\mu-\frac{\hbar\Omega}{qv_f}\right)\right]. \qquad (10.50)$$

The above equation can be rewritten as

$$G_2(\mathbf{q}) = \frac{Aq^2}{4\pi^2}\frac{\Theta(qv_f-\hbar\Omega)}{\sqrt{q^2v_f^2-(\hbar\Omega)^2}}\int_{\mathrm{Max}\left[1,\frac{2E_F-\hbar\Omega}{qv_f}\right]}^{\frac{2E_F+\hbar\Omega}{qv_f}} d\mu\sqrt{\mu^2-1}\Theta\left[\frac{2E_F+\hbar\Omega}{qv_f}-1\right]. \qquad (10.51)$$

Substituting the above expression back into the rate expression of Equation 10.39 gives

$$k_2 = \frac{e^2}{64\hbar\varepsilon^2\pi^3}\int_0^\infty dq q^3 e^{-2qz}\frac{\Theta(qv_f-\hbar\Omega)}{\sqrt{q^2v_f^2-(\hbar\Omega)^2}}\Theta\left[\frac{2E_F+\hbar\Omega}{qv_f}-1\right]\int_{\mathrm{Max}\left[1,\frac{2E_F-\hbar\Omega}{qv_f}\right]}^{\frac{2E_F+\hbar\Omega}{qv_f}}$$
$$\times d\mu\sqrt{\mu^2-1}\int_0^{2\pi} d\theta\left|\boldsymbol{\mu}_{eg}^D\cdot\left(i\hat{\mathbf{q}}+\hat{\mathbf{k}}\right)\right|^2. \qquad (10.52)$$

The θ integral can be performed easily, followed by an averaging over all possible orientations of the donor transition dipole to get

$$k_2 = \frac{e^2\mu_{eg}^2}{48\hbar\varepsilon^2\pi^2}\int_0^\infty dq q^3 e^{-2qz}\frac{\Theta(qv_f - \hbar\Omega)}{\sqrt{q^2v_f^2 - (\hbar\Omega)^2}}\Theta\left[\frac{2E_F + \hbar\Omega}{qv_f} - 1\right]\int_{\mathrm{Max}\left[1, \frac{2E_F - \hbar\Omega}{qv_f}\right]}^{\frac{2E_F + \hbar\Omega}{qv_f}} d\mu\sqrt{\mu^2 - 1}. \tag{10.53}$$

The integral over μ can be evaluated to get

$$k_2 = \frac{e^2\mu_{eg}^2}{96\hbar\varepsilon^2\pi^2}\int_0^\infty dq q^3 e^{-2qz}\frac{\Theta(qv_f - \hbar\Omega)}{\sqrt{q^2v_f^2 - (\hbar\Omega^2)}}\Theta\left[\frac{2E_F + \hbar\Omega}{qv_f} - 1\right] \tag{10.54}$$
$$\times\left(-r\sqrt{r^2-1} + s\sqrt{s^2-1} + \log\left[r + \sqrt{r^2-1}\right] - \log\left[s + \sqrt{s^2-1}\right]\right),$$

where $r = \mathrm{Max}\left[1, \frac{2E_F - \hbar\Omega}{qv_f}\right]$, and $s = \frac{2E_F + \hbar\Omega}{qv_f}$. The integrals over q in the expressions for k_1 and k_2 can be performed numerically, thus getting the total rate of transfer when the Fermi level in graphene is shifted into the conduction band.

10.6 LARGE z BEHAVIOR OF k_1 AND k_2

It is easy to analyze the large z behavior of k_1 and k_2 (see the Appendix for the detailed analysis). When $E_F < \frac{\hbar\Omega}{2}$ and $z > \frac{v_f}{2\Delta\varepsilon}$, with $\Delta\varepsilon = \frac{\hbar\Omega}{2} - E_F$ and $\hbar\Omega \gg \Delta\varepsilon$, the long-range behavior of k_1 is given by Equation 10.60 in the Appendix as

$$k_1 = \frac{e^2\mu_{eg}^2}{256\pi\Omega\hbar^2\varepsilon^2 z^4}. \tag{10.55}$$

In the case when $E_F > \frac{\hbar\Omega}{2}$ and $z > \frac{v_f}{4\Delta\epsilon}$, with $\hbar\Omega \gg \Delta\varepsilon$,

$$k_1 = \frac{e^2\mu_{eg}^2|\Delta\varepsilon|^{3/2}}{96\sqrt{2\Omega}\hbar^2\varepsilon^2\pi^{3/2}v_f^{3/2}z^{5/2}}e^{\frac{-4|\Delta\epsilon|z}{v_f}}, \tag{10.56}$$

as obtained in Equation 10.67 in the Appendix. The major contribution of the k_1 term to the rate comes only when $E_F > \frac{\hbar\Omega}{2}$ and it has a power law dependence (z^{-4}) on the distance. When $E_F < \frac{\hbar\Omega}{2}$, the contribution from k_1 decreases exponentially with z and hence is very small. Similarly, the long-range behavior of k_2 for both $E_F < \frac{\hbar\Omega}{2}$ and $E_F > \frac{\hbar\Omega}{2}$ is given by Equations 10.72 and 10.77 in the Appendix as

$$k_2 = \frac{1}{192\sqrt{z}}\left[\frac{e^4\mu_{eg}^4\hbar^3\Omega^5}{\pi^3\epsilon^4 v_f^7}\right]^{\frac{1}{2}} e^{\frac{-2z\hbar\Omega}{v_f}}\left(r\sqrt{r^2-1}-\log\left[r+\sqrt{r^2-1}\right]\right), \qquad (10.57)$$

where $r = 1 + \frac{2E_F}{\hbar\Omega}$. Therefore, the large z behavior of k_2 is exponential. Thus, in the case when $E_F < \frac{\hbar\Omega}{2}$, the rate of transfer to doped graphene has a power law dependence on the distance (arising from the k_1 term), whereas, when $E_F > \frac{\hbar\Omega}{2}$, the rate has an exponential dependence (arising due to both k_1 and k_2 terms). Therefore, as the Fermi level is increasingly shifted into the conduction band, there is a crossover of the distance dependence of the rate from power law to exponential and the crossover occurs over a region of E_F centered at $\frac{\hbar\Omega}{2}$.

10.7 RESULTS AND DISCUSSION

We performed numerical calculations for evaluating the rates of energy transfer from a fluorophore to doped graphene. Note that the numerical values of the rates k_1 and k_2 obtained using the equations derived earlier have been multiplied by a factor of 2 to account for the two inequivalent K-points of graphene. We take the emission energy of the fluorophore $\hbar\Omega$ to be 2.0 eV. Such low energy emission has been found in squarylium dyes [46]. We take $\mu_{eg} = 4.5\ D$. Figure 10.7 shows a plot of the logarithm of the rate as a function of the logarithm of the distance as the Fermi level is shifted increasingly into the conduction band in the range of 0.2–2.0 eV. For $E_F < 1.0$ eV, the log–log plot is linear, showing that the rate has a power law dependence (z^{-4}) on the distance. After around 1.0 eV, there is deviation from linearity in the log–log plot, and the dependence becomes exponential. In order to look into the crossover region more closely, in Figure 10.8, we show a plot of the rate as the Fermi level is moved into the conduction band in the range of 0.95–1.05 eV. This clearly shows that there is a crossover of the distance dependence of the rate from power law to exponential as the Fermi level is increasingly shifted into the conduction band. It should be possible to observe this effect experimentally.

Figure 10.9 shows a plot of the k_1 and k_2 terms and the total rate of transfer as a function of the Fermi energy of graphene at a fixed distance $z = 10$ Å. The contribution from the k_1 term decreases as the Fermi level is increasingly shifted into the conduction band, whereas that from the k_2 term increases. From the figure, it is clear that the total rate is governed mainly by the k_1 term up to the crossover point. Beyond

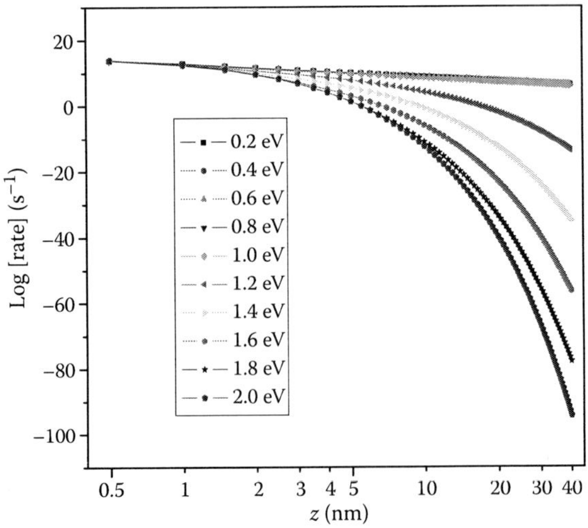

FIGURE 10.7 Rate of energy transfer as a function of distance, as the Fermi level is increasingly shifted into the conduction band. The emission energy of the fluorophore is taken to be $\hbar\Omega = 2.0$ eV. (Reprinted with permission from Swathi, R. S. and K. L. Sebastian, *J. Chem. Sci.*, 124, 233, 2012.)

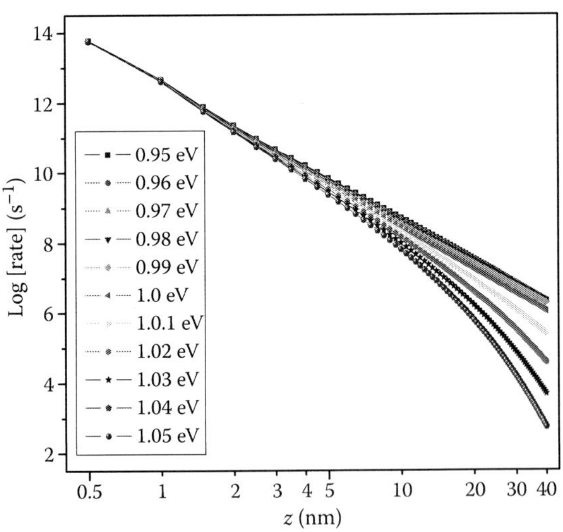

FIGURE 10.8 Rate of energy transfer as a function of distance, as the Fermi level is increasingly shifted into the conduction band (a closer look at the crossover region). The emission energy of the fluorophore is taken to be $\hbar\Omega = 2.0$ eV. (Reprinted with permission from Swathi, R. S. and K. L. Sebastian, *J. Chem. Sci.*, 124, 233, 2012.)

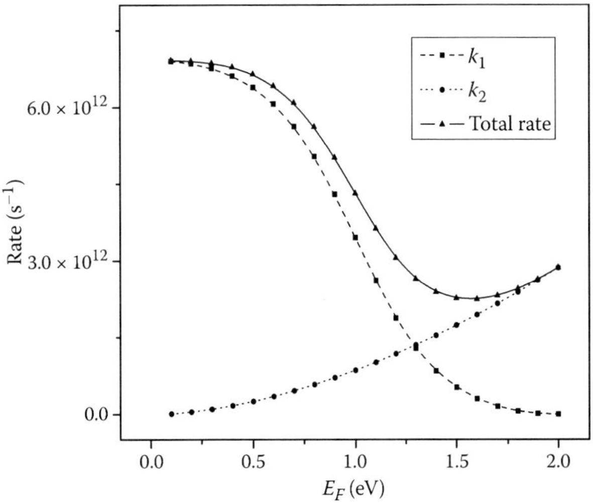

FIGURE 10.9 Rate of energy transfer as a function of the Fermi energy of graphene, at a distance of 10 Å. The figure shows plots of k_1 and k_2 separately, along with the total rate of transfer. (Reprinted with permission from Swathi, R. S. and K. L. Sebastian, *J. Chem. Sci.*, 124, 233, 2012.)

that, the contribution from k_1 term is small, and the total rate is governed by the k_2 term. This is clearly a density of states effect.

Finally, we mention a closely related work on energy transfer to graphene that appeared recently after we finished our work [28], wherein the authors consider the process of energy transfer to the electron–hole pairs and the plasmons of graphene [47]. The results obtained in [47] are in agreement with our calculations on energy transfer to the electron–hole pairs of undoped [26,27,29] and doped graphene [28,38]. However, they also study the possibility of collective excitations in graphene (plasmonic excitations), as a result of energy transfer. On the experimental side, nonzero Fermi energies have already been obtained for graphene on the modification of the carrier density by electrical or chemical means. Currently, attempts are being made to systematically study the effect of doping on fluorescence quenching by graphene [47]. By electrical gating, it is possible to control the flow of charge carriers to and away from graphene, which, in turn, affects the decay of a molecular emitter in proximity. It would be interesting, therefore, to envisage experiments wherein switching on and off of the molecular emission occurs as a result of the control of electrical gating in graphene. The decay of quantum emitters near doped graphene, causing surface plasmon excitations in graphene, has also been studied and is found to be rather promising in the area of graphene plasmonics [47].

10.8 SUMMARY

Motivated by experiments on fluorescence quenching of dyes tethered on CNTs and studies on the decay of excited states near metal surfaces, we have studied the process of EET from fluorophores to a sheet of graphene. Starting from the Fermi golden rule, we have developed an analytical model to evaluate the rate of transfer. The transition density of the fluorophore has been approximated as a point dipole, whereas the transition density of graphene has to be treated more rigorously. We have used the tight-binding model and the Dirac cone approximation for graphene. The long-range behavior of the rate of transfer has been found to have a $(distance)^{-4}$ dependence. Our numerical calculations on the study of energy transfer from pyrene to graphene have predicted a distance of ~300 Å up to which quenching by graphene can be significant. We have also studied EET from fluorophores to doped graphene. We have analyzed the distance dependence of the rate of transfer as the Fermi level of graphene is shifted away from the K-point into the conduction band. We have found a crossover of the dependence from power law (z^{-4}) to exponential as the Fermi level is increasingly moved into the conduction band. The point of crossover is at a shift of the Fermi level by $\hbar\Omega/2$. After our theoretical studies were reported, there have been a series of experimental studies on fluorescence quenching of dyes near graphene, with interesting applications for the process.

ACKNOWLEDGMENTS

K. L. Sebastian gratefully acknowledges the immense help that he received from Prof. B. M. Deb during his student days at the Indian Institute of Science. His work was supported by the Department of Science and Technology, Government of India, through the J. C. Bose Fellowship Program.

APPENDICES

APPENDIX 10.A BEHAVIOR OF RATE CONSTANT K_1

We first look at k_1. We consider two separate cases.

1. Case I: We consider the case where E_F is close to $\frac{\hbar\Omega}{2}$ but less than it. We put $\Delta\varepsilon = \frac{\hbar\Omega}{2} - E_F$. k_1 is then given by

$$k_1 = \frac{e^2\mu_{eg}^2}{48\hbar\varepsilon^2\pi^2}\int_0^\infty dq q^3 e^{-2qz}\frac{\Theta(\hbar\Omega - qv_f)}{\sqrt{(\hbar\Omega)^2 - q^2v_f^2}}\Theta\left[1+\frac{2\Delta\epsilon}{qv_f}\right]\int_{\mathrm{Max}\left[-1,-\frac{2\Delta\epsilon}{qv_f}\right]}^{1} d\upsilon\sqrt{1-\upsilon^2}. \tag{10.A1}$$

As $\Delta\varepsilon>0$, $\Theta\left[1+\frac{2\Delta\epsilon}{qv_f}\right]=1$. For $z>\frac{v_f}{2\Delta\varepsilon}$, the major contribution to the above integral is from $q\epsilon\left(0,\frac{\Delta\varepsilon}{v_f}\right)$. In this range,

$$\int_{\mathrm{Max}\left[-1,-\frac{2\Delta\varepsilon}{qv_f}\right]}^{1} d\upsilon\sqrt{1-\upsilon^2} = \int_{-1}^{1} d\upsilon\sqrt{1-\upsilon^2} = \frac{\pi}{2}. \tag{10.A2}$$

If $\frac{\hbar\Omega}{2} \gg \Delta\varepsilon$, then, in this range, $(\hbar\Omega)^2 - q^2v_f^2 \simeq (\hbar\Omega)^2$, and the integral in Equation 10.58 may be approximated as

$$k_1 = \frac{e^2\mu_{eg}^2}{96\pi\hbar^2\varepsilon^2\Omega}\int_0^\infty dq q^3 e^{-2qz} = \frac{e^2\mu_{eg}^2}{256\pi\Omega\hbar^2\varepsilon^2 z^4}. \tag{10.A3}$$

2. Case II: We now consider the case $E_F > \frac{\hbar\Omega}{2}$. Hence, $\frac{\hbar\Omega}{2} - E_F = -|\Delta\varepsilon|$. Then

$$k_1 = \frac{e^2\mu_{eg}^2}{48\hbar\varepsilon^2\pi^2}\int_0^\infty dq q^3 e^{-2qz}\frac{\Theta(\hbar\Omega - qv_f)}{\sqrt{(\hbar\Omega)^2 - q^2v_f^2}}\Theta\left[1-\frac{2|\Delta\varepsilon|}{qv_f}\right]\int_{\mathrm{Max}\left[-1,\frac{2|\Delta\varepsilon|}{qv_f}\right]}^{1} d\upsilon\sqrt{1-\upsilon^2}. \tag{10.A4}$$

For large z $\left(z \gg \frac{v_f}{4|\Delta\varepsilon|}\right)$, this may be approximated as

$$k_1 = \frac{e^2\mu_{eg}^2}{48\hbar\varepsilon^2\pi^2}\int_{\frac{2|\Delta\varepsilon|}{v_f}}^\infty dq q^3 e^{-2qz}\frac{\Theta(\hbar\Omega - qv_f)}{\sqrt{(\hbar\Omega)^2 - q^2v_f^2}}\int_{\frac{2|\Delta\varepsilon|}{qv_f}}^{1} d\upsilon\sqrt{1-\upsilon^2}. \tag{10.A5}$$

For z such that $\frac{4|\Delta\varepsilon|z}{v_f} \gg 1$, in the above integral over q, the contribution is from q in the vicinity of $\frac{2|\Delta\varepsilon|}{v_f}$. We now change the variable of integration from q to y defined by $q = \frac{2|\Delta\varepsilon|y}{v_f}$. Using the above transformation, k_1 is given by

$$k_1 = \frac{e^2\mu_{eg}^2|\Delta\varepsilon|^4}{3\hbar\varepsilon^2\pi^2v_f^4}\int_1^\infty dy y^3 e^{\frac{-4|\Delta\varepsilon|zy}{v_f}}\frac{\Theta(\hbar\Omega - 2|\Delta\varepsilon|y)}{\sqrt{(\hbar\Omega)^2 - 4|\Delta\varepsilon|^2y^2}}\int_{\frac{1}{y}}^{1} d\upsilon\sqrt{1-\upsilon^2}. \tag{10.A6}$$

Now, the major contribution to the above integral comes from values of $y \simeq 1$. If y is close to unity

$$\int_{\frac{1}{y}}^{1} d\upsilon\sqrt{1-\upsilon^2} \simeq \frac{2\sqrt{2}}{3}\left(\frac{y-1}{y}\right)^{3/2}. \tag{10.A7}$$

Hence,

$$k_1 \simeq \frac{2\sqrt{2}e^2\mu_{eg}^2|\Delta\varepsilon|^4}{9\hbar\varepsilon^2\pi^2 v_f^4}\int_1^{\infty} dy y^{3/2}(y-1)^{3/2} e^{\frac{-4|\Delta\varepsilon|zy}{v_f}} \frac{\Theta(\hbar\Omega - 2|\Delta\varepsilon|y)}{\sqrt{(\hbar\Omega)^2 - 4|\Delta\varepsilon|^2 y^2}}. \tag{10.A8}$$

For $\frac{\hbar\Omega}{2} \gg |\Delta\varepsilon|$ and y in the vicinity of 1, $\Theta(\hbar\Omega - 2|\Delta\varepsilon|y) = 1$, and $\sqrt{(\hbar\Omega)^2 - 4|\Delta\varepsilon|^2 y^2} \simeq \hbar\Omega$. The integral over y can now be evaluated to get

$$k_1 \simeq \frac{e^2\mu_{eg}^2|\Delta\varepsilon|^2}{48\sqrt{2}\Omega\hbar^2\varepsilon^2\pi^2 v_f^2 z^2} e^{\frac{-2|\Delta\varepsilon|z}{v_f}} K_2\left(\frac{2|\Delta\varepsilon|z}{v_f}\right). \tag{10.A9}$$

For large values of z, the asymptotic form of the Bessel function $K_2\left(\frac{2|\Delta\varepsilon|z}{v_f}\right)$ is given by $K_2\left(\frac{2|\Delta\varepsilon|z}{v_f}\right) \simeq \sqrt{\frac{\pi v_f}{4|\Delta\varepsilon|z}} e^{\frac{-2|\epsilon|z}{v_f}}$. Therefore, we get

$$k_1 \simeq \frac{e^2\mu_{eg}^2|\Delta\varepsilon|^{3/2}}{96\sqrt{2}\Omega\hbar^2\varepsilon^2\pi^{3/2} v_f^{3/2} z^{5/2}} e^{\frac{-4|\Delta\varepsilon|z}{v_f}}. \tag{10.A10}$$

Appendix 10.B Behavior of Rate Constant k_2

We now analyze k_2. As before, we consider two separate cases.

1. Case I: We consider the case $E_F < \frac{\hbar\Omega}{2}$. Therefore, $2E_F - \hbar\Omega < 0$, and hence, k_2 is given by

$$k_2 = \frac{e^2\mu_{eg}^2}{48\hbar\varepsilon^2\pi^2}\int_0^{\infty} dq q^3 e^{-2qz} \frac{\Theta(qv_f - \hbar\Omega)}{\sqrt{q^2 v_f^2 - (\hbar\Omega)^2}} \int_1^{\frac{2E_F+\hbar\Omega}{qv_f}} d\mu\sqrt{\mu^2 - 1}\Theta\left[\frac{2E_F + \hbar\Omega}{qv_f} - 1\right]. \tag{10.B1}$$

Using the two theta functions in the above expression, k_2 can be written as

$$k_2 = \frac{e^2\mu_{eg}^2}{48\hbar\varepsilon^2\pi^2} \int_{\frac{\hbar\Omega}{v_f}}^{\frac{2E_F+\hbar\Omega}{v_f}} dq \frac{q^3 e^{-2qz}}{\sqrt{q^2v_f^2 - (\hbar\Omega)^2}} \int_1^{\frac{2E_F+\hbar\Omega}{qv_f}} d\mu\sqrt{\mu^2-1}. \tag{10.B2}$$

We now make a change of the variable of integration from q to x defined by $q = \frac{\hbar\Omega}{v_f} + x$. Using this transformation, k_2 can be written as

$$k_2 = \frac{e^2\mu_{eg}^2}{48\hbar\varepsilon^2\pi^2} e^{\frac{-2z\hbar\Omega}{v_f}} \int_0^{\frac{2E_F}{v_f}} dx \frac{\left(\frac{\hbar\Omega}{v_f} + x\right)^3 e^{-2zx}}{\sqrt{x^2v_f^2 + 2\hbar\Omega v_f x}} \int_1^{\frac{2E_F+\hbar\Omega}{xv_f+\hbar\Omega}} d\mu\sqrt{\mu^2-1}. \tag{10.B3}$$

For large values of z, in the above integral over x, because of the presence of e^{-2zx} term, only small values of x are important. Therefore, for small values of x, the above expression can be simplified to get

$$k_2 = \sqrt{\frac{e^4\mu_{eg}^4\hbar^3\Omega^5}{\pi^4\varepsilon^4v_f^7}} \frac{e^{\frac{-2z\hbar\Omega}{v_f}}}{48\sqrt{2}} \int_1^{1+\frac{2E_F}{\hbar\Omega}} d\mu\sqrt{\mu^2-1} \int_0^{\frac{2E_F}{v_f}} dx \frac{e^{-2zx}}{\sqrt{x}}. \tag{10.B4}$$

The integral over μ can now be performed, and the upper limit in the integral over x can be extended to ∞ to get the following expression for k_2:

$$k_2 = \sqrt{\frac{e^4\mu_{eg}^4\hbar^3\Omega^5}{\pi^3\varepsilon^4v_f^7}} \frac{e^{\frac{-2z\hbar\Omega}{v_f}}}{192\sqrt{z}} \left(r\sqrt{r^2-1} - \log\left[r+\sqrt{r^2-1}\right]\right), \tag{10.B5}$$

where $r = 1 + \frac{2E_F}{\hbar\Omega}$.

2. Case II: We now consider the case $E_F > \frac{\hbar\Omega}{2}$. Therefore, $2E_F - \hbar\Omega > 0$, and hence, k_2 is given by

$$k_2 = \frac{e^2\mu_{eg}^2}{48\hbar\varepsilon^2\pi^2} \int_0^{\infty} dq q^3 e^{-2qz} \frac{\Theta(qv_f - \hbar\Omega)}{\sqrt{q^2v_f^2 - (\hbar\Omega)^2}} \int_{\mathrm{Max}\left[1, \frac{2E_F-\hbar\Omega}{qv_f}\right]}^{\frac{2E_F+\hbar\Omega}{qv_f}} d\mu\sqrt{\mu^2-1}\,\Theta\left[\frac{2E_F+\hbar\Omega}{qv_f} - 1\right]. \tag{10.B6}$$

The above equation can be simplified to get

$$k_2 = \frac{e^2\mu_{eg}^2}{48\hbar\varepsilon^2\pi^2} \int_{\frac{\hbar\Omega}{v_f}}^{\frac{2E_F+\hbar\Omega}{v_f}} dq \frac{q^3 e^{-2qz}}{\sqrt{q^2 v_f^2 - (\hbar\Omega)^2}} \int_{\mathrm{Max}\left[1,\frac{2E_F-\hbar\Omega}{qv_f}\right]}^{\frac{2E_F+\hbar\Omega}{qv_f}} d\mu\sqrt{\mu^2-1}. \quad (10.B7)$$

We now use the same procedure as was used for evaluating the integrals in Case I to get

$$k_2 = \sqrt{\frac{e^4\mu_{eg}^4\hbar^3\Omega^5}{\pi^4\varepsilon^4 v_f^7}} \frac{e^{\frac{-2z\hbar\Omega}{v_f}}}{48\sqrt{2}} \int_{\mathrm{Max}\left[1,\frac{2E_F}{\hbar\Omega}-1\right]}^{1+\frac{2E_F}{\hbar\Omega}} d\mu\sqrt{\mu^2-1} \int_0^{\frac{2E_F}{v_f}} dx \frac{e^{-2zx}}{\sqrt{x}}. \quad (10.B8)$$

On evaluating the integral over μ, we get

$$k_2 = \sqrt{\frac{e^4\mu_{eg}^4\hbar^3\Omega^5}{\pi^4\varepsilon^4 v_f^7}} \frac{e^{\frac{-2z\hbar\Omega}{v_f}}}{96\sqrt{2}} \left(-s\sqrt{s^2-1} + r\sqrt{r^2-1} + \log\left[s+\sqrt{s^2-1}\right]\right.$$

$$\left. - \log\left[r+\sqrt{r^2-1}\right]\right) \times \int_0^{\frac{2E_F}{v_f}} dx \frac{e^{-2zx}}{\sqrt{x}}, \quad (10.B9)$$

where $s = \mathrm{Max}\left[1, \frac{2E_F}{\hbar\Omega} - 1\right]$ and $r = 1 + \frac{2E_F}{\hbar\Omega}$. For E_F close to $\frac{\hbar\Omega}{2}$, $s = 1$. Using this and extending the upper limit of the integral over x to ∞ and then evaluating the integral, we get

$$k_2 = \sqrt{\frac{e^4\mu_{eg}^4\hbar^3\Omega^5}{\pi^3\varepsilon^4 v_f^7}} \frac{e^{\frac{-2z\hbar\Omega}{v_f}}}{192\sqrt{z}} \left(r\sqrt{r^2-1} - \log\left[r+\sqrt{r^2-1}\right]\right). \quad (10.B10)$$

REFERENCES

1. Valeur, B. *Molecular Fluorescence*. Wiley-VCH: Weinheim (2002).
2. Lakowicz, J. R. *Principles of Fluorescence Spectroscopy*. Springer: New York (2006).
3. May, V., and O. Kuhn. *Charge and Energy Transfer Dynamics in Molecular Systems*. Wiley-VCH: Weinheim (2004).

4. Förster, T. *Ann. Phys.* **2**, 55 (1948).
5. Stryer, L., and R. P. Haugland. *Proc. Natl. Acad. Sci. USA* **58**, 719 (1967).
6. Schuler, B., E. A. Lipman, and W. A. Eaton. *Nature* **419**, 743 (2002).
7. Zhao, R., and D. Rueda. *Methods* **49**, 112 (2009).
8. Wong, K. F., B. Bagchi, and P. J. Rossky. *J. Phys. Chem. A* **108**, 5752 (2004).
9. Sönnichsen, C., B. M. Reinhard, J. Liphardt, and A. P. Alivisatos. *Nat. Biotechnol.* **23**, 741 (2005).
10. Achermann, M., M. A. Petruska, S. Kos, D. L. Smith, D. D. Koleske, and V. I. Klimov. *Nature* **429**, 642 (2004).
11. Martinez, P. L. H., and A. O. Govorov. *Phys. Rev. B* **78**, 035314 (2008).
12. Alivisatos, A. P., D. H. Valdeck, and C. B. Harris. *J. Chem. Phys.* **82**, 541 (1985).
13. Drexhage, K. H. *J. Lumin.* **1**, 693 (1970).
14. Chance, R. R., A. Prock, and R. Silbey. *Adv. Chem. Phys.* **37**, 1 (1978).
15. Campion, A., A. R. Gallo, C. B. Harris, H. J. Robota, and P. M. Whitmore. *Chem. Phys. Lett.* **73**, 447 (1980).
16. Whitmore, P. M., A. P. Alivisatos, and C. B. Harris. *Phys. Rev. Lett.* **50**, 1092 (1983).
17. Persson, B. N. J., and N. D. Lang. *Phys. Rev. B* **26**, 5409 (1982).
18. Qu, L., R. B. Martin, W. Huang, K. Fu, D. Zweifel, Y. Lin, Y. P. Sun, C. E. Bunker, B. A. Harruff, J. R. Gord, et al. *J. Chem. Phys.* **117**, 8089 (2002).
19. Martin, R. B., L. Qu, Y. Lin, B. A. Harruff, C. E. Bunker, J. R. Gord, L. F. Allard, and Y. P. Sun. *J. Phys. Chem. B* **108**, 11447 (2004).
20. Ahmad, A., K. Kern, and K. Balasubramanian. *Chem. Phys. Chem.* **10**, 905 (2009).
21. Saito, R., G. Dresselhaus, and M. S. Dresselhaus. *Physical Properties of Carbon Nanotubes* (Imperial College Press: London, 1999).
22. Novoselov, K. S., A. K. Geim, S. V. Morozov, D. Jiang, Y. Zhang, S. V. Dubonos, I. V. Grigorieva, and A. A. Firsov. *Science* **306**, 666 (2004).
23. Neto, A. C., F. Guinea, and N. M. Peres. *Physics World* **1**, November (2006).
24. Geim, A. K., and P. Kim. *Scientific American* **90**, April (2008).
25. Katsnelson, M. I. *Materials Today* **10**, 20 (2007).
26. Swathi, R. S., and K. L. Sebastian. *J. Chem. Phys.* **129**, 054703 (2008).
27. Swathi, R. S., and K. L. Sebastian. *J. Chem. Phys.* **130**, 086101 (2009).
28. Swathi, R. S. *Ph.D. Thesis* (Department of Inorganic and Physical Chemistry, Indian Institute of Science: Bangalore, India, 2010).
29. Swathi, R. S., and K. L. Sebastian. *J. Chem. Sci.* **121**, 777 (2009).
30. Wallace, P. R. *Phys. Rev.* **71**, 622 (1947).
31. Sagar, A., K. Kern, and K. Balasubramanian. *Nanotechnology* **21**, 015303 (2010).
32. Xie, L., X. Ling, Y. Fang, J. Zhang, and Z. Liu. *J. Am. Chem. Soc.* **131**, 9890 (2009).
33. Kim, J., L. J. Cote, F. Kim, and J. Huang. *J. Am. Chem. Soc.* **132**, 260 (2010).
34. He, S., B. Song, D. Li, C. Zhu, W. Qi, Y. Wen, L. Wong, S. Song, H. Fang, and C. Fan. *Adv. Funct. Mater.* **20**, 453 (2010).
35. Chang, H., L. Tang, Y. Wang, J. Jiang, and J. Li. *Anal. Chem.* **82**, 2341 (2010).
36. Schedin, F., A. K. Geim, S. V. Morozov, E. W. Hill, P. Blake, M. I. Katsnelson, and K. S. Novoselov. *Nature Mater.* **6**, 652 (2007).
37. Bostwick, A., T. Ohta, T. Seyller, K. Horn, and E. Rotenberg. *Nature Phys.* **3**, 36 (2007).
38. Swathi, R. S., and K. L. Sebastian. *J. Chem. Sci.* **124**, 233 (2012).
39. Swathi, R. S., and K. L. Sebastian. *J. Chem. Phys.* **132**, 104502 (2010).
40. Frisch, M. J., G. W. Trucks, H. B. Schlegel, G. E. Scuseria, M. A. Robb, J. R. Cheeseman, J. A. Montgomery, Jr., T. Vreven, K. N. Kudin, J. C. Burant, et al. *Gaussian 03, Revision C.02*, Gaussian, Inc.: Wallingford, CT, 2004.
41. McCamant, D. W., P. Kukura, S. Yoon, and R. A. Mathies. *Rev. Sci. Instrum.* **75**, 4971 (2004).
42. Nestor, J., T. G. Spiro, and G. Klauminzer. *Proc. Natl. Acad. Sci. USA* **73**, 3329 (1976).

43. Matousek, P., M. Towrie, C. Ma, W. M. Kwok, D. Phillips, W. T. Toner, and A. W. Parker. *J. Raman Spectrosc.* **32**, 983 (2001).
44. Umapathy, S., A. Lakshmanna, and B. Mallick. *J. Raman Spectrosc.* **40**, 235 (2009).
45. Chen, Z., S. Berciaud, C. Nuckolls, T. F. Heinz, and L. E. Brus. *ACS Nano.* **4**, 2964 (2010).
46. Yanagi, K., K. Lakoubovskii, H. Matsui, H. Matsuzaki, H. Okamoto, Y. Miyata, Y. Maniwa, S. Kazaoui, N. Minami, and H. Kataura. *J. Am. Chem. Soc.* **129**, 4992 (2007).
47. Velizhanin, K. A., and A. Efimov. *Phys. Rev. B* **84**, 085401 (2011).

11 Third Law of Thermodynamics Revisited for Spin-Boson Model*

Sushanta Dattagupta and Aniket Patra

CONTENTS

11.1 INTRODUCTION

The laws of thermodynamics apply to the macroworld and are considered robust. They are expected to be valid for all material systems, as they arrived after years of experimentation. Indeed, the laws of thermodynamics are viewed to be so sacrosanct that they constitute the benchmark for all "microscopic" theories, based on either classical or quantum considerations. Such theories that fail to reproduce thermodynamics at the macrolevel need to be discarded!

Of the three laws of thermodynamics, the third law is the least discussed, though it is the most profound in some sense. This is so because, unlike the first two laws, the third law has direct contact with quantum mechanics. Thus, it is not surprising that the development, along with complete understanding, of the third law, was concomitant with the establishment of quantum theory itself in the first quarter of the last century.

* Dedicated to Professor B. M. Deb, who sets a high level of intellectual honesty.

In its simplest form, the third law of thermodynamics states that the entropy S and, from it, the derived heat capacity C (or the specific heat) must go to zero at absolute zero temperature. If, however, the system has defects, such as a disordered crystal, the entropy approaches a constant at $T = 0$, although the entropy per particle $s\left(\equiv S/N\right)$ does go to zero in thc so-called "thermodynamic limit" ($N \to \infty$).

What is then the connection with quantum mechanics, which is a zero-temperature theory in itself? A quantum system is characterized by its wave function but can admit several wave functions associated, however, with the same energy eigenvalue. The number of such distinct wave functions or state functions of the system is called degeneracy. The simplest textbook example of degeneracy is a 2- or 3-D harmonic oscillator. Another example is the hydrogen atom. When one views the degeneracy of the lowest energy level g in the context of the Boltzmann relation for the entropy, one immediately sees the relation between thermodynamics and quantum mechanics. Boltzmann hypothesized that the entropy S of a macrosystem can be written as

$$S = k_B \ln \Omega, \tag{11.1}$$

where k_B is the Boltzmann constant, and Ω is the number of microstates associated with the (fixed) energy E of the system. Then, at zero temperature, Equation 11.1 yields

$$S(T = 0) = k_B \ln g. \tag{11.2}$$

Clearly, the right-hand side of Equation 11.2 is a constant, but when divided by N, $s\left(\equiv S/N\right)$ vanishes in conformity of the third law, except, of course, if g increases faster than exponential in N! Such a situation may arise in glasses, which are characterized by an infinitely large number of energy minima in the ground state. The latter example, in apparent contradiction with the third law of thermodynamics, turns out to be a red herring, however, as glasses are "nonequilibrium" systems quite outside the purview of thermodynamics, in its standard formulation.

In Section 11.2, we will discuss the remarkable development of the third law of thermodynamics, which can indeed be viewed as one of the great triumphs of chemistry or what has more specifically come to be known as chemical physics. In the process of this development, at least four Nobel Prizes in Chemistry were awarded to Van't Hoff, Ostwald, Arrhenius, and Nernst. The clearest formulation of the third law was provided by Lewis, who however did not win the Prize! [1] We will discuss the ingenious experimental research of Lewis that led him to the most precise statement of the law. Our aim is to put forward a poser: What happens to the third law when it comes to a nano- or mesoscale system? The kind of quantum systems that we have in mind comprise, for example, a dot in which an isolated electron can move in a confined potential, an electron under an external magnetic field, constrained however by a parabolic well, or a qubit (a system of two quantum dots) in a quantum-dissipative environment. For such systems, the specific heat C does not quite follow the diktats of the third law. Where is then the resolution of this impasse? We argue that a small system, being tiny, is inevitably in contact with a larger system or the environment. Thus, a small system is never a closed one but is an *open system*, for which the thermodynamics has to be worked out anew. When this is done, we find a satisfactory resolution to the paradox of the specific heat *vis-à-vis* the third law. Our discussions will be based on the much-studied spin-boson problem, which will be introduced in Section 11.3 and which has come to be recognized as a standard paradigm of dissipative quantum systems [2]. In Section 11.4, we treat the specific heat of the spin-boson model in both the strong and weak damping regions. Finally, in Section 11.5, we present our concluding remarks.

11.2 BRIEF HISTORY OF THE THIRD LAW

The foundation of the third law is interestingly intertwined with another famous law of chemical thermodynamics, which is however embedded in kinetic considerations! The latter is what is called the Arrhenius law (1889), which yields the rate λ of a chemical reaction in terms of the activation energy E_a, which provides the difference in energy between the reactant and product states (Figure 11.1), i.e.,

$$\lambda = \lambda_0 \exp\left(-{E_a}/{k_B T}\right). \tag{11.3}$$

In the above equation, λ_0 is a "pre-exponential factor," which can be weakly temperature dependent. The formulation of the Arrhenius law was given a physical justification by Van't Hoff in the same year of 1889. (Incidentally, Van't Hoff was the first recipient of the Nobel Prize in Chemistry!) The pre-exponential factor λ_0 is related to the concept of an attempt frequency of a chemical reaction, particularly when the reaction occurs in the liquid state, which is in a state of perpetual Brownian motion. Indeed, the Arrhenius law is directly related to the concept of "moles" of chemical species that were expounded by another contemporary chemist, Ostwald (Figure 11.2).

FIGURE 11.1 Svante Arrhenius.

FIGURE 11.2 (Left) Jacobus Van't Hoff. (Right) Wilhelm Ostwald.

Ostwald, who initially had strong philosophical opposition to the atomic theory, was converted to the concept of moles as the fundamental constituent of matter, following Perrin's historical experiments on the Brownian motion [3].

As mentioned earlier, the clearest exposition of the third law of thermodynamics was provided by G. N. Lewis (Figure 11.3). Lewis analyzed a whole set of measurements of the rate of what is called first-order chemical reaction, which can be written, following a thermodynamic formulation of the Arrhenius law, as

$$K = K_0 \exp\left(-\frac{\Delta G}{RT}\right). \tag{11.4}$$

FIGURE 11.3 G. N. Lewis.

Here, ΔG is the differential Gibbs energy given by

$$\Delta G = \Delta H - T\Delta S, \tag{11.5}$$

where ΔH is the differential enthalpy, and ΔS is the differential entropy. We may rewrite Equation 11.5 as

$$\ln\left(\frac{K}{K_0}\right) = \frac{\Delta S}{R} - \frac{\Delta H}{RT}. \tag{11.6}$$

A plot of the left-hand side versus the inverse temperature $\left(= \frac{1}{T}\right)$ on a log scale yields a straight line, with an intercept $\left(= \frac{\Delta S}{R}\right)$ and a slope $\left(= -\frac{\Delta H}{R}\right)$, both being weakly temperature dependent. Lewis and his collaborators made very careful measurements of the slope, extrapolated to zero temperature, and came to the conclusion that not only ΔS reaches zero at $T = 0$ but S itself vanishes as long as the system has a ground state with only one configuration.

11.3 SPIN-BOSON MODEL

The spin-boson model can be motivated in terms of the physical problem of a quantum particle (e.g., an electron, muon, or proton) tunneling in an asymmetric double well (Figure 11.4). The relevant energy scales are the asymmetry $\hbar\varepsilon$, tunneling $\hbar\Delta_0$, and thermal k_BT, which can be all comparable but are much smaller than the barrier height E_0. Under this circumstance, the dynamics of the quantum particle moving in the double well can be described by what we shall refer to as the subsystem Hamiltonian

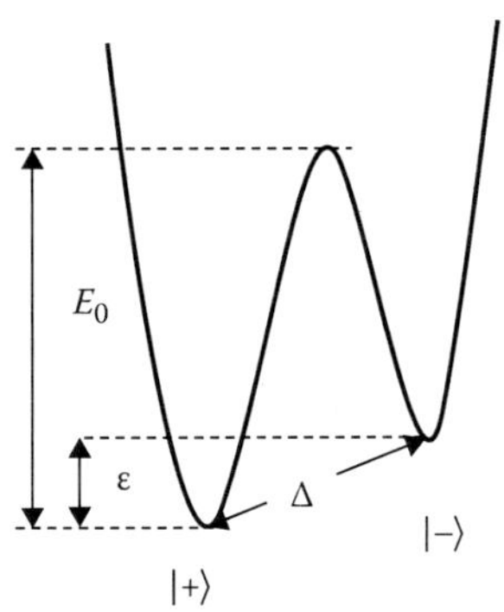

FIGURE 11.4 Double well.

$$\mathcal{H}_S = \frac{\hbar}{2}(\varepsilon\sigma_z - \Delta_0\sigma_x). \tag{11.7}$$

In the above expression, we have employed the Pauli matrices σ_z and σ_x and have decided to work in the representation in which σ_z is diagonal. Thus

$$\sigma_z|\mu\rangle = \mu|\mu\rangle,\ \mu = \pm, \tag{11.8}$$

where the two eigenstates $|+\rangle$ and $|-\rangle$ represent the left and right occupations of the particle, respectively (Figure 11.4). Appropriately then, the tunneling frequency Δ_0 couples to σ_x, which is totally off-diagonal in the σ_z-representation.

What about the environment? Think of the particle as a proton or a positive muon (μ^+) which, being positively charged, would carry with it the electron cloud of, for example, a metallic system like niobium (Nb) [4]. While the subsystem Hamiltonian $\mathcal{H}_S$ entails *coherent* tunneling with a (resultant) frequency Δ

$$\Delta = \sqrt{\left(\Delta_0^2 + \varepsilon^2\right)}, \tag{11.9}$$

the environmental coupling impedes the coherent evolution, leading to decoherence, dissipation, etc. [5]. At this stage then, it is pertinent to amplify these phrases: Decoherence in a quantum system occurs when the latter is observed over a time scale that is much longer than the quantal time scale τ_Q, i.e.,

$$\tau_Q = \hbar / k_B T. \tag{11.10}$$

The cause of this kind of decoherence is evidently thermal because of the alteration of the deBroglie wavelength of the wave packet associated with the quantum particle. Dissipation, on the other hand, arises from the exchange of energy between the system, which, in this case, comprise the tunneling particle and the environment of the electron cloud, or, for that matter, phonons that get excited because of the elastic distortion created by the particle (which are called interstitial sites). In either case, the coupling with the environment can be modeled as

$$\mathcal{H} = \frac{\hbar}{2}(\varepsilon\sigma_z - \Delta_0\sigma_x) + \sigma_z\sum_k g_k\left(b_k + b_k^\dagger\right) + \sum_k \hbar\omega_k b_k^\dagger b_k, \tag{11.11}$$

where $b_k\left(b_k^\dagger\right)$ are the bosonic operators, g_k is a coupling constant, and ω_k is the frequency of bosonic excitation. The last term in Equation 11.11 is then a free bosonic Hamiltonian, which lends time dependence to $b_k\left(b_k^\dagger\right)$ in the "interaction picture," thereby modulating the tunneling term in $\mathcal{H}_S$, again because σ_z does not commute with σ_x.

It turns out that both the electronic (fermionic) and the oscillator (bosonic) environments can be modeled by the last two terms of Equation 11.11 [6,7], provided that we properly interpret what is called the spectral density $J(\omega)$ of the bosonic excitations, which is defined by

$$J(\omega) = \frac{2}{\hbar^2}\sum_k g_k^2 \delta(\omega - \omega_k). \tag{11.12}$$

The frequency dependence of $J(\omega)$ is usually modeled as

$$J(\omega) = K\omega^s, \tag{11.13}$$

where $s = 1$ corresponds to "ohmic" dissipation, which is appropriate to a 1-D system of electrons (fermions). In contrast, $s = 3$ is relevant when the environment comprises a bunch of acoustic phonons [7].

Before concluding this section, it is important to point out that the "spin-boson Hamiltonian" has more general applicability than just dissipative tunneling. For instance, $\mathcal{H}_S$ can be thought of as a spin one-half particle subject to cross-magnetic fields, proportional to $-\varepsilon$ along the z-axis and Δ_0 along the x-axis. The coupling to bosonic fields can then be thought of in the context of spin-lattice relaxations, which is a well-known issue in the literature on magnetic resonance [8]. Alternatively, $\mathcal{H}_S$ can be envisaged to be the Hamiltonian of a spin-half Kondo impurity, which is influenced by conduction electrons, the dynamics of which is approximated by the bosonic modes [2]. In more recent times, the spin-boson Hamiltonian has been employed for the investigation of decoherence and dissipation of a qubit, which is a problem of great significance in quantum information processes [9].

11.4 SPECIFIC HEAT OF OPEN SYSTEMS

11.4.1 Preliminaries

Before we introduce the concept of an open system, it is useful to discuss the specific heat of the subsystem itself. The Hamiltonian $\mathcal{H}_S$ in Equation 11.7 does qualify for describing the prototype thermodynamics of a "small" quantum system, like a harmonic oscillator. What we have to do is to imagine $\mathcal{H}_S$ to be "weakly" coupled to a *classical* heat bath, with which the system undergoes exchange of energy. The consequent energy fluctuations provide the temperature of the system. All this can be put into statistical mechanical perspective in terms of the Gibbsian partition function

$$Z_S = Tr_S(e^{-\beta\mathcal{H}_S}) \tag{11.14}$$

and the concomitant Helmholtz free energy

$$F_S = -\frac{1}{\beta}\ln Z_S. \tag{11.15}$$

The trace denoted by Tr_S (…), which is the sum over the eigenstates of σ_z, can be worked out, yielding

$$Z_S = 2\cosh\left(\frac{\beta\hbar\Delta}{2}\right), \tag{11.16}$$

where Δ is defined in Equation 11.9. The heat capacity (or specific heat) defined by

$$C = k_B\beta^2 \frac{\partial^2}{\partial\beta^2}\ln Z_S \tag{11.17}$$

leads to the familiar Schottky relation [10]

$$C = \left(\frac{\hbar\beta\Delta}{2}\right)^2 \sec h^2\left(\frac{\hbar\beta\Delta}{2}\right). \tag{11.18}$$

A schematic of the specific heat, which is sketched in Figure 11.5, does indicate the validity of the third law, although the falloff at zero temperature is exponential, i.e.,

$$\lim_{T\to 0} C = 4(\hbar\beta\Delta)^2 e^{-2\beta\hbar\Delta}, \tag{11.19}$$

much like that of a quantum oscillator (or an Einstein oscillator). What is then an open system? When we "open" a new channel of the subsystem to an external environment, in this case by connecting the spin to the bosonic modes in the spin-boson system, as in Equation 11.11, we are led to an open system. In particular, if we trace out the bosonic degrees of freedom from the partition function or integrate them out from the equations of motion for spin operators, the thermodynamics and the

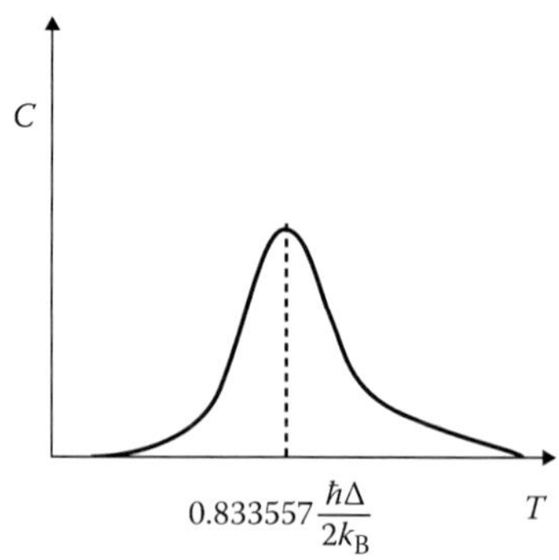

FIGURE 11.5 Schematic of the specific heat.

dynamics of the spins would appear to be dissipative. This is because the new channel opens up the possibility of exchange of energy, often irreversibly if the bosonic modes are infinite in number. The physics is much richer when the environment is quantal than in the case when it is classical. Now, temperature T will have to be thought of as an attribute of the bosonic environment, endowed to it by yet an external, classical, and invisible heat bath.

One may wonder if the epithet of an open system is a misnomer or not because, after all, the spin-boson Hamiltonian can be viewed as that of a composite quantum many-body system, which encompasses both the spin and the bosonic degrees of freedom! However, the point is that seldom is one interested in studying the subsystem over Poincare recurrence times for the entire global system, as described by Equation 11.11. The point is all the more self-evident when one thinks in terms of an actual experimental situation. For instance, in the case of neutron scattering of hydrogen, for which the spin-boson Hamiltonian is an appropriate model, the experimental time scale (i.e., the inverse of energy transfer in the inelastic scattering process divided by $\hbar$) is comparable to Δ_0^{-1} and ε^{-1}. Further, the probe, which, in this case, is the neutron, couples directly to the hydrogen and is "neutral" as far as the bosons are concerned. Hence, it is quite logical to integrate out the bosonic modes. Thus, the nomenclature of open systems is intertwined with how we interpret spectroscopic measurements.

What is then the correct definition of the specific heat for an open system? Recall that the density operator ρ for the system can be written as

$$\rho = \frac{e^{-\beta\mathcal{H}}}{Z_{tot}}, \tag{11.20}$$

where

$$Z_{\text{tot}} = Tr_{S+B}(e^{-\beta\mathcal{H}}). \tag{11.21}$$

In the coordinate representation, Equation 11.20 can be expressed as functional integrals

$$\rho_\beta(q'',\vec{x}'';q',\vec{x}') = \frac{1}{Z_{\text{tot}}} \int_{q(0)=q'}^{q(\hbar\beta)=q''} \mathcal{D}q(.) \int_{\vec{x}(0)=\vec{x}'}^{\vec{x}(\hbar\beta)=\vec{x}''} \mathcal{D}\vec{x}(.) \times e^{-\frac{S^{(E)}[q(.),\vec{x}(.)]}{\hbar}}, \tag{11.22}$$

where

$$S^{(E)}[q(.),\vec{x}(.)] = S_S^{(E)} + S_B^{(E)} + S_I^{(E)}. \tag{11.23}$$

Here, the qs represents the subsystem coordinates, whereas the $\vec{x}$s denotes the environment coordinates. $S^{(E)}[q(.),\vec{x}(.)]$ is the Euclidean action for the entire system (as described in Equation 11.11), which can be thought of as the summation of Euclidean actions for the subsystem, bath, and the interaction. The "reduced' density matrix is obtained from Equation 11.22 by integrating out the environmental coordinates, yielding

$$\rho_\beta^R(q'',q') = \int_{-\infty}^{+\infty} d\vec{x}\ \rho_\beta(q'',\vec{x};q',\vec{x}) = \frac{1}{Z_R} \int_{q(0)=q'}^{q(\hbar\beta)=q''} \mathcal{D}q(.) \times e^{-\frac{S_S^{(E)}[q(.),\vec{x}(.)]}{\hbar}} \times \mathcal{F}^{(E)}[q(.)], \tag{11.24}$$

where we have defined Z_R as

$$Z_R = \frac{\int_{-\infty}^{\infty} dq \int_{-\infty}^{\infty} d\vec{x}\rho_\beta(q,\vec{x};q,\vec{x})}{\int_{-\infty}^{\infty} d\vec{x}\rho_\beta^B(q,\vec{x};q,\vec{x})} = \frac{\int_{-\infty}^{\infty} dq \int_{-\infty}^{\infty} d\vec{x} \int_{q'=q''=q} \mathcal{D}q(.) \int_{\vec{x}'=\vec{x}''=\vec{x}} \mathcal{D}\vec{x}(.) \times e^{-\frac{S^{(E)}[q(.),\vec{x}(.)]}{\hbar}}}{\int_{-\infty}^{\infty} d\vec{x} \int_{\vec{x}'=\vec{x}''=\vec{x}} \mathcal{D}\vec{x}(.) \times e^{-\frac{S_B^{(E)}[q(.),\vec{x}(.)]}{\hbar}}} = \frac{Z_{tot}}{Z_B}, \tag{11.25}$$

the "influence functional" as

$$\mathcal{F}^{(E)}[q(.)] \equiv \frac{1}{Z_B} \int_{-\infty}^{\infty} d\vec{x} \int_{\vec{x}'=\vec{x}''=\vec{x}} \mathcal{D}\vec{x}(.) \times e^{-\frac{S_{B,I}^{(E)}[q(.),\vec{x}(.)]}{\hbar}}, \tag{11.26}$$

and

$$S_{B,I}^{(E)}[q(.),\vec{x}(.)] = S_B^{(E)} + S_I^{(E)}. \tag{11.27}$$

The influence functional $\mathcal{F}^{(E)}[q(.)]$ must reduce to unity if the interaction is switched off (i.e., $\mathcal{H}_I = 0$) because, in this limit, the reduced density operator

$$\rho^R = \frac{1}{Z_{\text{tot}}} Tr_B(e^{-\beta\mathcal{H}}) = Tr_B(\rho) \tag{11.28}$$

must equal

$$\rho^S = \frac{e^{-\beta\mathcal{H}_S}}{Tr_S(e^{-\beta\mathcal{H}_S})}. \tag{11.29}$$

This normalization ensures the presence of the prefactor Z_R^{-1} in Equation 11.24. Given the above considerations, the specific heat C can be expressed as [11–13]

$$C = k_B\beta^2 \frac{\partial^2}{\partial\beta^2} \ln Z_R. \tag{11.30}$$

Equations 11.29 and 11.30 form the basis of our further discussion of the specific heat.

11.4.2 Specific Heat in High Damping Limit

As discussed earlier, the effect of the interaction in the spin-boson model is to tweak the tunneling term, which can be immediately seen by examining the equation of motion of σ_x. This results in a "dressed" tunneling frequency, which is expected to be lower in its value (from Δ_0) if the damping is high. It is not surprising then that, in the most employed treatment of the spin-boson model in a functional integral method, i.e., the so-called dilute bounce gas approximation (DBGA), the effective tunneling frequency is assumed to be a small parameter [2]. We have shown earlier that a very transparent method to implement this scheme is to employ a "polaronic" transformation on Equation 11.11 with the aid of an operator [14]

$$S = e^{-\sigma_z \sum_q \frac{g_q}{\hbar\omega_q}\left(b_q - b_q^\dagger\right)}. \tag{11.31}$$

The transformed Hamiltonian $\tilde{\mathcal{H}}$ reads

$$\tilde{\mathcal{H}} = S\mathcal{H}S^{-1} = \tilde{\mathcal{H}}_S + \tilde{\mathcal{H}}_\mathcal{I} + \tilde{\mathcal{H}}_B = \frac{\hbar}{2}\varepsilon\sigma_z - \frac{\hbar}{4}\Delta_0(B_+\sigma_- + B_-\sigma_+) + \sum_q \hbar\omega_q b_q^\dagger b_q, \tag{11.32}$$

where

$$\begin{aligned}
\tilde{\mathcal{H}}_S &= \frac{\hbar}{2}\varepsilon\sigma_z, \\
\tilde{\mathcal{H}}_\mathcal{I} &= -\frac{\hbar}{4}\Delta_0(B_+\sigma_- + B_-\sigma_+), \\
\tilde{\mathcal{H}}_B &= \sum_q \hbar\omega_q b_q^\dagger b_q, \\
B_\pm &= e^{\pm 2\sum_q \frac{g_q}{\hbar\omega_q}\left(b_q - b_q^\dagger\right)},
\end{aligned} \tag{11.33}$$

and

$$\sigma_\pm = (\sigma_x \pm i\sigma_y). \tag{11.34}$$

The structure of the new interaction term, i.e., the second one in Equation 11.32, clarifies our earlier remark concerning the "dressing" of the tunneling frequency with the aid of the operators $B_\pm$. Thus, any perturbation analysis of the interaction term in Equation 11.32 is tantamount to assuming that the effective tunneling is weak yet treating the coupling to all orders, because the old coupling constant g_q is jacked up into the exponent.

We have shown before that second-order treatment of the interaction term $\tilde{\mathcal{H}}_\mathcal{I}$ in a resolvant operator expansion yields results that are equivalent to the DBGA. Here, we employ an alternative approach based on a cumulant expansion [15–17], hitherto unused (to the best of our knowledge) for calculating the specific heat.

We have

$$Z_{\text{tot}} = Tr_{S+B}\left(e^{-\beta\tilde{\mathcal{H}}_B}\mathcal{O}(\beta)\right) \tag{11.35}$$

where $\mathcal{O}(\beta)$ has the following definition:

$$\mathcal{O}(\beta) = \exp_-\left(-\int_0^\beta d\beta'\left\{\tilde{\mathcal{H}}_S + \tilde{\mathcal{H}}_\mathcal{I}(\beta')\right\}\right), \tag{11.36}$$

in which $\exp_-$ is the time-ordered (in the imaginary space) exponential and

$$\tilde{\mathcal{H}}_\mathcal{I}(\beta') = e^{\beta'\tilde{\mathcal{H}}_B}\tilde{\mathcal{H}}_\mathcal{I}e^{-\beta'\tilde{\mathcal{H}}_B}. \tag{11.37}$$

Stopping at the second order of $\tilde{\mathcal{H}}_\mathcal{I}$ in the argument of $\exp_-$, we obtain (from Equations 11.35 and 11.36)

$$Z_R = Tr_S\left[\exp\left\{-\int_0^\beta d\beta'\left\langle\tilde{\mathcal{H}}_S + \tilde{\mathcal{H}}_\mathcal{I}(\beta')\right\rangle_B^C + \int_0^\beta d\beta' \right.\right.$$
$$\left.\left.\times\int_0^{\beta'} d\beta''\left\langle\left(\tilde{\mathcal{H}}_S + \tilde{\mathcal{H}}_\mathcal{I}(\beta')\right)\left(\tilde{\mathcal{H}}_S + \tilde{\mathcal{H}}_\mathcal{I}(\beta'')\right)\right\rangle_B^C\right\}\right], \tag{11.38}$$

where

$$\left\langle\tilde{\mathcal{H}}_S + \tilde{\mathcal{H}}_\mathcal{I}(\beta')\right\rangle_B^C = \left\langle\tilde{\mathcal{H}}_S + \tilde{\mathcal{H}}_\mathcal{I}(\beta')\right\rangle_B = \tilde{\mathcal{H}}_S, \tag{11.39}$$

$$\begin{aligned}\left\langle\left(\tilde{\mathcal{H}}_S + \tilde{\mathcal{H}}_\mathcal{I}(\beta')\right)\left(\tilde{\mathcal{H}}_S + \tilde{\mathcal{H}}_\mathcal{I}(\beta'')\right)\right\rangle_B^C &= \left\langle\left(\tilde{\mathcal{H}}_S + \tilde{\mathcal{H}}_\mathcal{I}(\beta')\right)\left(\tilde{\mathcal{H}}_S + \tilde{\mathcal{H}}_\mathcal{I}(\beta'')\right)\right\rangle_B \\ &\quad -\left\langle\tilde{\mathcal{H}}_S + \tilde{\mathcal{H}}_\mathcal{I}(\beta')\right\rangle_B\left\langle\tilde{\mathcal{H}}_S + \tilde{\mathcal{H}}_\mathcal{I}(\beta'')\right\rangle_B \\ &= \left\langle\tilde{\mathcal{H}}_\mathcal{I}(\beta')\tilde{\mathcal{H}}_\mathcal{I}(\beta'')\right\rangle_B,\end{aligned} \tag{11.40}$$

and

$$\langle \ldots \rangle_B = Tr\left(\frac{e^{-\beta \tilde{\mathcal{H}}_B}}{Z_B} \ldots \right). \tag{11.41}$$

Here, we have used the fact that $\langle B_\pm \rangle_B = 0$ when we make use of the following expression for what is known as the "ohmic" spectral density:

$$J(\omega) = K\omega \exp\left(-\frac{\omega}{D} \right), \tag{11.42}$$

with D being a frequency cutoff. Using Equations 11.38 through 11.40, we obtain

$$Z_R = \frac{Z_{\text{tot}}}{Z_B} = \tilde{Z}_S \times \exp(I), \tag{11.43}$$

where

$$\tilde{Z}_S = 2\cosh\left(\frac{\beta\hbar\varepsilon}{2} \right), \tag{11.44}$$

and I is given by the following integral:

$$I = 4\left(\frac{\hbar^2 \Delta_0^2}{16} \right)\left(\frac{\pi}{\beta\hbar D} \right)^{2K} \int_0^{\beta} d\beta' \int_0^{\beta'} d\alpha \frac{1}{\left[\sin\left(\frac{\pi\alpha}{\beta} \right) \right]^{2K}}. \tag{11.45}$$

As we are considering the regime of high damping, K > 1. Upon partial integration, I can be written as

$$I = \int_0^{\beta} d\alpha(\beta - \alpha)\left(\frac{\pi}{\beta\hbar D} \right)^{2K} \left(\frac{\hbar^2 \Delta_0^2}{4} \right) \frac{1}{\left[\sin\left(\frac{\pi\alpha}{\beta} \right) \right]^{2K}}. \tag{11.46}$$

Since we are interested in checking the validity of the third law of thermodynamics, we have to eventually take the $\beta \to \infty$ limit. We can then replace $(\beta - \alpha)$ in the integrand of Equation 11.41 by β, resulting in

$$I = \int_0^{\beta} d\alpha\left(\frac{\pi}{\beta\hbar D} \right)^{2K} \left(\frac{\hbar^2 \Delta_0^2}{4} \right) \frac{\beta}{\left[\sin\left(\frac{\pi\alpha}{\beta} \right) \right]^{2K}}. \tag{11.47}$$

We have to remember though that, in Equation 11.42, we have already set a frequency cutoff, which, in turn, imposes an upper bound to β, which, in this case, is $\hbar D$. Therefore, in Equation 11.47, we use $\left(\frac{\pi\alpha}{\beta}\right)$ to be an expansion parameter in writing the denominator as a power series and then take the zero-temperature limit carefully, obtaining thereby

$$Z_R = \tilde{Z}_S \exp(f(K)\beta^{2(1-K)}), \tag{11.48}$$

where

$$f(K) = \left[\frac{1}{1-2K} + \frac{2K}{3!}\frac{\pi^2}{(3-2K)} - \frac{2K}{5!}\frac{\pi^4}{(5-2K)} + \frac{2K}{7!}\frac{\pi^6}{(7-2K)} - \ldots\right]. \tag{11.49}$$

Finally, using Equation 11.29, we get the following expression for specific heat:

$$C = 2k_B f(K)(1-K)(1-2K)\beta^{2(1-K)} + C_S, \tag{11.50}$$

where C_S is the Schottky correction term in the specific heat, which has already been discussed in Equations 11.18 and 11.19.

Note that K must be greater than 1 ($K > 1$) if the third law of thermodynamics is to remain valid. What it means is that, because the tunneling is assumed to be effectively weak, the approximate method presented above can be consistent only if K is appropriately large, at low temperatures. This conclusion is in conformity with the functional integral treatment of the specific heat [11].

11.4.3 Specific Heat in Weak Damping Limit

Given this issue in extending the results for the specific heat to K values less than one, we come to be aware of a problem with the DBGA or the equivalent polaronically transformed cumulant expansion method. As mentioned at the end of the previous section, the reasons are not difficult to decipher. It may be recalled that any perturbation treatment of the polaronically transformed interaction term willy-nilly presupposes tunneling to be infrequent, thus allowing the system to be "decohered" because of environmentally induced interruptions of the tunneling events. In other words, the method forces a lot more environmental excitations than what the system is ready for, especially at low temperatures. After all, $k_B T$ must be comparable to $\hbar\omega$, where ω is drawn out of the spectral function (i.e., the density of states) for the bosonic modes, for effective exchange of energy between the subsystem and the environment. Hence, it is not surprising that the approach in Section 11.4.2 may fail in dealing with the specific heat at very low temperatures (and more so at the absolute zero of temperature), wherein one needs a treatment in which Δ_0 is handled more accurately than what the method in Section 11.4.2 entails: concomitantly, the damping coefficient K ought to be extrapolated to low values. Such a weak-damping expansion is the subject of the discussion below.

We return to the original Hamiltonian of the spin-boson model

$$\mathcal{H} = \tilde{\mathcal{H}}_S + \tilde{\mathcal{H}}_\mathcal{I} + \tilde{\mathcal{H}}_B = \frac{\hbar}{2}(\varepsilon\sigma_z - \Delta_0\sigma_x) + \sigma_z \sum_k g_k \left(b_k + b_k^\dagger\right) + \sum_k \hbar\omega_k b_k^\dagger b_k, \tag{11.51}$$

where

$$\mathcal{H}_S = \frac{\hbar}{2}(\varepsilon\sigma_z - \Delta_0\sigma_x),$$

$$\mathcal{H}_\mathcal{I} = \sigma_z \sum_k g_k \left(b_k + b_k^\dagger\right) = \sigma_z V, \tag{11.52}$$

$$\mathcal{H}_B = \sum_k \hbar\omega_k b_k^\dagger b_k.$$

We employ the same cumulant expansion method; however, now the original interaction term (H_I in Equation 11.45) is treated perturbatively. The reduced partition function can be expressed as

$$Z_R = \frac{Z_{\text{tot}}}{Z_B} = Tr_S\left[\exp\left\{-\beta\mathcal{H}_S + \int_0^\beta d\beta' \int_0^\beta d\beta'' \langle V(\beta')V(\beta'')\rangle_B\right\}\right], \tag{11.53}$$

where

$$V(\beta') = e^{\beta'\mathcal{H}_B} V e^{-\beta'\mathcal{H}_B}. \tag{11.54}$$

Neglecting terms of the order of $1/D$ in $\langle V(\beta')V(\beta'')\rangle_B$ and performing a suitable rotation in the spin-space, we obtain

$$Z_R = Z_S \exp\left\{\int_0^\beta d\alpha \left(\left(\frac{K}{2\beta}\left(\frac{\pi^2}{3} + \frac{\pi^4}{15}\left(\frac{\alpha}{\beta}\right)^2 + \frac{2\pi^6}{189}\left(\frac{\alpha}{\beta}\right)^4 + \frac{\pi^8}{675}\left(\frac{\alpha}{\beta}\right)^6\right)\right.\right.\right.$$
$$\left.\left.\left. + \frac{2\pi^{10}}{10395}\left(\frac{\alpha}{\beta}\right)^8 + O\left(\left(\frac{\alpha}{\beta}\right)^{10}\right)\right)\right)\right\}, \tag{11.55}$$

where Z_S has been introduced in Equation 11.16. At this point, we carefully take the low temperature limit as discussed in Section 11.4.2 and get

$$Z_R = Z_S \exp\left[a_0 \frac{K}{\beta}\frac{\pi^2}{6} + a_2 \frac{K}{\beta^3}\frac{\pi^4}{30} + O\left(\frac{1}{\beta^5}\right)\right], \tag{11.56}$$

where

$$a_{2n} = \int_0^{\hbar D} d\alpha\, \alpha^{2n},\ n = 0,1,2,3\ldots. \tag{11.57}$$

Again, using Equation 11.29, we arrive at the following expression for specific heat for the system, which nicely satisfies the third law of thermodynamics:

$$C = k_B K\left[\frac{a_0}{\beta}\frac{\pi^2}{3} + \frac{a_2}{\beta^3}\frac{2\pi^4}{5} + O\left(\frac{1}{\beta^5}\right)\right] + C_S, \tag{11.58}$$

where C_S is again a Schottky correction term discussed in Equations 11.18 and 11.19.

11.5 CONCLUSIONS

Time-dependent phenomena, including dissipative quantum tunneling, have been extensively studied in the context of the spin-boson model. Surprisingly, thermodynamics have received comparatively less attention, although the specific heat has been adequately assessed within a path integral or functional integral approach. In this overview, we have introduced a much more straightforward methodology for calculating the specific heat with the aid of a cumulant expansion treatment. Our larger goal has been to critically examine the third law of thermodynamics for open systems using the paradigm of the spin-boson model. The profound implications of the third law have been summarized in both Sections 11.1 and 11.2. The historical perspective presented brings to the fore the monumental contribution of G. N. Lewis, who appears to have remained an unsung hero of chemistry, as the subject developed just prior to the advent of quantum mechanics [1].

The spin-boson model, its various ramifications, and analysis of its specific heat, when decoupled from its environment, have been the subject of Section 11.3. It is shown that the specific heat for the free Hamiltonian, *à la* Schottky, exhibits dominant exponential drop to zero temperatures, an aspect that is shared with the property of the Einstein oscillator. This aspect has been considered unsatisfactory *vis-à-vis* the third law that conjectures a power-law behavior with temperature, as far as the low-temperature specific heat is concerned [18].

The cumulant expansion scheme for the specific heat is dealt with in Section 11.4. For the sake of brevity, only the results for the much-studied ohmic dissipation are explicitly presented here. A polaronic transformation of the Hamiltonian enables us to treat wide regimes of coupling strength K (of the subsystem to the environment), although the method preordains the effective tunneling energy to be weak. As a result, the method suffers from a weakness, as discussed in Section 11.4.1, in that only K values larger than 1 retain physical validity. The cure for this weakness, which warrants a more accurate analysis of the tunneling energy, albeit restricted to small K values, has been the subject of Section 11.4.2. The results presented in Sections 11.4.1 and 11.4.2 put the spin-boson on firm footing in the context of the third law of thermodynamics.

Before concluding, it is interesting to note that environment-induced decoherence and dissipation yields a noninteger power law dependence of the specific heat on the temperature, with the power itself being dependent on the coupling strength K (for $K > 1$). This property is not only drastically different from the Schottky result for the free subsystem, underscoring the importance of the environment (or an open system), but is also different from that of the dissipative Einstein oscillator or dissipative cyclotron motion [12,13]. Both of the latter cases are characterized by linear dependence of the specific heat on the temperature, as far as low-temperature behavior is concerned. Interestingly, the linear dependence is restored when the damping is assumed to be weak, as shown in Section 11.4.2. Our results for the specific heat for the spin-boson model presented here may therefore be viewed as complementary to the recently studied examples of the free particle, the Einstein oscillator [12] and Landau diamagnetism [19], in the context of the third law of thermodynamics.

ACKNOWLEDGMENT

Aniket Patra thanks Patha Chakra (Study Circle), Visva-Bharati, for its kind hospitality during the time this work was carried out.

REFERENCES

1. Rao, C. N. R. *Public Lecture in the Indian Institute of Science Education and Research*. Kolkata, India. December 12, 2010.
2. Leggett, A. J., S. Chakravarty, M. P. A. Fisher, A. T. Dorsey, A. Garg, and W. Zwerger. 1987. *Rev. Mod. Phys.* **59**:1.
3. Einstein, A. 1956. *Investigations on the Theory of Brownian Motion*. Dover, NY.
4. Grabert, H., S. Linkwitz, S. Dattagupta, and U. Weiss. 1986. *Europhys. Lett.* **2**(**8**):631.
5. Dattagupta, S., and S. Puri. 2004. *Dissipative Phenomena in Condensed Matter*. Heidelberg: Springer-Verlag.
6. Chang, L. D., and S. Chakravarty. 1985. *Phys. Rev. B*. **31**:154.
7. Grabert, H., and H. Schober. 1994. *Hydrogen in Metals*. Vol. II, Ed. H. Wipf. Heidelberg: Springer-Verlag.
8. Abragam, A. 1961. *The Theory of Nuclear Magnetism*. London: Oxford University Press; Kubo, R. 1974. *Fluctuation, Relaxation and Resonance in Magnetic Systems*. Ed. By G. Hohler. Berlin: Springer Verlag; Slichter, C. P. 2010. Principles of Magnetic Resonance. *Springer Series in Solid State Science (Book 1)*.
9. Kamil E., and S. Dattagupta. 2012. *Pramana*. Issues-in-progress, pp. 9708.
10. Ashcroft, N. W., and N. D. Mermin. 1976. *Solid State Physics*. New York: Holt, Rinehart and Winston.

11. Weiss, U. 1998. *Quantum Dissipative Systems*. Singapore: World Scientific.
12. Dattagupta, S., Jishad Kumar, S. Sinha, and P. A. Sreeram. 2010. *Phys. Rev. E*. **81**:031136.
13. Jishad Kumar, T. M. Doctoral dissertation, Indian Institute of Science Education and Research. Kolkata, India. Unpublished.
14. Dattagupta, S., H. Grabert, and R. Jung. 1989. *J. Phys.* **1**:1405.
15. Kubo, R. 1962. *J. of the Phys. Soc. of Japan*. **17**:1100.
16. Dattagupta, S. 1987. *Relaxation Phenomena in Condensed Matter Physics*. Orlando: Academic Press.
17. Patra, A. 5th year integrated MS dissertation. Indian Institute of Science Education and Research. Kolkata, India. Unpublished.
18. Ingold, G. L., P. Hänggi, and P. Talkner. 2009. *Phys. Rev. E*. **79**:061105.
19. Kumar, J., P. A. Sreeram, and S. Dattagupta. 2009. *Phys. Rev. E*. **79**:021130.

12 Mechanism of Chemical Reactions in Four Concepts

María Luisa Cerón, Soledad Gutiérrez-Oliva, Bárbara Herrera, and Alejandro Toro-Labbé

CONTENTS

12.1 INTRODUCTION

The main objectives of research in chemistry in the last century or so are the determination of the structure and properties of molecular species, the characterization of its reactive propensity, and the elucidation of the mechanisms of the reactions in which they are involved. In the last few decades, chemistry has greatly evolved toward a less empirical science in which molecular structures, chemical reactivity, and reaction mechanisms are explained from the characterization of the electronic density. In this context, the conceptual density functional theory (CDFT) and the quantum theory of atoms in molecules (QTAIM) have been crucial for such a development.[1,2] On one hand, pioneer works by Parr, Geerlings, Chattaraj, and many others have set the foundation of CDFT, providing theoretical support for empirical concepts that chemists have used for years on an intuitive basis. Rigorous definitions of properties such as electronegativity,[3–5] hardness,[6] and electrophilicity,[7,8] together with easy ways to compute them and the formulation of reactivity principles and postulates that gives them an interpretative basis, have produced speedy development of the field of computational chemical reactivity. On the other hand, the works by Bader and collaborators on the characterization of the Laplacian of the electron density have provided new chemical insights on the electron density, thus introducing the concept of atoms in molecules that characterizes the behavior of atoms embedded in a molecular environment.[2,9] New definitions of chemical bond, together with the formulation of indicators of electronic distributions, came out from CDFT and QTAIM. Both theories have proven to be very useful and adequate in the characterization of the structure and properties of molecules. Although these theories are focused on the characterization of the structure and the intrinsic properties of isolated molecules, they have also been used to predict the specific way in which molecules interact to produce a chemical reaction.

Chemical reactions involve nuclear displacements and electronic rearrangements that bring the reacting species into products. These structural and electronic reorderings can be viewed as a sequence of chemical events, basically bond strengthening or formation, and bond weakening or breaking. Computation of reaction paths was possible only from the works by Fukui[10] and Schlegel[11] and the implementation of intrinsic reaction coordinate (IRC) routines in quantum chemistry codes. For years, the characterization of chemical reactions was limited to the analysis of energy profiles that gave thermodynamic and kinetic information, knowledge of the reaction and activation energies that produced crucial information about the energy involved in the process, and, through transition state theories, the time that reactants take to change into products. However, energy profiles do not give information about the mechanism of the reactions.

Details on the reaction mechanisms can be obtained through more advanced analysis of the energy profiles. In this context, the use of reaction path Hamiltonian, together with the so-called united reaction valley approach, which was proposed by Kraka et al.,[12,13] have been quite successful. In this approach, the reaction path is partitioned into various phases describing the most relevant changes of the system along the reaction coordinate. In the same line of thinking and at about the same time, the present authors formulated the reaction force concept,[14] focusing the attention not on the energy profile itself but on the derivative of the energy profile, thus defining the reaction force.[15–20] Interest on the reaction force is threefold: (1) It provides a framework to analyze chemical reactions by producing a rational partition of the reaction coordinate into three *reaction regions*: (a) reactants, (b) transition state, and (c) products; (2) a rational partition of the activation energy emerges from the reaction force analysis and permits identification of the physical nature of energy barriers and their classification in terms of structural or electronic predominance; (3) the derivative of the reaction force with respect to the reaction coordinate leads to the reaction force constant, a property that gives valuable information about the active modes that drives the reaction.

Since chemical reactions can be seen as a sequence of chemical events that involves the electronic activity taking place during the reaction, a descriptor of this activity becomes important to get insights into the reaction mechanism. The reaction electronic flux (REF), which was defined as the derivative of the electronic chemical potential with respect to the reaction coordinate, has been shown to be an adequate descriptor of the electronic activity.[21–25] This chapter is focused on the definitions, usefulness, and applications of the REF, the reaction force, the reaction force constant, and the partition of the activation energy.

The aim of this chapter is basically methodological, and we want to introduce and discuss the above-mentioned concepts that are designed as tools to advance toward a better understanding of the mechanisms of chemical reactions. In this context, we will discuss results that have been obtained through quantum chemical calculation and published elsewhere. Computation techniques and procedures will not be under discussion and the computational details will be given as needed. However, since the energy profile is on the focus of this chapter, it is necessary to mention that they were obtained using the IRC procedure[10,11] which is the minimum energy path that links transition states with reactants and products; in all cases the smallest step was used to ensure the quality of the numerical derivatives that led the reaction force and the reaction force constant.

12.2 FOUR CONCEPTS

12.2.1 Reaction Force

The minimum energy path $E(\xi)$, which links the transition state with the reactants and products, defines the IRC ξ. Figure 12.1 shows an energy profile of a generic elementary step of a chemical reaction along the IRC.

Quoted in the figure are the key quantities that determine $E(\xi)$: with the reaction energy ($\Delta E°$) giving the thermodynamic of the reaction and the activation energy ($\Delta E^{\neq}$) accounting for the kinetic information, both are crucial properties of the reaction that help characterize the chemical transformation. Also characterized on the figure are the respective activation and relaxation processes. Many chemical reactions require more than one elementary step to reach the products of the reaction; in such cases, the whole chemical process may be defined through a sequence of such elementary steps. The reaction force $F(\xi)$ is defined as[14–20]

$$F(\xi) = -\frac{dE}{d\xi}, \tag{12.1}$$

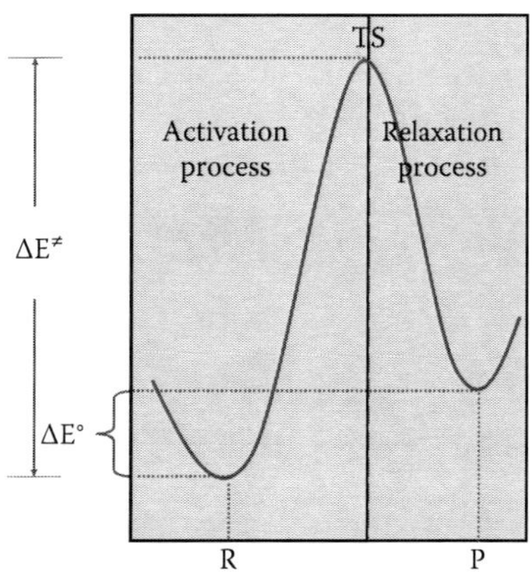

FIGURE 12.1 Energy profile of an elementary step.

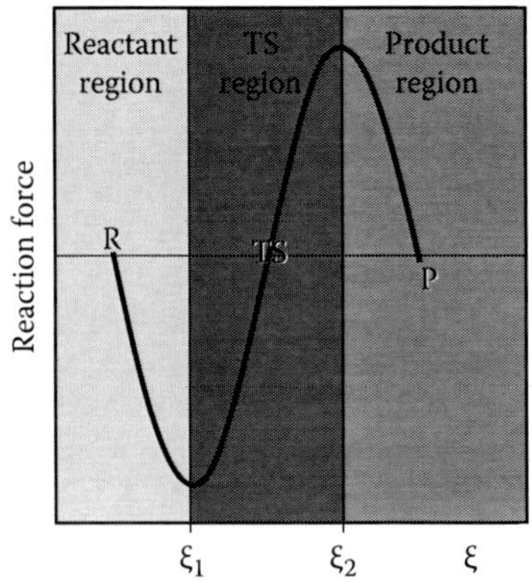

FIGURE 12.2 Reaction force profile along the reaction coordinate ξ. The three reaction regions where the specific mechanism takes place are also displayed in the figure.

so, as to the energy profile of Figure 12.1, it corresponds with the reaction force profile displayed in Figure 12.2. We first note that the reaction force profile presents two critical points, i.e., a minimum at ξ_1 and a maximum at ξ_2; these points allow proposing a rational partition of the reaction coordinate and defining three *reaction regions*: (1) the *reactant region*, which was defined from the reactant position at ξ_R to the force minimum at ξ_1, (2) the *transition state region*, which was defined from ξ_1 to ξ_2, and (3) the *product region*, which was defined from ξ_2 to ξ_P.

It is our experience in the study of many reactions, going from nucleophilic substitutions in catalytic reactions[22,26] to simple proton transfers in biological systems,[21] that, within the reactant region, the system prepares the reaction mainly through structural arrangements. At point ξ_1, the system can be seen as an *activated reactant* entering the transition state region and ready to continue with the reaction. The transition state region $\{\xi_1,\xi_2\}$ is mainly characterized by an intensive electronic activity; it is here where most bonds are formed and/or broken. At point ξ_2 the system is seen as an *activated product* ready to enter the product region, where it relaxes to the product structures. In summary, the reactant and product regions are mainly characterized by structural reordering, which activates the reactants or relaxes to the products; within the transition state region, most electronic activity takes place.

12.2.2 Partition of Activation Energy

From the perspective of the reaction force, the activation process involves, first, a structural reordering taking place principally within the reactant region, followed by electronic activity at the transition state region, leading to the formation and/or breaking of bonds. This can be rationalized through the following expression:

$$\Delta E^{\neq} = -\int_{\xi_R}^{\xi_0} F(\xi)d\xi > 0 = W_1 + W_2, \qquad (12.2)$$

where ξ_o denotes the position of the transition state structure, and $\{W_1,W_2\}$ are the reaction works defined within the respective reaction regions:

$$W_1 = -\int_{\xi_R}^{\xi_1} F(\xi)d\xi > 0 \quad W_2 = -\int_{\xi_1}^{\xi_0} F(\xi)d\xi > 0. \qquad (12.3)$$

Equation 12.2 gives very important information since it allows the identification of the physical nature of energy barriers in terms of the prevalence of structural or electronic works.

An interesting example illustrating the power of Equation 12.2 is the identification of the effect of a catalyst in a chemical reaction. A catalytic process that has been studied is the gas-phase keto-enol tautomerization of thymine, where a hydrogen atom is transferred from nitrogen to oxygen, which is bonded to the α-carbon. We have performed a calculation of the reaction in the presence and the absence of a Mg(II) ion.[18,19] The computed activation energies and the corresponding reaction works are displayed in Figure 12.3. We first note that the Mg(II) promotes the reaction by lowering the activation barrier from 49.6 to 44.2 kcal/mole, which is a decrease of about 5 kcal/mole with the whole catalytic effect concentrated in W_1 while W_2 remains practically constant. This result indicates that the action of the catalyst is focalized on the first part of the reaction, where structural reordering prevails over the electronic effects. This kind of result may be interesting to design catalyst acting on specific steps of the reaction.

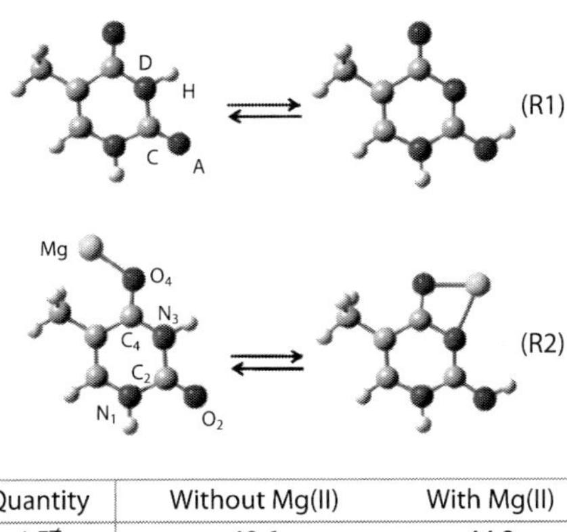

Quantity	Without Mg(II)	With Mg(II)
$\Delta E^{\neq}$	49.6	44.2
W_1	32.6	27.1
W_2	17.0	17.1

FIGURE 12.3 Keto-enol tautomerization reaction of thymine in gas phase with and without magnesium. The activation energies and reaction works are given in kilocalorie per mole and were computed at the DFT/B3LYP/6-311++G(d,p) level of theory.[18,19]

12.2.3 Reaction Force Constant

The reaction force constant has been introduced recently by Politzer et al.[27] to get insights on the reactive modes that might be driving the course of the reaction. It is defined as

$$\kappa(\xi) = -\frac{dF}{d\xi}. \tag{12.4}$$

The reaction force constant is positive in the reactant and product regions and negative in the transition state region, as shown in Figure 12.4. The negative $\kappa(\xi)$ indicates that the entire transition state region is a sequence of unstable configurations, so that the whole transition state region may be described as a chemically active or an electronically intensive stage of the reaction. On the other hand, the reaction force constant provides information about the active modes that might be driving the reaction at the different regions along the reaction coordinate. In the simple proton transfer reaction HONS→ONSH,[27] direct comparison of the reaction force constant with the force constants associated to the normal modes allows one to identify the modes that drive the course of the reaction. It goes as follows: the reaction is driven mainly by the in-plane NO stretch, which prevails over the other modes until entering the transition state region where activation of the other modes is completed, especially the H motion in the molecular plane that travels from the oxygen to the sulfur atom. At the product region, it is the NO stretching that leads the relaxation process.

Thus, the reaction force and reaction force constants indicate that the chemical process passes through a transition state region, which is defined as a sequence of unstable configurations having at least one imaginary frequency or a negative force constant. The above ideas are in line with experimental observations in the field of femtochemistry that, using ultrafast laser pulses on a femtosecond time scale, are able to *freeze* transient unstable configurations and identify a coalescence region of transition states and reactive intermediates.[28–30] In the context of femtochemistry,

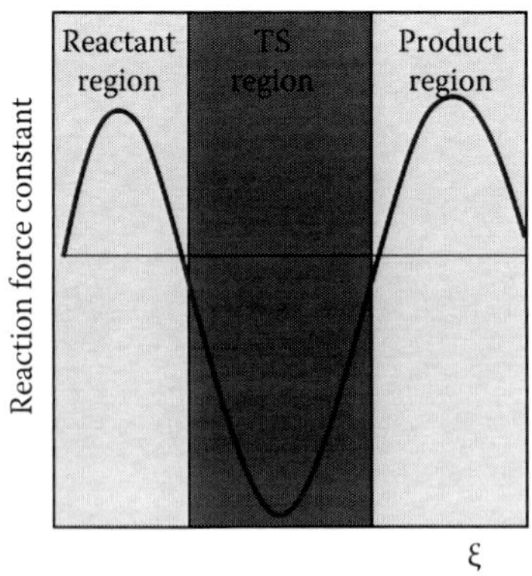

FIGURE 12.4 Reaction force constant for an elementary step.

the reaction force and reaction force constants identify the coalescence region with the transition state region, which is "populated" by activated reactants and activated products. The latter are characterized through at least one imaginary vibrational frequency and by a large electronic activity. We wish now to draw attention to the electronic activity that takes place during the progression from reactants to products.

12.2.4 Reaction Electronic Flux (REF)

Among the main developers of CDFT, Robert G. Parr,[1,5,6,8] Henry Chermette,[31] Paul Geerlings,[32] and Pratim K. Chattaraj[33] have brought key contributions to the field of chemical reactivity and provided rigorous definitions of chemical concepts that, in the past, were used and understood on an empirical basis.[31–36] One of these descriptors is the electronic chemical potential (μ)[5] that comes out as a Lagrange multiplier in the density functional theory (DFT) Euler–Lagrange equation for energy. It is very interesting to notice that this mathematical parameter is actually associated to electronegativity (χ), thus making the link between DFT and classical chemistry.[5,37,38] The electronic chemical potential is defined as

$$\mu = \left(\frac{dE}{dN}\right)_{\upsilon(\vec{r})} = -x, \tag{12.5}$$

where N and $v(r)$ are the total number of electrons of the system and the external potential, respectively. Physically, the electronic chemical potential characterizes the escaping tendency of electrons from an equilibrium distribution. Computationally, μ can be determined by making approximations aimed at avoiding the discontinuity of variable N in the derivative of the energy. The finite difference approximation and the Koopmans' theorem[39] lead to the following working formula:

$$\mu \simeq -\frac{1}{2}(IP + EA) \simeq \frac{1}{2}(\varepsilon_L + \varepsilon_H), \tag{12.6}$$

where IP is the first ionization potential, EA is the electron affinity, and ε_H and ε_L are the energies of the highest occupied molecular orbital (HOMO) and lowest unoccupied molecular orbital (LUMO), respectively. Since the above quantities can be determined easily from any quantum chemistry computational code, the evaluation of the chemical potential along the reaction coordinate is possible, and the profile $\mu(\xi)$ provides useful information about the electronic reordering that occurs during a chemical reaction. The quantity that more precisely describes the electronic activity taking place along the reaction coordinate is the REF, which is defined as[21–25]

$$J(\xi) = -\frac{d\mu}{d\xi}. \tag{12.7}$$

The physical meaning of the REF is obtained by making the analogy with classic thermodynamics and its relation with basic chemical events, namely, bond-formation and bond-breaking processes. Positive values of $J(\xi)$ account for spontaneous changes of the electron density that are led by bond-strengthening or bond-formation processes; negatives values of $J(\xi)$ witness nonspontaneous electronic changes that are controlled by bond-weakening or cleavage processes. Phenomenological insights on the REF can be obtained by partitioning it into two terms, namely, $J_p(\xi)$ and $J_t(\xi)$, accounting for the electron polarization and transfer effects,[23–25] i.e.,

$$J(\xi) = [J_p(\xi) + J_t(\xi)]. \tag{12.8}$$

The polarization flux describes how the electronic density of the fragments that define the system is affected by the presence of the other fragments. A key aspect of the theory is the calculation of this quantity. To do so, a rational fragmentation of the supramolecular system has to be done; in this context, the polarization flux can be written as a sum of the n_f contributions coming from the different fragments:

$$J_p(\xi) = \sum_{i=1}^{n_f} J_p^{(i)}(\xi) \tag{12.9}$$

with

$$J_p^{(i)}(\xi) = -\left(\frac{N_i}{N}\right)\frac{d\mu_i}{d\xi}, \tag{12.10}$$

where N_i and μ_i are the total number of electrons and the chemical potential of fragment i, respectively. We have proposed a computational procedure to determine the fragments' chemical potential. It takes advantage of the counterpoise method[40,41] that permits the determination of the electronic properties of fragments within a supermolecular system. This procedure is carried out all along the reaction coordinate. At each point, ξ calculations are performed on the fragmented system using the geometry it has in the supermolecule at that point and that was previously obtained through the IRC procedure. Having the total REF $J(\xi)$ and the polarization contribution $J_p(\xi)$ at hand, the REF accounting for the electronic transfer among the fragments is obtained through

$$J_t(\xi) = [J(\xi) - J_p(\xi)]. \tag{12.11}$$

The chemical potential is a global property of the system, which means that it has the same value everywhere in the system. However, we take advantage of producing an artificial partition of the supramolecular chemical potential that emerges when applying the equalization principle for the chemical potential[37,38] and can be expressed as follows:

$$\mu(\xi) = \sum_{i=1}^{n_f}\left(\frac{N_i}{N}\right)\mu_i^{\circ}(\xi), \tag{12.12}$$

TABLE 12.1
Chemical and Physical Partition of the REF for a Binary Reaction A + B

Physical/ Chemical	Fragment A	Fragment B	Supermolecule A:B
Polarization	$J_P^{(A)}(\xi) = -\left(\frac{N_A}{N}\right)\frac{d\mu_A}{d\xi}$	$J_P^{(B)}(\xi) = -\left(\frac{N_B}{N}\right)\frac{d\mu_B}{d\xi}$	$J_P(\xi) = J_P^{(A)}(\xi) + J_P^{(B)}(\xi)$
Transfer	$J_t^{(A)}(\xi) = -\left(\frac{N_A}{N}\right)\frac{d(\mu_A - \mu_A)}{d\xi}$	$J_t^{(B)}(\xi) = -\left(\frac{N_B}{N}\right)\frac{d(\mu_B - \mu_B)}{d\xi}$	$J_t(\xi) = J_t^{(A)}(\xi) + J_P^{(B)}(\xi)$
Total REF	$J^{(A)}(\xi) = J_P^{(A)}(\xi) + J_t^{(A)}(\xi)$	$J^{(B)}(\xi) = J_P^{(B)}(\xi) + J_t^{(B)}(\xi)$	$J(\xi) = J_P(\xi) + J_t(\xi)$

where $\mu_i^{\circ}(\xi)$ is the chemical potential of ith hypothetic fragment that constitutes the system such that $\mu_i^{\circ}(\xi) \equiv \mu(\xi)$. In this context,

$$J(\xi) = -\frac{d\mu}{d\xi} = -\sum_{i=1}^{n_f}\left(\frac{N_i}{N}\right)\frac{d\mu_i^{\circ}}{d\xi} = \sum_{i=1}^{n_f} J^{(i)}(\xi)\,. \tag{12.13}$$

It is necessary that the hypothetic fragmentation be consistent with the actual fragmentation performed to compute the polarization flux. Back to the electron transfer flux, using Equations 12.10 and 12.13, the electron transfer flux is given by

$$J_t(\xi) = \sum_{i=1}^{n_f}\left(\frac{N_i}{N}\right)\frac{d\mu_i}{d\xi} - \sum_{i=1}^{n_f}\left(\frac{N_i}{N}\right)\frac{d\mu_i^{\circ}}{d\xi} = \sum_{i=1}^{n_f}\left(\frac{N_i}{N}\right)\frac{d}{d\xi}\left[\mu_i(\xi) - \mu_i^{\circ}(\xi)\right] = \sum_{i=1}^{n_f} J_t^{(i)}(\xi). \tag{12.14}$$

Table 12.1 illustrates the results obtained for a binary reaction of type A + B. It can be observed that the REF admits both a chemical partition in terms of the fragments' contribution and a physical partition in terms of polarization and transfer contributions.

12.2.4.1 Characterization of Electronic Activity Using REF

We have studied the mechanisms of different chemical reactions. In this section, we provide two examples of the use of the REF as a descriptor of the electronic activity that takes place in the course of a chemical reaction.

12.2.4.1.1 Double-Proton Transfer in Adenine–Uracil

We have studied the double-proton transfer reaction in the adenine–uracil complex.[21] The calculations were performed at the HF/6-311G** level, and the REF was obtained by numerical differentiation of the chemical potential defined in terms of the frontier orbital energies, as indicated in Equation 12.6. The double-proton transfer in the complex adenine–uracil is a stepwise reaction in which the two protons are transferred asynchronously, so the reaction consists of two elementary steps. This gives rise to five reaction regions, which are delimited by the vertical lines displayed in Figure 12.5, which, in turn, shows the REF profile. The five reaction regions are reactant (R), transition state 1 (TS1), intermediate (I), transition state 2 (TS2), and product (P).

It can be observed that, within the reactant region, there is no electronic activity. This only shows up when entering the first transition state region, where the transfer of the first proton from uracil to adenine takes place. A positive peak indicates that the electronic activity, at this point, is spontaneous and that the REF is driven by bond-strengthening or bond-formation processes. Then, a zero-flux regime follows and remains all along the intermediate region. Entering the TS2 region, a negative peak emerges indicating that the second proton is transferred, this time from adenine to uracil. In this case, the electronic activity is nonspontaneous and is led by bond-weakening or bond-breaking processes.

It is interesting to quantify the observed electronic activity in terms of the energy involved in the process. To do so, we analyze simultaneously the HF/6-311G** energy profile with the REF profile. The energy-REF analysis is illustrated in Figure 12.6, where it can be observed that the reaction initiates with pure structural rearrangements from equilibrium. The right panel shows no electronic activity besides the basal one, and it becomes clear that the activation of the first proton transfer should be associated only to structural effects. The energy barrier for the first proton transfer is 15.52 kcal/mol. The partition of the activation energy given by Equation 12.2 indicates that 72% of it (W_1 = 11.20 kcal/mol) is due to structural arrangements. Within the intermediate region, the system spends 5.71 kcal/mol in structural arrangements to prepare the second proton transfer. The barrier for the second proton transfer measured from the intermediate region is 6.98 kcal/mol. In this case, 82% of the activation energy may be assigned only to structural rearrangements.

It is interesting to stress the highly localized electronic activity that shows up exclusively within the respective transition state regions. Roughly, electronic effects accounts for only 25% of the whole energetics of the reaction, which is exclusively activated by structural effects. Another interesting feature of the joint analysis of the energy and REF profiles concerns the mechanism of the reaction. It can be observed that the energetic profile is not clear about the nature of the transfer, concerted or stepwise. However, the REF profile is clearly consistent with a stepwise mechanism, a result that was later confirmed experimentally.[30]

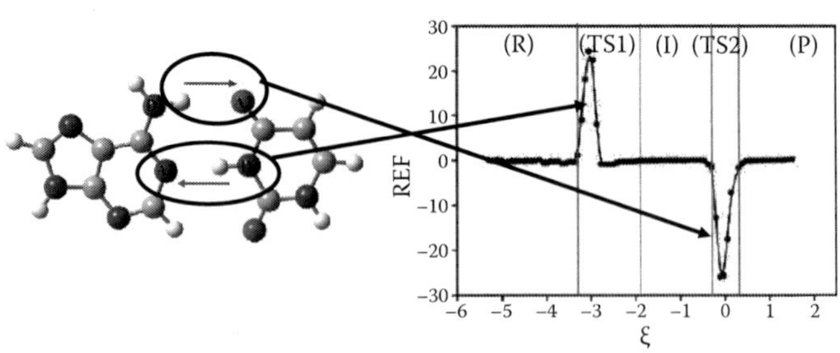

FIGURE 12.5 REF profile for the double proton transfer reaction in the adenine–uracil complex.

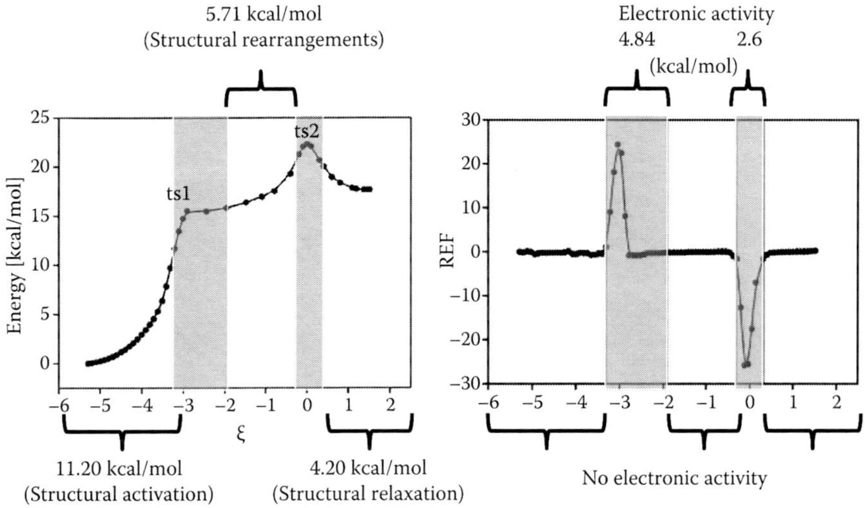

FIGURE 12.6 Energy cost of structural and electronic activity along the reaction coordinate for the double-proton transfer in the adenine–uracil complex.

12.2.4.1.2 Methanol Decomposition by Copper Oxide

In this section, we will review the reaction $CH_3OH + CuO \rightarrow CH_2O + H_2O + Cu$, which is methanol decomposition by copper oxide.[25,26] Results of the REF were obtained at the DFT/B3LYP/6-31G level using Lanl2DZ pseudopotential on copper. Figure 12.7 displays the REF profile, together with the structural schema of the reaction.

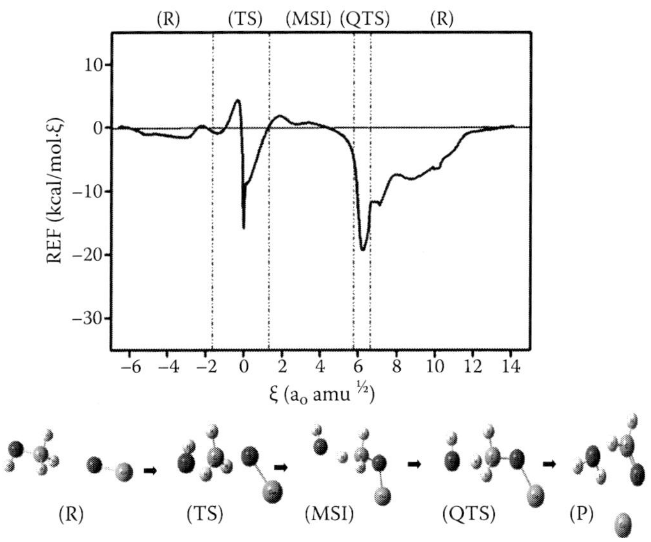

FIGURE 12.7 REF profile for the methanol decomposition reaction by copper oxide.

Again, this reaction occurs in two steps, giving rise to five reaction regions that are indicated on the figure. The reaction regions are defined as reactants (R), transition state (TS), metastable intermediate (MSI), and quasi-transition state (QTS, where structures present one imaginary frequency but not one corresponds to an energy maximum), and products (P). In terms of the electronic activity taking place during the process, the reaction starts with quite insignificant activity and only increases a little at the transition state, showing positive and negative peaks. Then, at the MSI region, the flux is again very small, but it becomes important when leaving the MSI region to enter the QTS one, with a large negative peak within the region witnessing a large electronic activity that remains until reaching the product of the reaction. It is interesting to characterize this electronic activity in terms of physical contributions, involving polarization and transfer, and chemically in terms of contributions coming from chemical fragments.[24,25]

To continue with the study, the system was separated into three chemical fragments that obey the structure of the transition state, as shown in the upper left side of Figure 12.8. The left panel of Figure 12.8 displays the REF and its decomposition into polarization and transfer contributions. Dashed lines indicate the splitting of polarization and transfer fluxes into chemical fragments' contributions shown in the right-hand panels.

It can be observed first that electron transfer effects are driving the reaction. The first peak displayed by the REF within the TS region is due to transfer, and

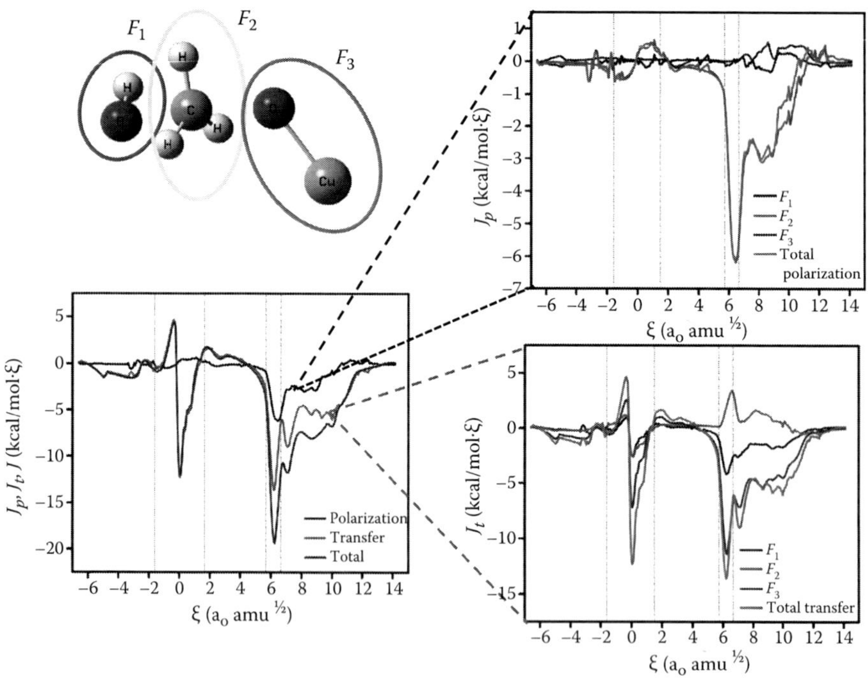

FIGURE 12.8 Physical and chemical partitions of the REF for the methanol decomposition reaction by copper oxide.

polarization flux is negligible. Complementary electronic population analysis indicated that the main electronic activity taking place in this region is copper reduction ($Cu^{2+} \rightarrow Cu^{+}$). In the second part of the REF profile, polarization effects are more noticeable, especially within the QTS region, but the transfer flux takes over mainly due to the reduction process $Cu^{+} \rightarrow Cu^{\circ}$.[25,26] On the other hand, the upper right-hand panel indicates that, chemically, the polarization effects are basically due to the CH_3 moiety. The other two fragments only show marginal contributions to $J_p(\xi)$. Concerning the activity associated to electronic transfer, Figure 12.8 shows that copper oxide contributes the most to it, and as already mentioned, this activity is mainly due to the successive reduction of copper.

In summary, methanol decomposition by copper oxide is initiated through structural reorderings that activate the electronic activity, which is principally manifested through a first reduction of copper. Then, the electronic activity decreases at the MSI region, but it becomes important when entering the QTS region. Here, polarization flux due to the CH_3 moiety, together with transfer flux mainly due to the second reduction of copper, drives the reaction until the products are reached. Structural relaxation in the product region appears to be coupled with the electronic activity. It is important to mention that the conclusions suggested by the REF analysis were totally confirmed by independent results provided by NBO electronic population analysis.[42]

12.3 CONCLUDING REMARKS

A complete methodology consisting of four key concepts in analyzing the mechanism of chemical reactions has been delivered. These concepts provide new important insights on the characterization of reaction mechanisms. The reaction force produces a good framework of reaction regions, in which different specific mechanisms might be operating. Most often, structural reorders occur within the reactants and product regions, whereas electronic activity mainly takes place within the transition state region. The reaction force also provides a rational partition of the activation energy that points to elucidating its physical nature. On the other hand, the reaction force constant grounds the existence of a transition state region defined as a sequence of unstable configurations, thus opening new interpretative possibilities in the whole field of reaction dynamics. Finally, the REF emerges as a very powerful tool to characterize the electronic activity taking place during the course of chemical reactions. It provides physical interpretations through polarization and transfer effects and identifies the specific role of the chemical species involved in the reaction.

ACKNOWLEDGMENTS

The authors wish to thank Professor Pratim K. Chattaraj for his kind invitation to contribute to this book honoring Professor Bidyendu Mohan Deb. It is a great pleasure for us to be part of this volume honoring such a great scientist. The authors wish to thank FONDECYT for the financial support through Projects N° 1090460, 1100881, and 1120093.

REFERENCES

1. Parr, R. G., and W. Yang. 1989. *Density Functional Theory of Atoms and Molecules*. Oxford: Oxford University Press.
2. Bader, R. F. W. 1990. *Atoms in Molecules: A Quantum Theory*. Oxford: Clarendon Press.
3. Mulliken, R. S. 1934. *J. Chem. Phys.* **2**:782–793.
4. Pauling, L. 1960. *The Nature of the Chemical Bond and the Structure of Molecules and Crystals: An Introduction to Modern Structural Chemistry.* Ithaca, NY: Cornell University Press.
5. Parr, R. G., R. A. Donnelly, M. Levy, and W. E. Palke. 1978. *J. Chem. Phys.* **68**:3801–3807.
6. Parr, R. G., and R. G. Pearson. 1983. *J. Am. Chem. Soc.* **105**:7512–7516.
7. Ingold, C. K. 1933. *J. Chem. Soc.* 1120.
8. Parr, R. G., L. V. Szentpaly, and S. B. Liu. 1999. *J. Am. Chem. Soc.* **121**:1922–1924.
9. Popelier, P. L. 2000. *Atoms in Molecules: An Introduction*. Upper Saddle River, NJ: Prentice Hall.
10. Fukui, K. 1970. *J. Phys. Chem.* **74**:4161–4163; Fukui, K. 1981. *Acc. Chem. Res.* **14**:363–368.
11. Gonzalez, C., and H.B. Schlegel. 1989. *J. Chem Phys.* **90**:2154–2161.
12. Kraka, E. 1998. *Encyclopedia of Computational Chemistry*. Hoboken, NJ: John Wiley. Vol. 4, pp. 2437–2463.
13. Kraka, E., and D. Cremer. 2010. *Acc. Chem. Res.* **43**:591–601.
14. Toro-Labbé, A. 1999. *J. Phys. Chem. A.* **103**:4398–4403.
15. Jaque, P., and A. Toro-Labbé. 2000. *J. Phys. Chem. A.* **104**:995–1003.
16. Toro-Labbé, A., S. Gutiérrez-Oliva, M. C. Concha, J. S. Murray, P. Politzer. 2004. *J. Chem. Phys.* **121**:4570–4575.
17. Gutiérrez-Oliva, S., B. Herrera, A. Toro-Labbé, and H. Chermette. 2005. *J. Phys. Chem. A.* **109**:1748–1751.
18. Rincón, E., P. Jaque, and A. Toro-Labbé. 2006. *J. Phys. Chem. A.* **110**:9478–9485.
19. Toro-Labbé, A., S. Gutiérrez-Oliva, J. S. Murray, and P. Politzer. 2007. *Mol. Phys.* **105**: 2619–2625.
20. Toro-Labbé, A., S. Gutiérrez-Oliva, P. Politzer, and J. S. Murray. 2009. *Chemical Reactivity Theory: A Density Functional View.* Chattaraj, P. K. (Editor). Boca Raton, FL: CRC Press. pp. 293–302.
21. Herrera, B., and A. Toro-Labbé. 2007. *J. Phys. Chem. A.* **111**:5921–5926.
22. Echegaray, E., and A. Toro-Labbé. 2008. *J. Phys. Chem. A.* **112**:11801–11807.
23. Vogt-Geisse, S., and A. Toro-Labbé. 2009. *J. Chem. Phys.* **130**:244308.
24. Duarte, F., and A. Toro-Labbé. 2011. *J. Phys. Chem. A.* **115**:3050–3059.
25. Cerón, M. L., E. Echegaray, S. Gutiérrez-Oliva, and B. Herrera, A. Toro-Labbé. 2011. *Science China: Chemistry*, **54**:1982–1988.
26. Cerón, M. L., B. Herrera, P. Araya, F. Gracia, and A. Toro-Labbé. 2011. *J. Mol. Model.* **17**:1625–1633.
27. Jaque, P., A. Toro-Labbé, P. Politzer, and P. Geerlings. 2008. *Chem. Phys. Lett.* **456**:135–140.
28. Zewail, A. H. 2000. *Pure Appl. Chem.*, **72**:2219–2231.
29. Zewail, A. H. 2000. *J. Phys. Chem. A* **104**:5660–5694.
30. Kwon, O. H., and A. H. Zewail. 2007. *Proc. Natl. Acad. Sci. USA*. **104**:8703–8708.
31. Chermette, H. 1999. *J. Comput. Chem.*, **20**:129–154.
32. Geerlings, P., F. De Proft, and W. Langenaeker. 2003. *Chem. Rev.* **103**:1793–1873.
33. Chattaraj, P. K., U. Sarkar, and D. R. Roy. 2006. *Chem. Rev.* **106**:2065–2091.

34. Sen, K. D. (Editor). 2001. *Reviews of Modern Quantum Chemistry: A Celebration of the Contributions of Robert G. Parr*. Singapore: World Scientific.
35. Toro-Labbé, A. (Editor). 2007. *Theoretical Aspects of Chemical Reactivity*. Elsevier.
36. Chattaraj, P. K. (Editor). 2009. *Chemical Reactivity Theory: A Density Functional View*. Boca Raton, FL: CRC Press.
37. Sanderson, R.T. 1951. *Science*, **114**:670–672.
38. Sanderson, R. T. 1971. *Chemical Bonds and Bond Energy*. New York: Academic Press.
39. Koopmans, T. 1934. *Physica*. **1**:104–113.
40. Boys, S., and F. Bernardi. 1970. *Molec. Phys*. **19**:553–566.
41. Simón, S., M. Durán, and J. J. Dannenberg. 1996. *J. Chem. Phys.* **105**:11024–11031.
42. Reed, A., L. Curtiss, and F. Weinhold. 1988. *Chem. Rev.*, **88**:899–926.

13 All-Atom Computation of Vertical and Adiabatic Ionization Energy of the Aqueous Hydroxide Anion

Jun Cheng and Michiel Sprik

CONTENTS

13.1 INTRODUCTION

The redox potentials measured by electrochemical experiment can be regarded as relative adiabatic ionization energies. Photoemission spectroscopy (PES) also probes the ionization of species, but the measured energy differences are absolute vertical ionization energies referred to vacuum. Correlating these two complementary ionization energies has been a long-standing ideal of fundamental physical electrochemistry.[1–4] Such comparisons have become much more precise due to recent progress in PES techniques for liquids and solutions.[5,6] Of special interest are small aqueous anions, such as hydroxide and halide anions. Work by the Berlin/Heidelberg group[5] has confirmed the ideas of Delahay about the ionization of these elementary aqueous anions.[1–3] Vertical electron detachment levels lie much deeper than the adiabatic redox level. The difference in the case of OH^- is as much as 2.9 eV. In fact, at –9.2 eV relative to vacuum, the OH^- vertical detachment level is separated by only 0.7 eV from the valence band maximum (VBM) of liquid water, as inferred from the PES threshold of liquid water (9.9 eV).[5]

Meaningful comparison of redox potentials and PES energies depends on more than accurate measurement of these quantities. What is needed in addition is a reliable estimate of the absolute value of the potential of the standard hydrogen electrode (SHE). Measurement of the absolute potential of the SHE requires the definition of a suitable reference state of the electron. Trasatti proposed to use an electron at rest just outside the surface of the electrolytic solution as reference.[7,8] With this choice, the absolute potential of the SHE can be interpreted as a work function, which makes it amenable to measurement by electrochemical techniques.[7–10] This definition is also compatible with the PES experiment. The value recommended in [7] is 4.44 V. This estimate seems to be accepted now in the electrochemical literature as correct within an error margin in the 20-mV range.[10] With this absolute SHE potential, the reduction potential of the $OH^\bullet/OH^-$ couple (+1.9 V vs. SHE)[11] translates to –6.3 eV relative to vacuum, which, compared to the –9.2 eV of the vertical detachment energy, leads to the 2.9 eV of relaxation energy mentioned above. The other way around, we can now assign a formal potential of 4.8 V vs. SHE to the vertical detachment level of OH^-, which is an extraordinarily high positive potential on the electrochemical scale.

The impressive improvement in accuracy of experiment-derived solution phase energy level diagrams has set uncompromising benchmarks for electronic structure calculation. As is already evident from the huge relaxation energies, solvent effects are crucial. The standard approach in quantum chemistry to account for solvation is to use an implicit (continuum) solvent model augmented, where necessary, with an appropriate number of explicit water molecules.[12–14] This approach is remarkably successful in reproducing reduction potentials. Combined with the Marcus theory of electron transfer,[15,16] the continuum theory is also capable of a semiquantitative explanation of the large relaxation energies and, hence, the PES energies, as was already pointed by Delahay.[2]

The challenge was also taken up by the density functional theory–based molecular dynamics (DFTMD) community. DFTMD is an all-atom method treating solute and solvent at the same level of theory. DFTMD allows for the sampling of solvent fluctuations and coupling to the extended electronic states of the solvent, effects that are excluded from implicit solvent schemes. However, the elimination of these constraints proved to be problematic for the description of the free radicals formed by oxidation of stable closed-shell anions. This is the conclusion of a number of DFTMD investigations of the aqueous solvation of the hydroxyl ($OH^•$) radical.[17–22] All of these simulations employed the generalized gradient approximation (GGA). The $OH^•$ radical showed a preference for hemibonded coordination to a special water molecule. While the hemibonded dimer is probably a minority species,[23] the GGA has a definite tendency to overstabilize this conformation. This is generally blamed on the delocalization error in the GGA (for a dissenting view, see [21]).

The question of the energy level diagram of the $OH^•/OH^-$ couple was addressed in a DFTMD study by our group reported in [22]. Similar methods have also been applied to aqueous transition metal complexes.[24,25] We found that application of the GGA leads to serious underestimation of the reduction potential and the vertical ionization potential (IP) of OH^- (aq). The reduction potential computed using Becke–Lee–Yang–Parr (BLYP)[26,27] was too small by at least 0.5 eV. We attributed this error to overestimation of the stability of the oxidized state (the radical) as a result of hemibonding. Finite-system-size errors, according to our argument in [22], mainly affect the vertical IP, increasing the discrepancy with experiment for this quantity to 1 eV. We also observed a pronounced asymmetry in the position of the vertical detachment level of OH^- (minus the IP) and the attachment level of $OH^•$ relative to the computed redox level. This effect was seen as evidence for nonlinearity in the response of the solvent to the change of oxidation state of the solute.

The $OH^•/OH^-$ energy levels in [22] were computed directly with respect to SHE using a computational hydrogen electrode scheme developed in our group. The key step in this scheme is reversible insertion of a proton in the aqueous DFTMD model system. The calculation of [22] was an early application of this scheme, which was adjusted and refined in subsequent more technical publications.[28,29] The latest version of the scheme finds a redox potential of 1.3 V *versus* SHE in the BLYP approximation compared to 1.9 V of experiment.[30] A far more serious shortcoming of [22] is, in hindsight, the omission of the calculation of the vertical IP of liquid water. This calculation has been completed very recently and turned out to be quite revealing.[31] The vertical IP of water according to BLYP was computed to be 3.5 eV smaller than measured by experiment. The error in the position of the VBM of liquid water exceeds by far the underestimation of the adiabatic or vertical IP of OH^-(aq). This observation made us reconsider our view about the main origin of the failure of the GGA to describe the ionization of the hydroxide ion, which must be sought, we now believe, in spurious interaction with the misplaced valence band of water.

The topic of this chapter is defined by this change of mind. After a revision of our molecular dynamics hydrogen electrode scheme, starting from its foundation in the formal theory of absolute electrode potentials, we revisit the calculation of the $OH^•/OH^-$ energy level diagram in [22]. The diagram will be extended with the band edges of liquid water, as obtained in [31], which will be the basis of our analysis of the flawed performance of the GGA for the computation of ionization potentials in solution. Further support for our argument is provided by the results of DFTMD calculations of the same system using a hybrid functional (HSE06[32]) containing a fraction of the exact exchange. All these data will be combined in a level diagram analyzed and discussed in Section 13.6. We also comment on the role of all-atom calculations in general and DFTMD in particular in the numerical study of redox properties.

13.2 CHEMICAL REACTIONS

The reversible potential *vs* SHE for the reduction of a neutral free radical $X^{\bullet}$ to the anion X^- in aqueous solution is defined by the free-energy change of the reaction

$$X^{\bullet}(s) + \frac{1}{2}H_2(g) \rightarrow X^-(s) + H^+(s) \tag{13.1}$$

where s is a generic notation for the aqueous phase and g is that for the gas phase. The electron accepted by the radical is supplied by the oxidation of a gas-phase H_2 molecule. Adopting microscopic units, the standard free-energy change $\Delta_{red}G^\circ$ of reaction 1 is specified in terms of an energy per particle measured in electron volts. Dividing by the charge of an electron ($-e_0$), we obtain the standard SHE potential

$$U^\circ_{X^{\bullet}/X^-}(\text{she}) = -\frac{\Delta_{red}G^\circ}{e_0}. \tag{13.2}$$

Free radicals such as $OH^{\bullet}$ are highly reactive. Reaction 1 is therefore strongly exergonic, and U° is large and positive ($U^\circ = 1.9$ V for the reduction of the aqueous hydroxyl radical[11]).

The anion X^- could also have been produced by acid dissociation of the hydride XH, establishing a link between redox and acid base chemistry. This link is best visualized by the proton-coupled electron transfer triangle familiar from physical organic chemistry.[33] The triangle for XH is given in Figure 13.1. Starting with the acid dissociation (horizontal arrow in Figure 13.1)

$$XH(s) \rightarrow X^-(s) + H^+(s) \tag{13.3}$$

and then transferring an electron from the conjugate base (X^-) to the aqueous proton, reducing it to (half) a H_2 molecule (vertical arrow)

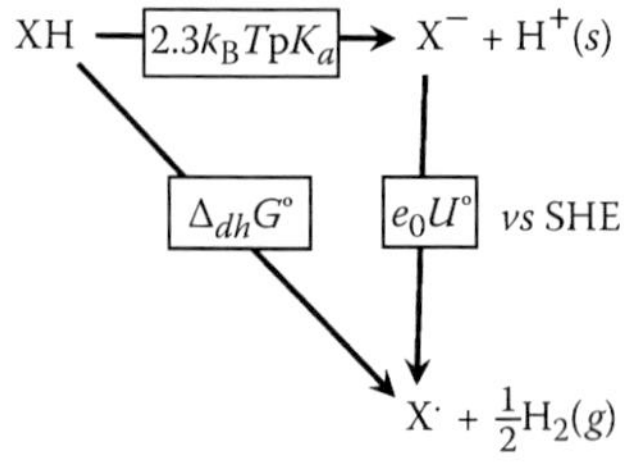

FIGURE 13.1 Relation between the dehydrogenation free energy $\Delta_{dh}G^\circ$ of a hydride XH, its pK_a and the standard reduction potential U° of the conjugate base X^-.

$$X^-(s) + H^+(s) \rightarrow X^\bullet(s) + \frac{1}{2}H_2(g) \tag{13.4}$$

should give the same product as direct dehydrogenation of XH (diagonal arrow)

$$XH(s) \rightarrow X^\bullet(s) + \frac{1}{2}H_2(g). \tag{13.5}$$

The free energy of solvation of the proton is the common reference for the equilibrium constant of both reactions 3 and 4. The reverse of reaction 4 is reaction 1, defining the redox potential of the $X^\bullet/X^-$ couple (Equation 13.2). Expressing the equilibrium constant of reaction 3 in terms of the pK_a of XH, Hess law requires that

$$\Delta_{dh}G^\circ_{XH} = 2.30k_BTpK_{aXH} + e_0U^\circ_{X^\bullet/X^-}(\text{she}) \tag{13.6}$$

where $\Delta_{dh}G^\circ_{XH}$ is the standard free-energy change of reaction 5.

The triangle relation of Equation 13.6 is a strong and most useful constraint for computation. The methods for calculation of pK_a and redox potentials must be consistent and satisfy Equation 13.6 by construction. Quantum chemistry schemes based on implicit solvent models normally meet this consistency requirement.[12,13,34] For all-atom schemes, this is much harder to achieve. This is ultimately why we decided to compute pK_a by completely removing the acid proton from solution[28,29] rather than transferring it to the solvent, which would be the natural approach in DFTMD.[35,36]

13.3 ABSOLUTE REDOX POTENTIALS

Equation 13.2 is the fundamental equation of equilibrium electrochemistry. It is derived by equating electrical and chemical work or alternatively from Nernst equation.[9,37–39] It states that the standard free-energy change $\Delta_{red}G^\circ$ of reaction 1 can be directly measured as the open-circuit potential $U^\circ_{X^\bullet/X^-}$ (SHE) of an electrochemical cell with $X^\bullet$ reduction taking place at the cathode and H_2 oxidation at the anode. The other way around, we can predict open-circuit potentials from the standard chemical potentials of the species involved in the cell reaction. Electron transfer inevitably creates or eliminates ions. The standard chemical potentials of ions, in essence, their solvation free energies, are listed in thermodynamic tables relative to the aqueous proton.[12,13,40] This is the free energy of the product ion pair of reaction 1 and is all that is needed to determine the reaction free energy.

To relate the open-circuit potential to the ionization energies measured in PES experiments, redox potentials must be converted to an absolute electrode potential referred to vacuum. With the same vacuum reference for the $X^\bullet/X^-$ and the H^+/H_2 couple, this leads to separation of the potential *vs* SHE in two absolute potentials

$$U^\circ_{X^\bullet/X^-}(\text{she}) = U^\circ_{X^\bullet/X^-}(\text{abs}) - U^\circ_{H^+/H_2}(\text{abs}). \tag{13.7}$$

The absolute potentials on the right-hand side are identified with single electrode potentials according to the definition of Trasatti.[7,8] These potentials can be expressed in terms of electronic and ionic work functions. The following two sections are a summary of the heuristic derivation of these expressions, as presented by Fawcett in [9].

13.3.1 Electrochemical Potential

The thermodynamic potential controlling electrochemical equilibrium is the electrochemical potential $\tilde{\mu}_i^\alpha$ defined for component i in phase α by

$$\tilde{\mu}_i^\alpha = \frac{\partial G^\alpha}{\partial n_i}. \tag{13.8}$$

Here, n_i is the quantity (mass) of component i in phase α with Gibbs free energy G^α. Electrochemical potentials are separated in a chemical and electrical contribution

$$\tilde{\mu}_i^\alpha = \mu_i^\alpha + q_i\phi^\alpha. \tag{13.9}$$

μ_i^α is the chemical potential of component i in phase α, q_i is its charge, and φ^α is the inner potential of the phase, which is also referred to as Galvani potential. In absolute microscopic units, n_i in Equation 12.8 is the number of particles of species i, and q_i is the charge of a single particle. The activity dependence of the chemical potential in these units is given by the expression

$$\mu_i^\alpha = \mu_i^{\alpha,\circ} + k_B T \ln a_i^\alpha \tag{13.10}$$

where a_i^α is the activity of i in phase α. In ideal solutions, $a_i^\alpha = \rho_i^\alpha / c^\circ$, where ρ_i^α is the number density and $c^\circ = 1$ mol dm^{-3} is the standard concentration. $\mu_i^{\alpha,\circ}$ is the corresponding standard chemical potential at temperature T.

Equations 13.8 and 13.9 only make sense if phase α is finite and conducting. Only for finite bodies can we define the increase in free energy due the addition of macroscopic amounts of *charged* particles, as implied by Equation 13.8. Excess charge in conductors accumulates at the surface. The potential inside a conductor is constant and given by φ^α. The Galvani potential accounts therefore for charge imbalances at the surface only. These also include possible surface dipoles. To make this explicit, the Galvani potential is further resolved in a surface dipole potential χ^α and the outer potential ψ^α, i.e.,

$$\phi^\alpha = \psi^\alpha + \chi^\alpha. \tag{13.11}$$

The outer potential ψ^α, or Volta potential, is measured at a point in vacuum "just outside" the surface of phase α. This point is chosen close enough to the surface that (almost) all the work to bring a test charge from infinity toward a surface with a net charge density and dipole moment has been done but still at a sufficiently long distance away to be out of reach of image forces.[38] This is the common reference point for the work function W_i^α of a charged particle i, specifying the reversible work needed to take it out of phase α.

The relation between chemical potentials and work functions is at the very heart of the theory of electrode potentials. The electrochemical potential $\tilde{\mu}_i^\alpha$ is the negative of the reversible work for transferring a particle from solution all the way to infinity. A separate potential is introduced for the work of bringing the particle to just outside the surface of a phase. This is the real potential defined as

$$\alpha_i^\alpha = \tilde{\mu}_i^\alpha - q_i\psi^\alpha = \mu_i^\alpha + q_i\chi^\alpha. \tag{13.12}$$

The work function W_i^α is the difference between the standard real potential $\alpha_i^{\alpha,\circ}$ and the standard chemical potential $\mu_i^{g,\circ}$ of species i in the gas phase[9]

$$W_i^\alpha = \mu_i^{g,\circ} - \alpha_i^{\alpha,\circ}. \tag{13.13}$$

The work function should be distinguished from the solvation free energy

$$\Delta_s G_i^{\alpha,\circ} = \mu_i^{\alpha,\circ} - \mu_i^{g,\circ} \tag{13.14}$$

which is based on a comparison of chemical potentials. The difference is a change of sign and the contribution of the surface potential

$$\Delta_s G_i^{\alpha,\circ} = -\left(W_i^\alpha + q_i\chi^\alpha\right). \tag{13.15}$$

The solvation free energy is the negative of the work function for neutral solutes ($q_i = 0$) but not for ionic solutes.

After this preparation, we are now ready to write down one of the central equations of this review, the expression for the real potential of the aqueous proton. Rearranging Equation 13.13, the standard real potential can be written as

$$\alpha_i^{\alpha,\circ} = -W_i^\alpha + \mu_i^{g,\circ}. \tag{13.16}$$

The form of Equation 13.16 is familiar from the solution chemistry of neutral species. The chemical potential in solution is separated in a contribution from solvation (the excess chemical potential) and a gas-phase term describing the chemical reactivity of the species.[41] The chemical reference state for hydrogen is the hydrogen molecule. $\mu_i^{g,\circ}$ in Equation 13.16 is therefore the free energy for the formation of a proton from H_2 in the gas phase, i.e.,

$$\frac{1}{2}\mathrm{H_2(g)} \rightarrow \mathrm{H^+(g)} + e^-(\mathrm{vac}). \tag{13.17}$$

The standard free-energy change of reaction 17 can be computed from the free energy for dissociating a H_2 molecule in $H^\bullet$ atoms and the ionization potential of $H^\bullet$. Denoting this energy by $\Delta_f G^{g,\circ}_{H^+}$ and substituting in Equation 13.16. we have

$$\alpha^{\circ}_{H^+} = -W_{H^+} + \Delta_f G^{g,\circ}_{H^+} \tag{13.18}$$

where we have suppressed the superindex for the aqueous phase. The value of $\Delta_f G^{g,\circ}_{H^+}$ at $T = 298$ K is given in thermodynamic tables of gas-phase chemistry as 15.72 eV. However, in solution chemistry, it is convenient to use the same reference for activity in solution and the gas phase.[12,40] This makes the work function (Equation 13.13) and solvation free energy (Equation 13.14) independent of the definition of standard state. The gas-phase activity is therefore represented relative to standard concentration and not standard pressure ($p^\circ = 1$ bar). At $T = 298$ K, the difference, the so-called standard state compression term, is given by $k_B T \ln(RTc)^\circ/p^\circ) = 82$ meV. The $^\circ$ symbol is officially reserved for the pressure-based reference of gas-phase chemistry.[40] Rather than introducing yet another superscript, we will stick with our unconventional notation. This means that $\Delta_f G^{g,\circ}_{H^+}$ will have to be adjusted to 15.81 eV. The discussion of the value of W_{H^+} will be deferred to Section 13.3.4.

We close this section with a reminder of a fundamental issue in electrochemistry: Not all the quantities in Equations 13.8 through 13.13 are accessible to measurement by electrochemical or thermodynamic methods.[9,37–39] Only the electrochemical potential $\left(\tilde{\mu}_i^\alpha\right)$, the work function ($W_i^\alpha$) or equivalently the real potential (α_i^α) and the Volta potential (ψ_α) are. Equations 13.9, 13.11, and 13.13 are therefore formal resolutions. It is not possible to assign actual values to the separate terms, the chemical potential (μ_i^α), the Galvani potential (φ^α), nor the surface potential (χ^α), without making "extrathermodynamic" assumptions. These quantities must therefore be considered "unphysical," at least from the point of view of thermodynamics.[42] This statement, which is called the "Gibbs–Guggenheim Principle" in [42], is often met with disbelief from theoretical and computational chemists, particularly in the case of the chemical potential (Equation 13.10). The standard chemical potential $\mu_i^{\alpha,\circ}$ is essentially the (absolute) solvation free energy $\Delta_s G_i^\alpha$ of species i. One would hope that a molecular simulation contains all information needed to compute $\Delta_s G_i^\alpha$. Indeed, there seems to be a way around this thermodynamic verdict for computation and also mass spectroscopic.[9,10,12,13,34,43] This continues to be, however, hazardous territory, particularly for DFT calculations in periodic systems.[44–46]

13.3.2 Potentials as Work Functions

The Nernst equation for the reversible potential of $X^\bullet/X^-$ *vs* SHE is obtained by resolving the reaction free energy in Equation 13.2 in the chemical potentials of the components

$$e_0 U_{X^\bullet/X^-}(\text{she}) = \left(\mu^{\circ}_{X^\bullet} - \mu^{\circ}_{X^-} + k_B T \ln \frac{a_{X^\bullet}}{a_{X^-}}\right) - \left(\mu^{\circ}_{H^+} - \frac{1}{2}\mu^{\circ}_{H_2}\right) \tag{13.19}$$

where we have assumed standard conditions for the hydrogen electrode. Chemical potentials for the same half-reaction have been grouped together in Equation 13.19. Adding the surface potential of water denoted by χ^S to the terms inside each of the brackets leads to an expression for U of the form of Equation 13.7, with the potentials given by

$$e_0 U_{X^\bullet/X^-}(\text{abs}) = \mu^\circ_{X^\bullet} - \left(\mu^\circ_{X^-} - e_0\chi^S\right) + k_B T \ln \frac{a_{X^\bullet}}{a_{X^-}} \quad (13.20)$$

$$e_0 U^\circ_{H^+/H_2}(\text{abs}) = \mu^\circ_{H^+} + e_0\chi^S - \frac{1}{2}\mu^{g,\circ}_{H_2}. \quad (13.21)$$

The potentials of Equations 13.20 and 13.21 depend on the properties of one electrode only and can be regarded as single-electrode potentials. A similar claim could already be made for the quantities between brackets in Equation 13.19. The potentials of Equations 13.20 and 13.21, however, are work functions and can be interpreted as absolute electrode potentials, as anticipated by the notation. This is why the surface potential was added in. With this additional term, we can replace $\mu^\circ_{H^+}$ by the real potential $\alpha^\circ_{H^+}$ combining Equation 13.21 with Equation 13.12, i.e.,

$$e_0 U^\circ_{H^+/H_2}(\text{abs}) = \alpha^\circ_{H^+} - \frac{1}{2}\mu^{g,\circ}_{H_2}. \quad (13.22)$$

In Section 13.3.1, we have already derived an expression for the real potential of the aqueous proton. This is Equation 13.18. This involved choosing a reference state for the proton, for which we took the regular reference state for hydrogen in the gas phase, namely, molecular hydrogen. To be consistent, we must set therefore $\mu^{g,\circ}_{H_2} = 0$ in Equation 13.22. Then, with Equation 13.18, for the real potential, Equation 13.22 leads to

$$e_0 U^\circ_{H^+/H_2}(\text{abs}) = \Delta_f G^{g,\circ}_{H^+} - W_{H^+}. \quad (13.23)$$

As explained in [9], ion work functions are uniquely defined thermochemical constants that can be measured using electrochemical cells with air gaps. Both terms in Equation 13.23 are therefore known from experiment.

The potential for the X$^\bullet$/X half-reaction of Equation 13.20 can be similarly expressed in real potentials. The result is

$$e_0 U_{X^\bullet/X^-}(\text{abs}) = \alpha^\circ_{X^\bullet} - \alpha^\circ_{X^-} + k_B T \ln \frac{a_{X^\bullet}}{a_{X^-}}. \quad (13.24)$$

The first two terms can be interpreted as an electronic work function. Using Equation 13.16, we can write

$$\alpha^{\circ}_{X^{\cdot}} - \alpha^{\circ}_{X^{-}} = \mu^{g,\circ}_{X^{\cdot}} - \mu^{g,\circ}_{X^{-}} + W_{X^{-}} - W_{X^{\cdot}} \tag{13.25}$$

The right-hand side of Equation 13.25 represents a thermodynamic cycle consisting of desolvating the anion X^-, ionizing it in the gas phase and reinserting the product radical $X^{\cdot}$ in solution. This is the definition of the adiabatic ionization potential (AIP) of an aqueous species. However, instead of removing it all the way to infinity, the electron is left at the reference point used for work functions. With this electrochemical definition of ionization, the absolute standard potential for the $X^{\cdot}/X^-$ couple is, according to Equation 13.24, simply equal to the AIP, i.e.,

$$U^{\circ}_{X^{\cdot}/X^{-}}(\text{abs}) = \frac{\text{AIP}_{X^{-}}}{e_0}. \tag{13.26}$$

Substitution in Equation 13.19 with Equation 13.23 for the potential *vs* SHE gives

$$e_0 U^{\circ}_{X^{\cdot}/X^{-}}(\text{she}) = \text{AIP}_{X^{-}} + W_{H^+} - \Delta_f G^{g,\circ}_{H^+}. \tag{13.27}$$

The first term AIP_{X^-} specifies the work needed to extract an electron bound in X^-. Comparing Equation 13.27 through Equation 13.19, we see that the standard reaction free energy has been reinterpreted in terms of work functions, which is what will enable us to relate $U^{\circ}_{X^{\cdot}/X^{-}}$ to the PES experiment. The same relation is also the key to computation of $U^{\circ}_{X^{\cdot}/X^{-}}$.

13.3.3 Energy Level Picture

The absolute electrode potential of the $X^{\cdot}/X^-$ couple was identified in Section 13.3.2 with the AIP of the anion (Equation 13.26). This relation is as profound as it is intuitive. Equation 13.7 (or equivalently Equation 13.27) transforms reversible potentials to the vacuum scale, enabling us to compare electrochemical energies to ionization energies determined by PES. The ionization energies measured by PES are, however, vertical IPs. A vertical ionization potential will be referred to simply as IP. The relation of an AIP to the corresponding IP is explained by the Marcus–Hush theory.[15,16,47] The atomistic statistical mechanical formulation was worked out by Warshel[48,49] and others.[50,51] This formalism also applies to the general case of nonlinear solvent response. In view of the small size of the ion, breakdown of the linear approximation to the response of the solvent to a change in the charge of the aqueous hydroxide anion is a possibility, as we speculated in our previous DFTMD study of this system.[22] We therefore briefly revisit the nonlinear electron transfer theory, broadly following [51] (see [4] for a more spectroscopic perspective).

The theory can be summarized in two key relations linking the adiabatic ionization energy and the vertical detachment and attachment energies. To emphasize the generality of these relations, we change the notation and consider the oxidation of a species R to O, with R standing for the reduced state and O standing for the oxidized state. The vertical ionization potential (electron detachment energy) of R will be indicated by IP_R, and the vertical electron affinity (electron attachment energy) of O by EA_O. The adiabatic IP of R equals the adiabatic EA of O. To keep notation consistent with this symmetry, the redox free energy will be referred to as ΔA. These three quantities are related as

$$IP_R = \Delta A + \lambda_O \tag{13.28}$$

$$EA_o = \Delta A - \lambda_R \tag{13.29}$$

where λ_O and λ_R are the reorganization energies in response to the vertical oxidation of R and vertical reduction of O, respectively (see Figure 13.2). λ_O and λ_R are positive energies. IP_R and EA_O are normally positive as well, which means that ΔA is too.

Equations 13.28 and 13.29 seem self-evident when drawn as transitions between the total energy curves (Born–Oppenheimer surfaces), as shown in Figure 13.2 (see also [4]). However, the energy curves in Figure 13.2 are potentials of mean force or diabatic free-energy curves. ΔA, λ_O, and λ_R are therefore free energies. It is not immediately clear that free-energy curves can be treated similar to energy curves in the gas phase. Equations 13.28 and 13.29 are however rigorous and can be derived by taking the energy gap as reaction coordinate.[48–51] The equations familiar from the Marcus theory are recovered by assuming that the free-energy curves are quadratic (linear response regime). Again, as a consequence of the special properties of potentials of mean force for energy gaps, the second derivatives must be identical (oxidation state independent), and hence,

$$\lambda_o = \lambda_R \equiv \lambda. \tag{13.30}$$

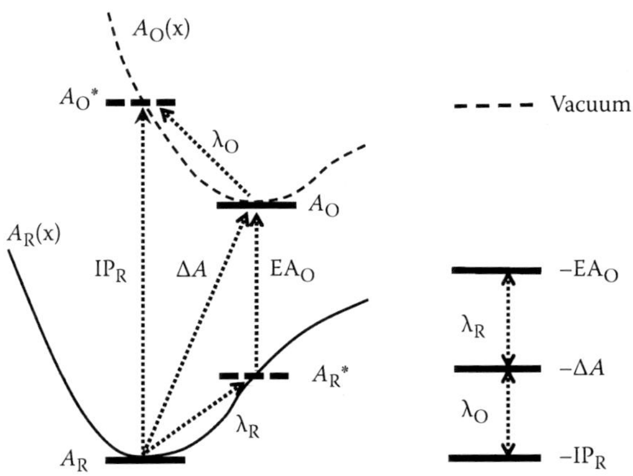

FIGURE 13.2 Free-energy curves $A(x)$ of the reduced (R) and oxidized species (O) plotted against an unspecified reaction coordinate x. Nonequilibrium states created by vertical electron detachment (oxidation) or attachment (reduction) are marked by a superscript *. The corresponding energy level scheme is indicated on the right.

Substituting in Equations 13.28 and 13.29 and adding and subtracting give

$$\Delta A = \frac{1}{2}(\mathrm{IP_R} + \mathrm{EA_O}) \tag{13.31}$$

$$\lambda = \frac{1}{2}(\mathrm{IP_R} - \mathrm{EA_O}). \tag{13.32}$$

These fundamental relations have important implications for energy level diagrams, such as will be constructed later. $-\mathrm{IP_R}$ and $-\mathrm{EA_O}$ are interpreted as orbital energies in a one-electron picture. $-\mathrm{IP_R}$ plays the role of the HOMO energy of R, and $-\mathrm{EA_O}$ is identified with the LUMO energy of O. These orbital energies will be indicated by ε_R and ε_O, respectively. Note that this is strictly only an analogy because the total energy differences and orbital energies are usually not the same in practical DFT. (In the case of the electron affinity, they are not even equal in principle.) $-\Delta A$ is interpreted as a thermodynamic energy level, effectively a chemical potential μ. Equations 13.28 and 13.29 impose a definite ordering of the levels, namely, $\varepsilon_R < \mu < \varepsilon_o$. In the linear regime, μ is positioned exactly midway in between the HOMO of R and LUMO of O (Equation 13.31). The R and O levels are separated by twice the reorganization energy (Equation 13.32, see also Figure 13.2).

For electrochemical applications, it is often convenient to represent the donor (HOMO) and acceptor (LUMO) levels, together with the chemical potential (μ) relative to the SHE. Equation 13.27 tells us how to do this for the chemical potential $\mu = -\Delta A = -\mathrm{AIP_R}$. According to Equation 13.28, the difference between μ and ε_R is a relaxation energy, i.e., a quantity not involving electron transfer (non-Faradaic), which is therefore invariant under a change in energy reference. The same applies to ε_O in virtue of Equation 13.29. The "SHE potential" equivalent of ε_R and ε_O is therefore obtained by replacing the AIP in the equation for the reversible potential by the vertical IP and EA, respectively, giving

$$e_0 U_R(\mathrm{she}) = \mathrm{IP_R} + W_{H^+} - \Delta_f G^{g,\circ}_{H^+} \tag{13.33}$$

$$e_0 U_O(\mathrm{she}) = \mathrm{EA_O} + W_{H^+} - \Delta_f G^{g,\circ}_{H^+}. \tag{13.34}$$

13.3.4 Experimental Absolute SHE

The question of the numerical value of the absolute hydrogen electrode potential is obviously an important one that has given rise to some controversy.[8] A key constraint for the definition of the absolute hydrogen electrode potential in Section 13.3.2 was that it can be determined by experiment.[7,8] Equation 13.23 relates $U^\circ_{H^+/H_2}(\mathrm{abs})$ to the work function of the proton. The relevant experimental data are listed in Table 13.1.

TABLE 13.1
Experimental Data for the Determination of the Absolute SHE Potential according to Equation 13.23[a]

$\frac{1}{2}H_2(g) \rightarrow H^+(g) + e^-(vac)$	$\Delta_f G^{g,\circ}_{H^+} = 15.81\text{ eV}$
Work function of H^+ (aq)[7]	$W_{H^+} = 11.36\text{ eV}$
Solvation free energy of H^+ (aq)[43]	$\Delta_s G^{\circ}_{H^+} = -11.53\text{ eV}$
Surface potential of water (from Equation 13.15)	$\chi^S = +0.17\text{ V}$
Recommended value for absolute SHE[7]	$U^{\circ}_{H^+/H_2}(\text{abs}) = 4.44\text{ V}$

Note: The standard state for H^+ is ideal gas at 1 mol dm^{-3}. The standard state for e^- is vacuum at rest.

[a] Also listed is the value for the absolute SHE, as recommended in [7]. The estimate for the surface potential has been obtained using the formalism of Section 13.3.1.

This is not the only possibility for obtaining an estimate of the absolute SHE potential from electrochemical experiment. As shown in [7], the absolute SHE potential can also be determined from measurements of the potential of zero charge and contact potential (Volta potential difference) of the interface of an inert metal electrode, such as Hg with an electrolytic solution. Surveying the experimental data available at the time Trasatti decided that $U^{\circ}_{H^+/H_2}(\text{abs}) = 4.44$ V is the best estimate.[7] This value was accepted by IUPAC as their recommended value. Fawcett has recently reviewed the procedure, also taking into account experimental data from mass spectroscopy experiments.[43] His conclusion was that only a small adjustment was necessary (from 4.44 to 4.42 V).

The mass spectroscopy data of [43] are of special interest because what is measured by this technique is the chemical potential $\mu^{\circ}_{H^+}$ rather than the real potential $\alpha^{\circ}_{H^+}$. (From a formal perspective, this statement has the status of an "extrathermodynamic" assumption.[9,12,13,34]) Recall from Equation 13.12 that $\alpha^{\circ}_{H^+}$ and $\mu^{\circ}_{H^+}$ are related to each other by the surface potential χ^S of water. A similar relation holds for the solvation free energy and the work function (Equation 13.15). Applying this relation to the experimental estimates in Table 13.1 yields the $\chi^S = 170$ meV value given in the table. 170 meV is within 50 meV consistent with estimates of χ^S from other sources.[10] It should be mentioned that there are good arguments that 4.44 V (or 4.42 V) is not an appropriate value for use in quantum chemistry calculations.[12,13,34] For the conversion of redox potentials computed using implicit solvent models, a value of $U^{\circ}_{H^+/H_2}(\text{abs})$ based on the chemical potential rather than the real potential is more consistent (see also [10]).

13.4 FULL ENERGY PERTURBATION

In Section 13.3.2, we derived a formal expression for a reversible potential *vs* SHE. To implement this expression in a calculation, we must find a way to compute the

AIP of the reduced state of the electroactive solute with the work function of the proton as energy reference. Electronic energy levels obtained from vertical ionization or orbital energies are aligned relative to the SHE using again the work function of the proton as energy reference (Equations 13.33 and 13.34). For extended model systems as used here, consistency between ionization energies and proton work function is absolutely crucial. In particular, it is not possible to just use the experimental value for the proton work function (Table 13.1). To ensure consistency, it is imperative to compute the work function of the proton in the same DFTMD supercell as used for the electronic energy calculation. The method we employ is a DFTMD implementation[52] of Warshel's microscopic version[48,49] of the Marcus theory of electron transfer.[16] Recently, this scheme has been generalized to reversible insertion of protons for the calculation of pK_a.[53] The method has been presented in detail in [28] and [29]. Here, we summarize the expressions used in calculations and refer to the two technical papers just mentioned for explanation.

13.4.1 Thermodynamic Integration

Free energies in electronic structure calculation are computed from total energy differences. The scheme used in our approach to convert total energy differences into a free-energy difference is based on thermodynamic integration of energy gaps. Thermodynamic integration is a free-energy perturbation (FEP) method. Reactants R are transformed into products P using an auxiliary (mapping) Hamiltonian $\mathcal{H}_\eta$, which is a linear combination of the Hamiltonian for the reactants and that for the products $\mathcal{H}_\eta = (1 - \eta)\mathcal{H}_R + \eta\mathcal{H}_P$. Here, η is a coupling parameter that is gradually increased from 0 (reactants) to 1 (products). Intermediate values $0 < \eta < 1$ correspond to hybrid systems that are a mixture of R and P. These systems have no physical counterpart in experiment but may be simulated by determining the atomic forces from the mapping Hamiltonian $\mathcal{H}_\eta$. Canonical averages over MD configurations obtained with Hamiltonian $\mathcal{H}_\eta$ will be denoted by brackets $\langle \ldots \rangle_\eta$. The total free-energy change ΔA is calculated from the integral of the vertical energy gap ΔE with respect to the coupling parameter

$$\Delta A = A(\eta = 1) - A(\eta = 0) = \int_0^1 d\eta \langle \Delta E \rangle_\eta. \tag{13.35}$$

The gap ΔE is defined as the total energy difference between the initial and final configurations, from which an electron e^-, a proton H^+, or both have been removed while keeping all other atoms fixed.

The coupling parameter integral Equation 13.35 converts vertical energy gaps (ΔE) into adiabatic gaps (ΔA). Vertical energy gaps for electron removal have an experimental counterpart. There is indeed a close correspondence between the thermodynamic integration scheme and the generalized Marcus theory of Section 13.3.3. We will demonstrate this for our model redox reaction: the reduction of a neutral radical $X^\bullet$ to the anion X^-. The reduced state R in this reaction is X^-, and the oxidized state O is $X^\bullet$. Conforming with the notation of [28] and [29], the vertical energy gap

between the oxidized and reduced states (solute plus solvent) will be indicated by $\Delta_{ox}E_{X^-}$, with the subscript specifying the solute. Equation 13.35 gives the corresponding adiabatic energy gap

$$\Delta_{ox}A_{X^-} = \int_0^1 d\eta \left\langle \Delta_{ox}E_{X^-} \right\rangle_\eta . \tag{13.36}$$

The integrand at the reactant end, $\left\langle \Delta_{ox}E_{X^-} \right\rangle_{\eta=0}$, is the vertical IP of X^- averaged over the thermal fluctuations of the solvent. The average gap $\left\langle \Delta_{ox}E_{X^-} \right\rangle_{\eta=0}$ computed by the DFTMD simulation would be equal to the experimental IP for the ideal case of the exact functional, if it were not for the uncertainty in the reference for electrostatic potentials in systems under periodic boundary conditions. The bias V_0 in the reference for DFTMD potentials as a result of this uncertainty was discussed in some detail in [28] and [29]. It acts as an offset $-e_0V_0$

$$\langle \Delta_{ox}E \rangle_{\eta=0} = \mathrm{IP}_{X^-} - e_0V_0 \tag{13.37}$$

where IP_{X^-} now stands for the true unbiased IP for a given functional. For stable condensed phase systems, $V_0 > 0$. Removing electrons from the periodic model system costs less energy than from the same system with an interface to vacuum.

The gap at the product end $\langle \Delta_{ox}E \rangle_{\eta=1}$ of the integral Equation 13.36 is a similarly biased estimate of the vertical electron affinity $\mathrm{EA}_{X^\bullet}$ of the oxidation product, radical $X^\bullet$, i.e.,

$$\langle \Delta_{ox}E \rangle_{\eta=1} = \mathrm{EA}_{X^\bullet} - e_0V_0. \tag{13.38}$$

Thermodynamic integration interpolates between these two extremes. In the limit of linear solvent response, the trapezium rule is exact, and the integral is the mean of the end points. The linear response approximation for the oxidation free energy is therefore determined by the IP of the reduced state and EA of the oxidized state[48,49,52]

$$\Delta_{ox}A_{X^-} = \frac{1}{2}\left(\mathrm{IP}_{X^-} + \mathrm{EA}_{X^\bullet}\right) - e_0V_0. \tag{13.39}$$

This equation reproduces Equation 13.31 subject to an offset that cannot be ignored (see below). In the nonlinear generalization of Equation 13.39, the $\frac{1}{2}(\mathrm{IP} + \mathrm{EA})$ term is replaced by the AIP, and we can write

$$\mathrm{AIP}_{X^-} = \Delta_{ox}A_{X^-} + e_0V_0 \tag{13.40}$$

with $\Delta_{ox}A_{X^-}$ given by Equation 13.36. Relaxation energies are estimated in the linear approximation using Equation 13.32 or in the nonlinear case by subtracting the adiabatic IP from the vertical IP and EA according to Equations 13.28 and 13.29.

The AIP is directly the absolute electrode potential of a redox couple (Equation 13.26). Unfortunately, it is not possible to use Equation 13.40 because V_0 is, in general, not known. The point of the hydrogen insertion method is that we do not need to know V_0 when computing the electrode potential relative to the SHE because the work function of the proton in periodic systems is also biased by V_0 but with the opposite sign. The challenge is therefore to find a procedure for the insertion/removal of a proton based on Equation 13.35 consistent with the FEP scheme for electron removal. This procedure should be equally suitable for the computation of acidity constants and was in fact originally developed for that purpose.[53] In the next section, we therefore will review first the proton insertion method for computation of pK_a.

Finally, we mention in passing that the free energies computed from thermodynamic integration are not strictly equivalent to the free energies of Section 13.3.3. The latter are the minima of the potentials of mean force, whereas thermodynamic integration yields the thermodynamic free energies of the diabatic states. The difference vanishes for quadratic free-energy curves. (Recall from the discussion in Section 13.3.3 that the harmonic potentials of mean force in the energy gap must have the same second derivative.) The small discrepancy for nonlinear systems will be ignored.

13.4.2 Acidity Constants

Acid dissociation is a homogeneous reaction transferring protons from an acid to a solvent molecule. The hydrogen electrode also involves proton transfer, but the exchange is between two phases, namely, the solution and the gas phase. As explained in Section 13.2, Hess law effectively forces us to also treat acid dissociation by reversible insertion of protons in solution. This can be achieved by a thermodynamic cycle transferring the acid proton from the acid to the point just outside the solution and reinserting it again as a solvated proton. Such a scheme was proposed in [53] and worked out in detail in [28] and [29]. The pK_a of the acid dissociation reaction $XH \rightarrow X^- + H^+$ is obtained from the following expression:

$$\begin{aligned} 2.30\, k_B T \mathrm{p}K_{a\mathrm{XH}} &= \mathrm{ADP}_{\mathrm{XH}} - W_{\mathrm{H}^+} \\ &\quad + k_B T \ln\left[c^\circ \Lambda^3_{\mathrm{H}^+}\right]. \end{aligned} \tag{13.41}$$

The first term, $\mathrm{ADP}_{\mathrm{XH}}$, is the adiabatic deprotonation energy of XH with the point just outside the solution interface as reference, and W_{H^+} is again the work function of the aqueous proton. The last term of Equation 13.41 accounts for the translational free energy generated by the dissociation.[29] This contribution has been approximated by the chemical potential of a free gas phase proton at standard concentration (c°). Λ_{H^+} is the thermal wavelength of the proton. This approximation is justified because of the small mass of the proton compared to the conjugate base X^-. Under ambient condition, the translational contribution amounts to a correction of $k_B T \ln\left[c^\circ \Lambda^3_{\mathrm{H}^+}\right] = -0.19\,\mathrm{eV}$ or −3.2 pK units.

The deprotonation free energy is estimated from a thermodynamic integral Equation 13.35 as

$$\Delta_{dp}A_{XH} = \int_0^1 d\eta \langle \Delta_{dp}E_{XH} \rangle_\eta \tag{13.42}$$

where the energy gap in the integrand now corresponds to the energy for vertical deprotonation of XH. The adiabatic deprotonation energy ADP_{XH} of Equation 13.41 is obtained from $\Delta_{dp}A_{XH}$ of Equation 13.42 according to[29]

$$ADP_{XH} = \Delta_{dp}A_{XH} - e_0V_0 - \Delta_{zp}E_{H(X)}. \tag{13.43}$$

Equation 13.43 is similar to Equation 13.40. However, the bias due to the shift in the electrostatic reference appears with the opposite sign because the particle that is removed has a positive charge. We have also added a zero-point motion correction. This is necessary because the DFTMD simulation as it is usually applied treats all nuclei as classical particles. The approach adopted in [29] is to subtract the zero-point energy of the acid proton attached to the base in full solution. All other protons are classical particles. This effective acid proton zero-point motion is indicated by $\Delta_{zp}E_{H(X)}$ in Equation 13.43. $\Delta_{zp}E_{H(X)}$ is estimated from the peaks in the (classical) velocity autocorrelation function of the acid proton as obtained from the DFTMD trajectory.

Vertical detachment or attachment of protons is not feasible in experiment. Still, we can calculate the vertical deprotonation energy $\Delta_{dp}E$ in Equation 13.42 by switching off the charge of the acid proton, transforming it instantaneously in a "dummy." This process can be viewed as the fictitious discharge reaction $XH \rightarrow Xd^-$, where d is a dummy proton d has no Coulombic interactions with the electrons and nuclei in the system. Instead d is subject to a harmonic restraining potential V_{restr}, keeping it close to the equilibrium position of the H^+ nucleus in the protonated system. This construction ensures that the deprotonation is reversible. Letting the system relax and switching the charge of the dummy back on, we obtain the protonation energy of the species to which the dummy is bound. The reasons this procedure can be carried out in practice are special to the geometry of hydrogen bonds.[28,53] Perturbations by a properly tuned restraining potential are minimal but can still have some effect on the deprotonation free energy. An approximate analytic expression for these perturbations has been worked out in the appendix of [29]. It is shown there that, provided the minimum and spring constants of the restraining potential are matched to the geometry and vibrational frequencies of the real acid proton, the error is small, compared to other sources of uncertainty in the calculation. Corrections for the restraining potential have therefore been omitted in Equation 13.43.

The expression for pK of Equation 13.41 also contains the work function of the proton. Computation of W_{H^+} is the key step in the proton insertion scheme. It is estimated by the deprotonation free energy of the hydronium cation (H_3O^+). The H_3O^+ ion is modeled by a (flexible) pyramidal structure stabilized by the restraining potential. Thus, identifying W_{H^+} with $ADP_{H_3O^+}$, we apply Equation 13.43 and find

$$W_{H^+} = \Delta_{dp}A_{H_3O^+} - e_0V_0 - \Delta_{zp}E_{H^+(OH_2)}. \tag{13.44}$$

However, the hydronium ion is not a regular acid. The trigonal H_3O^+ geometry on which our estimate for the work function (solvation free energy) is based is only one of the configurations that the solvated proton can adopt. The implications of this restriction are quantified below and discussed in detail in [29]. Substituting Equations 13.43 and 13.44 into Equation 13.41, we obtain for the pK of XH, i.e.,

$$\begin{aligned} 2.30 k_B T \mathrm{p}K_{a\mathrm{XH}} &= \Delta_{\mathrm{dp}} A_{\mathrm{XH}} - \Delta_{\mathrm{dp}} A_{\mathrm{H_3O^+}} \\ &\quad - \left(\Delta_{\mathrm{zp}} E_{\mathrm{H(X)}} - \Delta_{\mathrm{zp}} E_{\mathrm{H^+(OH_2)}} \right) \\ &\quad + k_B T \ln \left[c^\circ \Lambda_{\mathrm{H^+}}^3 \right]. \end{aligned} \tag{13.45}$$

Assuming that differences in zero-point motion of hydronium and acid can be ignored, pKa is the difference in deprotonation free energies adjusted by a constant. The constant, –3.2 units, is effectively the pKa of the constrained hydronium ion, because, in this case, the deprotonation integrals and zero-point motion terms rigorously cancel. The model H_3O^+ is therefore 1.5 pK units more acid than the pKa = –1.74 of the hydronium in the Brønsted theory.

13.4.3 Redox Potentials

Redox potentials are computed using Equation 13.27. The electronic and proton work functions are determined by evaluating the thermodynamic integrals for adiabatic oxidation of X^- (Equation 13.36) and reversible deprotonation of the hydronium ion (Equation 13.42, with XH = H_3O^+). Substitution of Equations 13.40 and 13.44 into Equation 13.27 gives

$$\begin{aligned} e_0 U^\circ_{\mathrm{X^\bullet/X^-}}(\mathrm{she}) &= \Delta_{\mathrm{ox}} A_{\mathrm{X^-}} + \Delta_{\mathrm{dp}} A_{\mathrm{H_3O^+}} \\ &\quad - \Delta_f G^{g,\circ}_{\mathrm{H^+}} - \Delta_{\mathrm{zp}} E_{\mathrm{H^+(OH_2)}}. \end{aligned} \tag{13.46}$$

In contrast to the expression for pK_a (Equation 13.45), the zero-point motion term in the work function of the proton (Equation 13.44) is not canceled out. The value that we use for $\Delta_{\mathrm{zp}} E_{\mathrm{H^+(OH_2)}}$ is 0.35 eV. This is a rough estimate that we determined by summing the zero-point motion of three vibrational modes with frequencies obtained from the stretching, bending, and librational peaks in the classical velocity autocorrelation function of a proton in liquid water. A correction of –0.35 V amounts to a substantial decrease in reduction potential. The $\Delta_f G^{g,\circ}_{\mathrm{H^+}}$ term in Equation 13.46 is given the experimental value of 15.81 eV (see Section 13.3.1).

To complete the free-energy triangle of Figure 13.1, we need an equivalent expression for the dehydrogenation free energy $\Delta_{\mathrm{dh}} G^\circ_{\mathrm{XH}}$ of XH, i.e., the free energy of reaction 5. Rather than relying on Hess law, we prefer an independent estimate obtained by simultaneously removing a proton and electron. In the notation we have introduced, the corresponding thermodynamic integral is written as

$$\Delta_{\mathrm{dh}} A_{\mathrm{XH}} = \int_0^1 \mathrm{d}\eta \langle \Delta_{\mathrm{dh}} E_{\mathrm{XH}} \rangle_\eta. \tag{13.47}$$

The vertical energy gap $\Delta_{\mathrm{dh}} E_{\mathrm{XH}}$ is the difference in total energy of $X^{\bullet}$ (product) relative to XH (reactant), with the energy of the solvent included. Because the net charge of the system remains the same, a bias in the electrostatic reference is of no concern. We must however again account for the free energy for the creation of one more translational degree of freedom, as well as the zero-point motion energy of the acid proton adding corrections similar to Equations 13.41 and 13.43, respectively. This gives

$$\begin{aligned} \Delta_{\mathrm{dh}} G^{\circ}_{\mathrm{XH}} &= \Delta_{\mathrm{dh}} A_{\mathrm{XH}} - \Delta_f G^{g,\circ}_{\mathrm{H}^+} \\ &\quad - \Delta_{\mathrm{zp}} E_{\mathrm{H(X)}} + k_{\mathrm{B}} T \ln\left[c^{\circ} \Lambda^3_{\mathrm{H}^+}\right]. \end{aligned} \tag{13.48}$$

The combined effect of the last two terms in Equation 13.48 is a downward adjustment of 0.54 eV, which is definitely not negligible. This correction is the sum of the corrections terms in Equation 13.45 for the pK and Equation 13.46 for the redox potential. Because the thermodynamic corrections add up, Hess law (Equation 13.6) requires that the thermodynamic integrals satisfy the triangle relation of Figure 13.1, i.e.,

$$\Delta_{\mathrm{dh}} A_{\mathrm{XH}} = \Delta_{\mathrm{dp}} A_{\mathrm{XH}} + \Delta_{\mathrm{ox}} A_{\mathrm{X}^-}. \tag{13.49}$$

Equation 13.49 is a most powerful condition to assess the statistical accuracy of the DFTMD sampling. The experience of the applications to date shows that the dehydrogenation free energy matches the free energy of sequential deprotonation and oxidation within an uncertainty of 100 meV for simulation runs of a duration accessible to DFTMD($\approx$ 10 ps).

13.4.4 Half-Reactions and Model Systems

Equation 13.46 is the working equation of our computational hydrogen electrode method. It was obtained by adding Equation 13.40 for the AIP to Equation 13.44 for the work function of the proton. The offset term $e_0 V_0$ canceled because it was assumed that an electron and proton are inserted/removed from the same model system. To obtain the results discussed in Section 13.5, however, we used separate DFTMD model systems for the ionization of the electroactive solute and the deprotonation of the reference hydronium. This approach was referred to as a half-reaction scheme in previous publications. V_0 is strongly dependent on system composition, and cancellation is no longer guaranteed in a half-reaction scheme. A similar argument applies to the application of a half-reaction scheme for computation of the acidity constant of a water molecule discussed in Section 13.5. The calculation is based on Equation 13.45, but the proton is removed from a water molecule in pure water and inserted again in pure water. The DFTMD model system was the familiar minimal periodic cubic cell of length 9.86 Å containing 32 water molecules corresponding to ambient density. One proton or hydrogen atom more or less represents only a minimal change in composition. Variation in V_0 for these very similar model systems is small and can be ignored. (We refer to [28] and [29] for a more quantitative justification of this claim.)

Electrostatic interactions in DFTMD are evaluated using Ewald summation methods. Our model systems are small, and either reactant or product state carries a net charge. Finite-system-size errors are therefore a concern. Finite-system-size errors for the cubic supercells used in the half-reaction scheme are fortunately of secondary importance (see also Section 13.5.1). The supercell contains at most only a single ion. The interactions between this ion, its periodic images, and homogeneous counter charge inserted by the Ewald sum (the Wigner potential) are screened by polarization induced in the solvent. As it turns out, in an MD cell of cubic symmetry, this leads to an almost perfect cancellation of errors.[54] The net charge of a supercell also made the total energy sensitive to shifts in the electrostatic reference. This effect, however, should be carefully distinguished from finite-system-size errors, which vanish in the limit of infinite system size. V_0, in contrast, approaches a constant for systems larger than a certain minimum size, which depends on the composition. This is an intrinsic periodic boundary effect that cannot be eliminated by increasing the system size.

13.5 IONIZATION OF AQUEOUS OH^-

In this section, we present the results of the application of the DFTMD-FEP method to the aqueous $OH^{\bullet}/OH^-$ redox reaction and the band edges of liquid water. All energies are represented as potentials *vs* SHE in a level diagram that is analyzed in Section 13.6.

13.5.1 Adiabatic Ionization Potential

The computation of the reversible potential of $OH^{\bullet}/OH^-$ are carried out in conjunction with the calculation of the acidity and the dehydrogenation free energy of a water molecule. The advantages of coupling these calculations were explained in Section 13.2. The results for the free energies were first reported in [30]. They are summarized in Table 13.2. The DFTMD simulation was carried out using the CP2k package.[55,56] Technical details of the electronic structure calculation (basis sets, etc.) can be found in the original papers.[29,30] The table lists the thermodynamic integrals $\Delta_{ox}A$ of Equation 13.36, $\Delta_{dp}A$ of Equation 13.42, and $\Delta_{dh}A$ of Equation 13.47, and the reaction free energies ΔG derived from these integrals using Equations 13.45, 13.46, and 13.48 (differences in zero-point corrections have been ignored). ΔA for oxidation and dehydrogenation was determined from the vertical energy gaps of five to six different values of the coupling parameter, each averaged over DFTMD trajectories of 5–10 ps. For deprotonation, only three integration points were used ($\eta = 0.0, 0.5, 1.0$), but the runs were considerably longer (40–60 ps for BLYP). The three thermodynamic integrals computed in this approximation satisfy Hess law (Equation 13.49) within 0.1 eV, which can therefore be regarded as an empirical estimate of the statistical accuracy of the calculation.

TABLE 13.2
Standard Free-Energy Changes for the Dehydrogenation of a Water Molecule in Bulk Solution (Energies in Electron Volts)

	BLYP		HSE06		
	ΔA	ΔG	ΔA	ΔG	Exp.
$H_2O \rightarrow OH^- + H^+$	16.34	0.7	16.29	0.7	0.83
$OH^- + H^+ \rightarrow OH^\bullet + \frac{1}{2}H_2$	2.10	1.3	2.52	1.7	1.9
$H_2O \rightarrow OH^\bullet + \frac{1}{2}H_2$	18.54	2.1	18.89	2.5	2.72
$H_3O^+ \rightarrow H_2O + H^+$	15.35	–0.19	15.29	–0.19	–0.10

Note: Thermodynamic integrals (Equation 13.35) are denoted by ΔA, with the reaction free energies calculated using these integrals by ΔG. The BLYP and HSE06 results were taken from [30]. Experimental redox energies are from [11]. The data in the last row are the deprotonation integral and acidity of the hydronium ion (see Section 13.4.2 and [29]).

The standard state for H_2O is the liquid. Conversion of the acidity constant of H_2O to pK units therefore yields pK_W. This is the convention used in Table 13.2. In practice, this amounts to decreasing the free energies for deprotonation and dehydrogenation by a further $k_BT\log([H_2O]/c^\circ) = 104$ me V. This energy, equivalent to –1.74 pK units, is, by definition, the Brønsted acidity of the hydronium. Both the BLYP and HSE06 calculations gave $pK_W = 12$, two pK units less than the experimental value of 14. The error in the oxidation free energies is significantly larger. The BLYP redox potentials are underestimated by as much as 0.6 V. The exact exchange component in the HSE06 functional reduces the error, but the redox free energies are still too low. Similar discrepancies were found in [28] and [29]. We are confident that the error can be attributed to shortcomings in the DFT functional and not to finite system size or periodic boundary effects. Limitations in system size could affect the free energy for the oxidation of the OH^- anion, which is a charge recombination reaction. However, the dehydrogenation reaction (the diagonal in Figure 13.1), which involves only neutral species, suffers from the same problem. We will return to this issue in Section 13.6.

The last entry in Table 13.2 gives $\Delta_{dp}A_{H_3O^+}$, which is the integral for deprotonation of the constrained hydronium used in the calculation of the work function of the proton. The value of this quantity, which is central in the hydrogen insertion method, was taken from [29]. Recall that, according to Equation 13.45, the pKa of H_3O^+ is a constant whatever $\Delta_{dp}A_{H_3O^+}$ happens to be. Of course, the work function of the proton depends on $\Delta_{dp}A_{H_3O^+}$ as is clear from Equation 13.44. Unfortunately, the unknown bias in the electrostatic reference (V_0) makes it impossible to determine the absolute value of W_{H^+}. However, as explained, only relative values are needed for computation of redox potential *vs* SHE and acidity pK_a. Shifts in the average electrostatic potential in a phase cancel. This also applies to the contribution of the surface potential χ^S, which differentiates between the solvation free energy and the work function (Equation 13.15). For the calculation of homogeneous redox potentials *vs* SHE, all these electrochemical subtleties are not relevant. Neither are they relevant in the measurement of redox potentials *vs* SHE. The distinction between the real and chemical potentials becomes important for the comparison to PES energies (see Section 13.6). The electrochemical formalism of Sections 13.3.1 and 13.3.2 is also crucial for the computation of potentials of electrodes, which are not under the control of equilibrium charge exchange with the electrolyte.[9,31]

13.5.2 Vertical Ionization Potential

The calculation of vertical ionization potentials is an intrinsic part of the calculation of adiabatic IPs in our FEP scheme. The vertical IP is the $\eta = 0$ point in the coupling parameter integral (Equation 13.37). Replacing the oxidation integral in Equation 13.46 by the average gap at $\eta = 0$ should give us therefore an estimate of the equivalent potential U_R (SHE) of the electron detachment level, as defined in Equation 13.33. The vertical IP is, by construction, consistent with the AIP, and the difference can be directly interpreted as the reorganization energy λ_O in the oxidized state (Equation 13.28). Similarly, the potential equivalent of the electron attachment level U_O (SHE) of Equation 13.38 is obtained from the point at $\eta = 1$ of the oxidation integral in Equation 13.46. The difference with the full integral is the reorganization free energy λ_R of Equation 13.29.

The results for U_R (SHE) for OH^- (aq) and U_O (SHE) for $OH^\bullet$ are listed in Table 13.3 under the column labeled ΔE. The columns labeled ε are alternative estimates obtained using the orbital energies, instead of the total energy difference. Substituting

TABLE 13.3
Vertical Ionization Potential of OH^-, Vertical Electron Affinity of $OH^\bullet$, and VBM and CBM of Pure Liquid Water Represented as Potentials *vs* SHE in V

		BLYP		HSE06		
		ΔE	ε	ΔE	ε	Exp.
OH^-	HOMO	2.1	1.41	3.29	2.99	4.8
$OH^\bullet$	LUMO	−0.9	0.59	−0.62	−0.09	−1.0[a]
H_2O	VBM	2.31	1.97	3.65	3.56	5.5
	CBM		−2.60		−3.24	−3.2

Note: The potentials in the column labeled ΔE have been computed from the total energy differences. Estimates computed from orbital energies (HOMO and LUMO, respectively) are listed under the heading ε. For the source of the experimental data, see text.

[a] Linear extrapolation, assuming $\lambda_R = \lambda_O = 2.9$ eV (see Figure 13.2).

the negative of the energy of the HOMO of OH^- in Equation 13.33 aligns this level relative to the SHE. The LUMO of $OH^{\bullet}$ is aligned using a similar procedure. The last column in Table 13.3 gives the corresponding experimental values represented again as potentials *vs* SHE. The transformation to the electrochemical scale was carried out by subtracting the experimental absolute potential of the SHE, given Table 13.1 from the PES data and a change of sign. The vertical attachment level of $OH^{\bullet}$ is not strictly an experimental value but an extrapolation. Assuming that the solvent response is linear, we estimate the reorganization energy on the reduced surface according to Equation 13.30 and obtain the vertical electron affinity using Equation 13.29 (see also Figure 13.2). Note that, by referring experimental ionization energies to the SHE rather than calculated ionization energies to vacuum, we avoid the question raised in [34] about the vacuum level for computation.

13.5.3 Band Edges of Liquid Water

From an electronic structure perspective, water can be regarded as a disordered wide gap insulator.[57] The fundamental gap as given in [57] is 8.7 eV. This value has become the consensus experimental reference for computational work on the electronic structure of liquid water.[58–60] Adding to the VBM at 9.9 eV below vacuum[5,6] places the conduction band minimum (CBM) at –1.2 eV relative to vacuum. It should be mentioned that the position and even definition of the CBM of water are the subject of some debate.[4,57] Coe et al. in [4] argue in favor of a CBM at –0.5 V. Converting the spectroscopic levels to SHE potentials, we find U_{VBM} (SHE) = 5.5 V and U_{CBM} (SHE) = –3.2 V (see also Table 13.3). On the electrochemical scale, these are extreme potentials, which implies that electrochemical activation of liquid water is not easy.

All-atom methods, at a purely technical level, make no difference between the IP of a solvated species and the pure liquid. The method applied in Section 13.5.2 should therefore be equally suitable to align the energy levels of the pure solvent to the SHE. Vertical energy gaps were sampled from a DFTMD trajectory of the cubic 32 H_2O molecule system, which was the basis for the OH^-, $OH^{\bullet}$, and H_3O^+ model systems (Section 13.4.4). The calculation was however less demanding. Fluctuations in the energy gap for the pure liquid are small compared to the fluctuations in the energy gaps of solutes. The result for the VBM obtained by comparing the energy for the removal of an electron to the free energy for the deprotonation of the hydronium can be found in Table 13.3. Alternatively, we could have used the average energy of the HOMO of the liquid as an estimate of the VBM. This crucial energy level will be indicated by ε_{HOMO}. The result is also listed in Table 13.3. For the CBM, we have only calculated the estimate based on ε_{LUMO}, which is the average energy of the LUMO of the liquid.

13.6 ENERGY LEVEL DIAGRAM

DFTMD simulation enables us to align the energy levels of solutes and solvent. Using the data of Tables 13.2 and 13.3, we have constructed such a level diagram for the $OH^{\bullet}/OH^-$ couple. The result is shown in Figure 13.3. The levels of the solvent, as published in [31], will be discussed first.

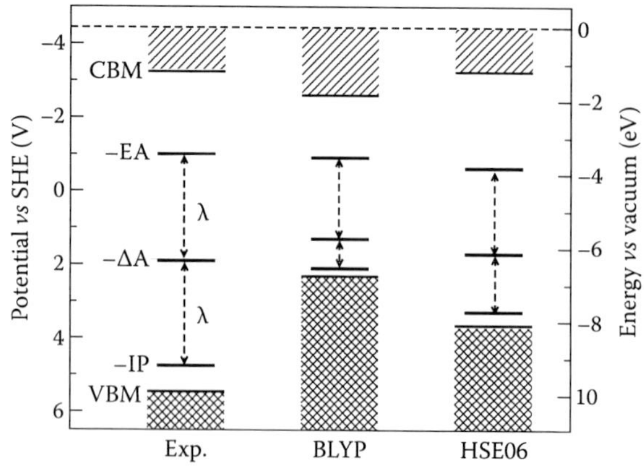

FIGURE 13.3 Energy levels of the $OH^{•}/OH^{-}$ couple aligned with the band edges of liquid water. The data for the vertical IP of OH^{-}, electron affinity EA of $OH^{•}$, VBM, and CBM of water are from Table 13.3. The $OH^{•}$ reduction potential (ΔA) is taken from Table 13.2. All levels, except the CBM, of water are computed from the total energy differences.

13.6.1 Solvent Levels

The most conspicuous feature of the level diagram is probably the contrast in the position of the VBM and CBM of liquid water. The GGA is notorious for underestimating band gaps in solids. The discrepancy becomes worse for larger band gaps.[61] Our results for water are in line with this expectation. The Kohn–Sham band gap ($\varepsilon_{LUMO} - \varepsilon_{HOMO}$) in the BLYP approximation is 4.6 eV too small, consistent with what has been observed in previous investigations.[58–60] However, our method also gives the absolute level positions. What this additional information reveals is that the underestimation of the band gap is predominantly caused by a failure to account for the energies of the occupied orbitals. The effective potential of the HOMO is 3.5 V less positive, which places the VBM 3.5 eV above the VBM obtained from PES. The error in positioning the LUMO, 1 eV below the CBM in experiment, is comparatively smaller. Application of HSE06 raises the LUMO to about the correct level. In addition, the HOMO level improves. It moves to more positive potentials but not as much as is needed. The difference with experiment is still 2 V.

The relation between vertical IP and EA and HOMO and LUMO energy is a fundamental issue in DFT.[62,63] For isolated molecules, it is well known that the absolute value of ε_{HOMO} can be several electron volts less than the ΔSCF IP using the same GGA functional. These two quantities should be identical in exact DFT (ionization theorem).[62] Could this be the reason for the unrealistically high position of the HOMO in the GGA? The results in Table 13.3 suggest that this is not the case. The IP result is equally bad. The gap between ε_{HOMO} and -IP, as computed for BLYP, is 0.7 eV, much less than the discrepancy with experiment. For HSE06, -IP and ε_{HOMO} are even closer.

A possible explanation for this, for a chemist, perhaps somewhat surprising observation can be found in [63]. There, it is shown that, for perfectly ordered solids,

ε_{HOMO} must be equal to -IP because of the extended nature of the orbitals in the energy bands. This remains true, even if the value is wrong. Our calculations suggest therefore that the behavior of liquid water is closer to that of a solid than that of a molecule. We note in this context that the experimental IP of liquid water (9.9 eV) is 2.7 eV smaller than the IP of a gas-phase water molecule (12.6 eV, as quoted in [62]), which gives an indication of the band dispersion in liquid water.

13.6.2 Solute Levels

The vertical levels of the solute follow the trend of the VBM and CBM of water. The vertical detachment level of OH^- ends up too high compared to experiment, whereas the vertical attachment level of $OH^•$ is in fair agreement with experiment. The suggestion that the vertical detachment level of OH^- is pushed up by the VBM of water is hard to avoid. The attachment level of $OH^•$ is much further away from the solvent VBM and, therefore, far less sensitive to perturbation by the VBM. We now believe that the proximity to the VBM of the solvent is the main reason for the underestimation of the IP of OH^-. This error is carried over to the AIP, i.e., the redox potential, as can be seen from Equation 13.31. With EA_O only marginally affected, the error in the redox potential (ΔA) is approximately half of the error in IP_R. Underestimation of the redox potential in the GGA is therefore mainly a generic solvent effect and not due to some self-interaction error specific to the open-shell character of $OH^•$, as we argued in [22]. Accordingly, the improved accuracy of HSE06 must be attributed to the increase in the gap between the IP of OH^- and water. The destabilization of hemi-bonded coordination of the OH^- radical is only a secondary effect.

The $OH^•/OH^-$ level diagram in the GGA shows a pronounced asymmetry in the position of detachment and attachment levels relative to the adiabatic level. Interpreting the gaps as reorganization energies following the nonlinear generalization of the Marcus theory of Section 13.3.3, this would lead to the conclusion that $\lambda_R \approx 3\lambda_O$. In [22] the asymmetry was explained as the result of nonlinear solvent relaxation. In light of the analysis of the underestimation of the redox potential, we must abandon this line of argument. The asymmetry, we now believe, is simply a reflection of the difference in energy separation of the vertical detachment and attachment levels from the VBM of the solvent. It is therefore a spurious electronic effect. The most direct evidence is the result for HSE06. While the ionization potentials are still somewhat underestimated, the addition of exact exchange effectively restores the symmetry of the level diagram. Relaxation of the solvent is essentially linear again.

A final more technical point is the analysis of the orbital energies. The result for the HOMO of OH^- is consistent with the IP for both BLYP and HSE06 (see again Table 13.3). The HOMO of OH^- in the GGA is located at 0.6 eV above the HOMO of water. The same gap is found for HSE06. The relative positions of the HOMO of OH^- and the HOMO of liquid water are therefore in good agreement with the PES estimate (0.7 eV), despite the large error in the absolute position. The discrepancy between HOMO and minus the IP for BLYP is 0.7 eV, which decreases to 0.3 for HSE06. Compared to isolated molecules, the deviations from the ionization theorem in solution must be considered rather modest.[62] The BLYP LUMO of $OH^•$ lies 1.5 eV below the vertical attachment level. For HSE06, the difference is 0.5 eV. Introduction of exact exchange moves the LUMO in the direction of the vertical attachment level (minus the electron affinity), and the results for the aqueous OH radical seem to be in accordance with this expectation.

More interesting is the comparison of the position of the LUMO of the oxidized state and the HOMO of the reduced state. For BLYP, the LUMO of $OH^{\bullet}$ lies 0.8 eV above the HOMO of OH^-. For an isolated molecule at fixed geometry, the LUMO of the oxidized state normally lies below the HOMO of the reduced state. This suggests that the reversal of the level ordering in the gas phase is due to the reorganization of the solvent. For HSE06, the gap has increased to 3.1 eV. The effect of exact exchange is rather extreme and probably enhanced by the condensed phase environment.

13.7 CONCLUSION AND OUTLOOK

The key point made in this chapter concerns the electronic levels of solvated species in the energy gap of the solvent. These levels are necessarily constrained by the band edges of the solvent. The separation between the vertical energy levels of the solute and the band edges of the solvent is therefore a factor in the calculation of redox potentials. If the vertical detachment energy of the reduced state is close to the valence band of the solvent, underestimation of the ionization potential of the solvent must lead to underestimation of the ionization potential of the solute. The vertical detachment level of the hydroxide anion is, according to the PES experiment, only 0.7 eV above the valence band edge of water.[5] The ionization potential of the hydroxide anion and the valence band of water are therefore correlated, and errors in the calculation of the VBM will affect the accuracy of calculations of the reduction potential of $OH^{\bullet}/OH^-$. As we have shown in a recent publication,[31] the ionization potential of liquid water is seriously underestimated in the GGA. Coupling between the localized state of the hydroxide and the extended states of water is inevitable and is, in our view, the origin of the underestimation of the reduction potential of $OH^{\bullet}$, as found in our previous DFTMD study of the ionization of the hydroxide anion reported in [22].

These observations are not new. A similar electronic correlation mechanism has been proposed as explanation for the underestimation of charge state transition energies of point defects in semiconductors.[64,65] Further support for the interaction with the solvent band structure is provided by the comparison to implicit solvent models, which omit such interactions. Errors in the calculation of reduction potentials are significantly smaller for the very same GGA functionals (see, for example, [14]). The unfavorable comparison to implicit solvent models raises the question why all-atom methods are used if the huge increase in computational costs only leads to deterioration of accuracy. The answer must be that there are situations where an all-atom approach is required. The reactivity and transport of excess electrons in water almost certainly involve conducting states as intermediates. Localization of holes in water is still subject to debate but could possibly play a role in transport[21] and reaction kinetics.[66] Electrochemical interfaces are another example of systems for which interactions between localized and extended states are important. The DFTMD/

FEP scheme has already been applied to solid–water interfaces with encouraging results.[30,31,67–70] Indeed, we see electrochemistry as the main area for application of the methods reviewed in this chapter.[31]

ACKNOWLEDGMENTS

Marialore Sulpizi, Joost VandeVondele, and Jochen Blumberger are acknowledged for their crucial contributions to the development and validation of the FDFTMD/FEP methods applied in this work. Jun Cheng is grateful for the financial support of Emmanuel College Cambridge and the Engineering and Physical Sciences Research Council, U.K. We also acknowledge support from the UKCP consortium for access to HECToR, U.K.'s high-end computing resource funded by the Research Councils.

REFERENCES

1. Von Burg, K., and P. Delahay. 1981. *Chem. Phys. Lett.* **78**:287–290.
2. Delahay, P., and K. Von Burg. 1981. *Chem. Phys. Lett.* **83**:250–254.
3. Delahay, P. 1982. *Acc. Chem. Res.* **15**:40–45.
4. Coe, J. V., A. D. Earhart, M. H. Cohen, G. J. Hoffman, H. W. Sarkas, and K. H. Bowen. 1997. *J. Chem. Phys.* **107**:6023–6031.
5. Winter, B., M. Faubel, I. V. Hertel, C. Pettenkofer, S. E. Bradforth, B. Jagoda-Cwiklik, L. Cwiklik, and P. Jungwirth. 2006. *J. Am. Chem. Soc.* **128**:3864–3865.
6. Seidel, R., S. Thümer, and B. Winter. 2011. *J. Phys. Chem. Lett.* **2**:633–641.
7. Trasatti, S. 1986. *Pure & Appl. Chem.* **58**:955–966.
8. Trasatti, S. 1990. *Electrochim. Acta.* **35**:269–271.
9. Fawcett, W. R. 2004. *Liquids, Solutions, and Interfaces*. Oxford: Oxford University Press.
10. Fawcett, W. R. 2008. *Langmuir.* **24**:9868–9875.
11. Stanbury, D. M. 1989. *Adv. Inorg. Chem.* **33**:69–138.
12. Kelly, C. P., C. J. Cramer, and D. Truhlar. 2006. *J. Phys. Chem. B.* **110**:16066–16081.
13. Kelly, C. P., C. J. Cramer, and D. Truhlar. 2007. *J. Phys. Chem. B.* **111**:408–422.
14. Jinnouchi, R., and A. B. Anderson. 2008. *J. Phys. Chem. C.* **112**:8747–8750.
15. Marcus, R. A. *J. Chem. Phys.* **1956**, *24*, 966–978.
16. Marcus, R. A. *Rev. Mod. Phys.* **1993**, *65*, 599–610.
17. Vassilev, P., M. J. Louwerse, and E. J. Baerends. 2004. *Chem. Phys. Lett.* **398**:212–216.
18. Vassilev, P., M. J. Louwerse, and E. J. Baerends. 2005. *J. Phys. Chem. B.* **109**:23605–23610.
19. Kalack, J. M., and A. P. Lyubartsev. 2005. *J. Phys. Chem. A.* **109**:378–386.
20. VandeVondele, J., and M. Sprik. 2005. *Phys. Chem. Chem. Phys.* **7**:1363–1367.
21. Codorniu-Hernández, E., and P. G. Kusalik. 2012. *J. Am. Chem. Soc.* **134**:532–538.
22. Adriaanse, C., M. Sulpizi, J. VandeVondele, and M. Sprik. 2009. *J. Am. Chem. Soc.* **131**:6046–6047.
23. Chipman, D. M. 2011. *J. Phys. Chem. A.* **115**:1161–1171.
24. Seidel, R., M. Faubel, B. Winter, and J. Blumberger. 2009. *J. Am. Chem. Soc.* **131**:16127–16137.
25. Moens, J., R. Seidel, P. Geerlings, M. Faubel, B. Winter, J. Blumberger. 2010. *J. Phys. Chem. B.* **114**:9173–9182.
26. Becke, A. D. 1988. *Phys. Rev. A.* **38**:3098.

27. Lee, C., W. Yang, and R. Parr. 1988. *Phys. Rev. B.* **37**:785.
28. Cheng, J., M. Sulpizi, M. Sprik. 2009. *J. Chem. Phys.* **131**:154504.
29. Costanzo, F., R. G. Della Valle, M. Sulpizi, and M. Sprik. 2011. *J. Chem. Phys.* **134**:244508.
30. Cheng, J., M. Sulpizi, J. Joost VandeVondele, and M. Sprik. 2012. *ChemCatChem.* **4**:636–640.
31. Cheng, J., and M. Sprik. 2012. *Phys. Chem. Chem. Phys.* **14**:11245–11267.
32. Krukau, A. V., O. A. Vydrov, A. F. Izmaylov, and G. E. Scuseria. 2006. *J. Chem. Phys.* **125**:224106.
33. Warren, J. J., T. A. Tronic, and J. M. Mayer. 2010. *Chem. Rev.* **110**:6961–7001.
34. Isse, A. A., and A. Gennaro. 2010. *J. Phys. Chem. B.* **114**:7894–7899.
35. Davies, J. E., N. L. Doltsinis, A. J. Kirby, C. D. Roussev, and M. Sprik. 2002. *J. Am. Chem. Soc.* **124**:6594.
36. Leung, K., I. M. B. Nielson, and L. J. Criscenti. 2009. *J. Am. Chem. Soc.* **131**:18358–18365.
37. Bard, A. J., and L. R. Faulkner. 2001. *Electrochemical Methods* (2nd ed); New York: John Wiley & Sons.
38. Bagotsky, V. S. 2006. *Fundamentals of Electrochemistry*. Hoboken, NJ: John Wiley & Sons.
39. Hamann, C. H., A. Hamnett, and W. Vielstich. 2007. *Electrochemistry*. Weinheim: Wiley-VCH.
40. Marcus, Y. 1987. *J. Chem. Soc. Faraday Trans. 1.* **83**:339–349.
41. Beck, T. L., M. E. Paulitis, and L. R. Pratt. 2006. *The potential distribution theorem and models of molecular solutions*. Cambridge: Cambridge University Press.
42. Pethica, B. A. 2007. *Phys. Chem. Chem. Phys.* **9**:6253–6262.
43. Tissandier, M. H., K. A. Cowen, W. Y. Feng, E. Gundlach, M. H. Cohen, A. D. Earhart, and J. V. Coe. 1998. *J. Phys. Chem. A.* **102**:7787–7794.
44. Hunt, P., and M. Sprik. 2005. *Comp. Phys. Comm.* **6**:1805.
45. Leung, K. 2010. *J. Phys. Chem. Lett.* **1**:496–499.
46. Kathmann, S. M., J. Kuo, C. J. Mundy, G. K. Schenter. 2011. *J. Phys. Chem. B.* **115**:4369–4377.
47. Hush, N. S. 1961. *Trans. Faraday Soc.* **57**:577–580.
48. King, G., and A. Warshel. 1990. *J. Chem. Phys.* **93**:8682.
49. Lee, F. S., Z. T. Chu, M. B. Bolger, and A. Warshel. 1992. *Protein Engineering.* **5**:215.
50. Tachiya, M. 1993. *J. Phys. Chem.* **97**:5911–5916.
51. Zhou, H.-X., and A. Szabo. 1995. *J. Chem. Phys.* **103**:3481–3494.
52. Blumberger, J., I. Tavernelli, M. L. Klein, and M. Sprik. 2006, *J. Chem. Phys.* **124**:064507.
53. Sulpizi, M., and M. Sprik. 2008. *Phys. Chem. Chem. Phys.* **10**:5238–5249.
54. Ayala, R., and M. Sprik. 2008. *J. Phys. Chem. B.* **112**:257–269.
55. The CP2K developers group, http://cp2k.berlios.de. 2008.
56. VandeVondele, J., M. Krack, F. Mohamed, M. Parrinello, T. Chassaing, and J. Hutter. 2005. *Comp. Phys. Comm.* **167**:103–128.
57. Bernas, A., C. Ferradini, and J.-P. Jay-Gerin. 1997. *Chem. Phys.* **222**:151–160.
58. Prendergast, D., J. C. Grossman, and G. Galli. 2005. *J. Chem. Phys.* **123**:014501.
59. Lu, D., F. Gygi, and G. Galli. 2008. *Phys. Rev. Lett.* **100**:147601.
60. Garbuio, V., M. Cascella, L. Reining, R. D. Sole, and O. Pulci. 2006. *Phys. Rev. Lett.* **97**:137402.
61. Marsman, M., J. Paier, A. Stroppa, and G. Kresse. 2008. *J. Phys.: Condens. Matter.* **20**:064201.
62. Chong, D. P., O. V. Gritsenko, and E. J. Baerends. 2002. *J. Chem. Phys.* **116**:1760–1772.
63. Mori-Sánchez, P., A. J. Cohen, and W. Yang. 2008. *Phys. Rev. Lett.* **100**:146401.
64. Komsa, H.-P., P. Broqvist, and A. Pasquarello. 2010. *Phys. Rev. B.* **81**:205118.

65. Alkauskas, A., P. Broqvist, and A. Pasquarello. 2011. *Phys. Status Solidi B.* **248**:775–789.
66. Marsalek, O., C. G. Elles, P. A. Pieniazek, E. Pluharova, J. VandeVondele, S. E. Bradforth, P. Jungwirth. 2011. *J. Chem. Phys.* **135**:224510.
67. Cheng, J., and M. Sprik. 2010. *Phys. Rev. B.* **82**:081406(R).
68. Cheng, J., and M. Sprik. 2010. *J. Chem. Theor. Comp.* **6**:880–889.
69. Gaigeot, M.-P., M. Sprik, and M. Sulpizi. 2012. *J. Phys.: Condens. Matter.* **24**:124106.
70. Sulpizi, M., M.-P. Gaigeot, and M. Sprik. 2012. *J. Chem. Theor. Comp.* **8**:137–147.

14 Vibrational Spectral Diffusion and Hydrogen Bonds in Normal and Supercritical Water

Amalendu Chandra

CONTENTS

14.1 INTRODUCTION

Water exhibits many peculiar properties that are generally attributed to its hydrogen-bonded network structure.[1,2] Some of these peculiar properties include negative volume of melting, density maximum, high dielectric constant, high mobility of protons and hydroxide ions, etc. It is also a ubiquitous solvent that easily dissolves ions under ambient conditions. However, water above its critical temperature (647 K) and pressure (22.1 MPa) behaves very differently from normal water.[3–5] For example, unlike water under ambient conditions, supercritical water behaves more like a nonpolar solvent that prefers to dissolve organic solutes rather than ionic solutes. More importantly, the density of supercritical water can be tuned from liquidlike to gaslike densities without any liquid–gas phase transition, leading to huge changes in its solvation properties. The altered solvent properties of supercritical water are believed to arise from changes in the hydrogen bond characteristics of the solvent as one moves from ambient to supercritical conditions.

Understanding the hydrogen bond properties of water has been a grand challenge for liquid state theories and molecular simulations. The first simulation of liquid water[6] was carried out more than four decades ago by using the method of Monte Carlo,[7] where the energies of different configurations were calculated from empirical interaction potentials. However, the many-body interactions and their participation in chemical and physical processes pose a challenge to the predefined empirical potentials of water, which usually rely on pair interactions. In addition, essentially all of the available empirical potentials are optimized for ambient water, and they are not readily applicable to the high-temperature supercritical phase. The technique of *ab initio* molecular dynamics solves the above problems of using empirical interaction potentials by incorporating a full quantum mechanical description of the interparticle interactions. In particular, the Car–Parrinello and Born–Oppenheimer molecular dynamics methods have enabled simulations of liquid water entirely from first principles.[8,9] These simulations treat electronic degrees of freedom through the quantum density functional theory and the ionic motion classically at a finite temperature, although other quantum mechanical methods for electronic structure calculations can also be used. Thus, no predefined interaction potential is used in this scheme, so that it is ideally suited to study complex liquids, such as water. Understandably, the *ab initio* molecular dynamics simulations are computationally much more expensive than classical simulations, which use empirical interaction potentials. However, recent advances in more efficient numerical algorithms combined with significant increase in high-performance computer power have made *ab initio* simulation a viable method for studying finite-temperature chemical and physical processes in water and other solvents from first principles.

A water molecule can engage very effectively in hydrogen bond formation with other water molecules. It has been found that the majority of water molecules form four hydrogen bonds oriented tetrahedrally on average, as shown in Figure 14.1. In fact, the microscopic structure of ambient water is characterized by the tetrahedral ordering of hydrogen bonds between water molecules, which is subject to fluctuations due to the thermal motion of atoms. The dynamics of these hydrogen bond fluctuations can be captured experimentally through time-dependent vibrational spectroscopy.[10] In these experiments, the dynamics of vibrational frequency fluctuations,

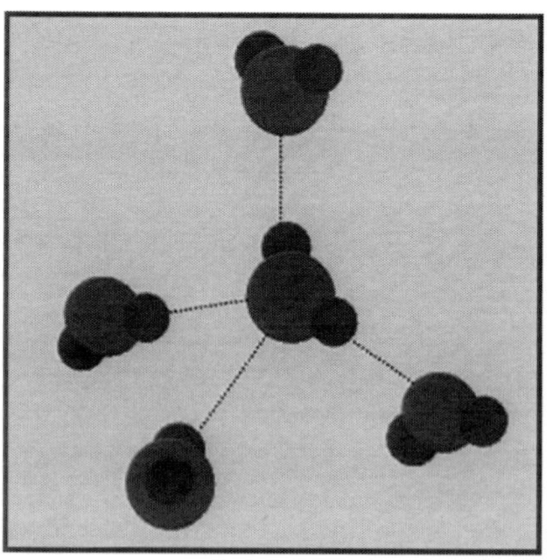

FIGURE 14.1 Tetrahedral hydrogen-bonded coordination of a water molecule in the liquid phase.

the so-called vibrational spectral diffusion, are determined and then mapped to the dynamics of hydrogen bond fluctuations. Recent experiments have reported two time scales for the dynamics of vibrational spectral diffusion in ambient water: (1) short-time decay in the range of 100–200 fs and (2) a longer-time decay in the range of 1–2 ps.[11–13] The altered solvent properties of supercritical water are thought to be related to the changes of the intermolecular structure at elevated temperatures, especially the change in the hydrogen bond network. This structural change can have a substantial effect on the hydration properties and dynamics of supercritical aqueous solutions. For example, a recent study on vibrational spectral diffusion on supercritical water reported a time scale of less than 300 fs,[14] which is much faster than the time scale of 1–2 ps found for ambient water.

In this chapter, we briefly review some of our recent studies on structure and dynamics of normal and supercritical water based on first principles simulations. The work reviewed here was based on the methods of *ab initio* molecular dynamics[8,9] for trajectory generation and time series analysis[15] for frequency calculations. We consider normal water at room temperature and also supercritical water at three different densities ranging from 1.0 to 0.35 g cm^{-3} at a temperature of 673 K. More details of the work reviewed here can be found in [16,17]. The next section of this chapter contains a brief account of the *ab initio* simulations and time series analysis. The results of the hydrogen-bonded structures and their relations to the vibrational frequencies and also the dynamics of the hydrogen bond and vibrational frequency fluctuations in normal and supercritical water are discussed in Section 14.4. This chapter is concluded in Section 14.5.

14.2 *AB INITIO* MOLECULAR DYNAMICS SIMULATIONS AND TIME SERIES ANALYSIS

In *ab initio* molecular dynamics simulations, the forces on the nuclei are obtained directly from an electronic structure calculation performed "on the fly" via adiabatic dynamics principle.[8] In this scheme, the force $F_i()$ on particle i at time t is given by

$$F_i(t) = -\nabla < \psi_e(t)|H_e|\psi_e(\mathrm{t})>, \quad (14.1)$$

where ψ_e and H_e are the electronic wave function and the electronic Hamiltonian operator (including the nuclear repulsion terms) of the system under investigation, respectively. In the current study, the *ab initio* molecular dynamics simulations were performed by using the Car–Parrinello molecular dynamics (CPMD) method.[8] This method combines a quantum mechanical treatment of the electrons within the framework of Kohn–Sham formulation of the density functional theory[18] and a classical treatment of the nuclei. The forces on the nuclei are calculated on the fly as the simulation progresses. The CPMD code[19] was used to carry out the *ab initio* simulations of normal and supercritical water at varying density. All hydrogen atoms were assigned the mass of deuterium to reduce the influence of quantum effects on the calculated properties. In addition, the choice of D_2O in place of H_2O ensured that electronic adiabaticity and energy conservation were maintained throughout the simulations. Further details of the simulations can be found in [16,17].

A major objective of carrying out the *ab initio* simulations was to find the correlations of hydrogen bonds and spectral observables like vibrational stretch frequencies. For this purpose, a time series analysis was performed by using the wavelet method. The details of this method can be found elsewhere.[15–17,20] Here, we only discuss the basic idea. A time-dependent function $f(t)$ is expressed in terms of base functions, which are constructed as translations and dilations of a mother wavelet, i.e.,

$$\psi_{a,b}(t) = a^{-1/2}\,\psi\left(\frac{t-b}{a}\right). \tag{14.2}$$

The coefficients of the expansion are given by the wavelet transform of $f(t)$, which is defined as

$$L_{\psi}\, f(a,b) = a^{-1/2} \int_{-\infty}^{\infty} f(t)\, \bar{\psi}\left(\frac{t-b}{a}\right) \tag{14.3}$$

for $a > 0$ and b is real. The mother wavelet was taken to be of Morlet–Grossman form.[21] The wavelet transform is a function of the variables a and b. The inverse of the scale factor a is proportional to the frequency, and thus, the wavelet transform at each b gives the frequency content of $f(t)$ over a time window about b. For calculations of OD stretch frequencies, the time-dependent function $f(t)$ for a given OD bond was constructed to be a complex function with its real and imaginary parts corresponding to the instantaneous OD distances and the corresponding momentum along the OD bond. The analysis was then performed for all the OD modes in a simulation system.

14.3 DISTRIBUTIONS OF VIBRATIONAL FREQUENCIES AND HYDROGEN BONDS

The hydrogen bonds are known to cause a red shift in the stretch frequencies of water molecules. The stretch frequency increases as the associated hydrogen bond becomes longer or weaker. Hence, studies of such correlations between the length of hydrogen bonds and the stretch frequencies of water and how such correlations change as one moves to supercritical state from ambient state would reveal important information about the changes of hydrogen-bonded structures and their implications on experimental observables such as vibrational spectroscopy. The results of such frequency–structure correlations are shown in Figures 14.2 and 14.3 for normal and supercritical water, respectively. In these figures, the contour plots of the conditional probability of observing a particular frequency for a given D··O distance are shown. Here, the oxygen of D··O is the one nearest to the D atom, i.e., it is the hydrogen bond distance in the event of the chosen OD mode being hydrogen bonded. Figure 14.2 shows the results for a density of 1.0 g cm^{-3} for ambient water, and Figure 14.3 shows the results for two lower densities of 0.7 and 0.35 g cm^{-3} for supercritical water.

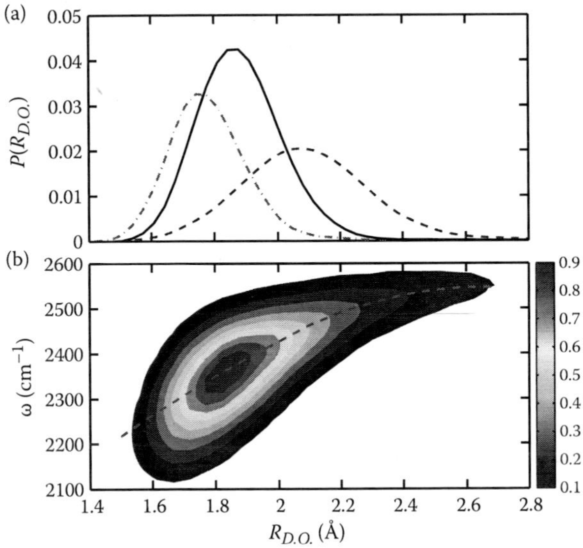

FIGURE 14.2 (a) Distribution of the D··O distance for OD frequencies at (black) average value and (dashed-dotted) –100 cm^{-1} and (dashed) +100 cm^{-1} of the average value. (b) Joint probability distribution of the OD frequency and D··O distance. (Reproduced from Mallik, B. S., A. Semparithi, and A. Chandra, *J. Phys. Chem. A*, 112, 5104, 2008. With permission.)

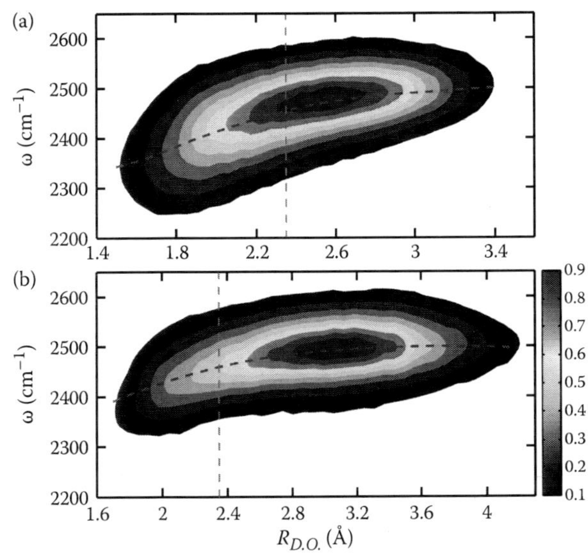

FIGURE 14.3 Joint probability distributions of OD frequency and D··O distance at 673 K form water densities of (a) 0.7 and (b) 0.35 g cm^{-3}, respectively. (Reproduced from Mallik, B. S., and A. Chandra, *J. Phys. Chem. A*, 112, 13518, 2008. With permission.)

On average, the stretch frequency of OD modes is seen to increase with the D··O distance. Similar results were also reported by others for ambient water.[22,23] However, as can be seen from Figure 14.3, the rate of increase of OD stretch frequency with D··O distance is found to gradually decrease as the density is lowered. In addition, with decreasing density, the maximum probability moves to a larger distance–higher frequency value. The distributions for ambient and supercritical water also reveal a clear change in the shape with increase in temperature. In the supercritical state, the distribution is extended to a larger D··O distance, which can be attributed to weaker and less number of hydrogen bonds at the elevated temperature. The vertical dashed line in these figures represent the dividing surface between the hydrogen-bonded and nonhydrogen-bonded states as given by the geometric criterion used to define the hydrogen bonds.[16,17] It can be seen that, while the major part of the probability distribution falls on the left side of the dividing line for the ambient water, an opposite behavior is observed for supercritical water at lower densities. For low-density systems, the major part of the frequency–distance probability distribution falls in the nonhydrogen-bonded region. The number of hydrogen bonds per water was found to be 3.6 for ambient water and 2.8, 2.3, and 1.4 for supercritical water at densities of 1.0, 0.7, and 0.35 g cm^{-3}, respectively.[17] This means that the hydrogen bonds play less significant roles in determining the properties of supercritical water at low densities, unlike ambient water where hydrogen bonds play dominant roles in determining the solvent properties of liquid water.

14.4 DYNAMICS OF VIBRATIONAL SPECTRAL DIFFUSION AND HYDROGEN BONDS

The dynamics of fluctuating frequencies in an aqueous medium can be described in terms of the time correlation function of fluctuations of such frequencies around their average values. In fact, the frequency time correlation function serves as the key dynamical quantity in the studies of vibrational spectral diffusion. This correlation function is defined as

$$C_{\omega}(t) = \frac{\langle \delta\omega(t)\delta\omega(0)\rangle}{\langle \delta\omega(0)^2\rangle}. \tag{14.4}$$

The average of Equation 14.4 is over the initial time and over all the OD groups present in a system. The results of the dynamics of frequency time correlation are shown in Figure 14.4 for both ambient and supercritical water.[17] The frequency time correlations reveal fast decay in a short time, which is then followed by a relatively slower decay. In addition, the dynamics is found to be significantly faster for supercritical water than that for ambient water. Generally, two time scales are found to be present in the decay of the frequency correlations: (1) a short time scale of about 100 fs and (2) a slower time scale of about 2 ps for ambient water and about 150–600 fs for supercritical water. For ambient water, the short- and long-time decays have been attributed to the dynamics of intact hydrogen-bonded pairs and breaking dynamics

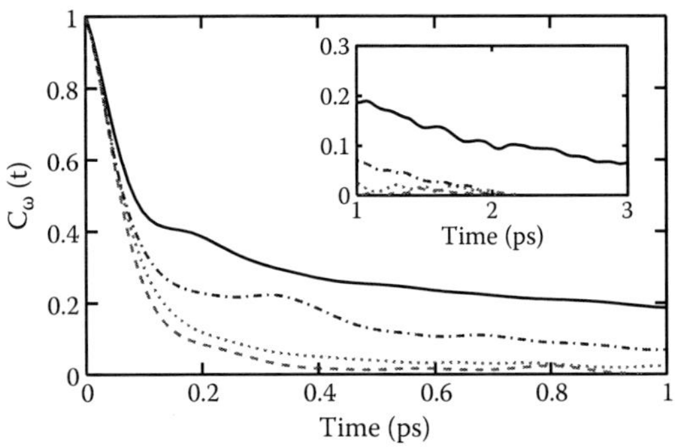

FIGURE 14.4 Time correlation functions of OD fluctuating frequencies of normal and supercritical heavy water. (Reproduced from Mallik, B. S., and A. Chandra, *J. Phys. Chem. A*, 112, 13518, 2008. With permission.)

of hydrogen bonds, respectively.[16] We note that such biphasic decay was also found in other recent theoretical studies of vibrational spectral diffusion, which used classical and quantum–classical models of water.[22,23] The reformation of hydrogen bonds is a rather fast event in ambient water, and it can also contribute to the short-time decay of the frequency fluctuations. These theoretical results are in general agreement with the findings of recent experiments on vibrational spectral diffusion in aqueous medium.[11–13]

The faster dynamics of spectral diffusion in supercritical water is also in agreement with available experimental results.[14] More importantly, theoretical analysis revealed that, for supercritical water, the slower component of the spectral diffusion does not necessarily capture the breaking dynamics of hydrogen bonds.[17] Rather, an interplay between the dynamics of hydrogen bonds, free OD groups, and rotation of OD bonds was found to determine the times scales of spectral diffusion in supercritical water in a rather subtle manner. At higher density of supercritical water, the slower component of spectral diffusion is still determined by the lifetimes of hydrogen bonds. However, the lifetime of dangling OD groups determines the slower component at the lower density. The reverse was found to hold for the faster component of the spectral diffusion in supercritical water of varying densities.

14.5 CONCLUSIONS

The *ab initio* molecular dynamics technique provides a powerful method in studying the properties of chemical systems under varying thermodynamic conditions without having to employ any empirical interaction potentials. In this chapter, a brief review has been made on our recent studies on water dynamics by using this method combined with a time series analysis. We have discussed the frequency–structure correlations of water molecules in both supercritical and normal water. Our calculations reveal that hydrogen bonds still persist to some extent in the supercritical state. However, the quantitative details of hydrogen bonding depend on the density. At lower density, the number of hydrogen bonds decreases so much that the nonhydrogen-bonded water molecules play more dominant roles in determining the properties of supercritical water.

The dynamics of vibrational frequency fluctuations or the spectral diffusion in (deuterated) water is described in terms of the vibrational frequency time correlation function of OD stretch modes of (deuterated) water molecules. For ambient water, the frequency time correlation function shows a short-time decay with a time scale of about 100 fs corresponding to the motion of intact hydrogen bonds and a long-time decay with a time scale of about 2 ps corresponding to an average lifetime of hydrogen bonds. The short-time dynamics of the frequency correlation can also have some contributions from hydrogen bond reformation events, which also occur at a rather fast time scale in ambient water. The vibrational spectral diffusion in supercritical water occurs with a much shorter time scale compared to that in ambient water. In the supercritical state, the frequency time correlation function decays with two time scales: one around 100 fs or less, and the other in the region of 150–600 fs. Unlike ambient water, for supercritical water, the slower component of the spectral diffusion does not necessarily correspond to the hydrogen bond dynamics at all densities. Rather, the time scales of spectral diffusion in supercritical water are determined in a rather subtle manner by an interplay between the dynamics of hydrogen bonds and free OD groups and also the fast rotation of OD bonds. We finally note that the theoretical methodology discussed here has also been used in recent years to study vibrational spectral diffusion in other aqueous systems such as aqueous ionic solutions, water with an organic solute, and also interfacial systems.[24–26]

ACKNOWLEDGMENTS

We gratefully acknowledge financial support from the Department of Science and Technology and Council of Scientific and Industrial Research, Government of India.

REFERENCES

1. *Water: A Comprehensive Treatise*, edited by Franks, F. New York: Plenum. Vol. 1–8, pp. 1972–1979.
2. Clark, G. N. I., C. D. Cappa, J. D. Smith, R. J. Saykally, and T. Head-Gordon. 2010. *Mol. Phys.* **108**:1415.
3. Soper, A. K., F. Bruni, and M. A. Ricci. 1997. *J. Chem. Phys.* **106**:247.
4. Osada, M., K. Toyoshima, T. Mizutani, K. Minami, M. Watanabe, T. Adschiri, and K. Arai. 2003. *J. Chem. Phys.* **118**:4573.
5. Takebayashi, Y., S. Yoda, T. Sugeta, and K. Otake. 2004. *J. Chem. Phys.* **120**:6100.
6. Barker, J. P., and R. O. Watts. 1969. *Chem. Phys. Lett.* **3**:144.
7. Allen, M. P., and D. J. Tildesley. 1987. *Computer Simulation of Liquids*. Oxford, NY.
8. Car, R., and M. Parrinello. 1985. *Phys. Rev. Lett.* **55**:2471.
9. Marx, D., and J. Hutter. 2009. *Ab Initio Molecular Dynamics: Basic Theory and Advanced Methods*. Cambridge: Cambridge University Press.
10. Bakker, H. J., and J. L. Skinner. 2010. *Chem. Rev.* **110**:1498 and references therein; Nibbering, E. T. J., and T. Elsaesser. 2004. *Chem. Rev.* **104**:1887 and references therein.

11. Bakker, H. J., M. F. Kropman, A. W. Omta, and S. Woutersen. 2004. *Physica Scripta.* **69**:C14; Woutersen, S., and H. J. Bakker. 1999. *Phys. Rev. Lett.* **83**:2077.
12. Fecko, C. J., J. D. Eaves, J. J. Loparo, A. Tokmakoff, and P. L. Geissler. 2003. *Science.* **301**:1698; Fecko, C. J., J. J. Loparo, S. T. Roberts, and A. Tokmakoff. 2005. *J. Chem. Phys.* **122**:054506.
13. Park, S., and M. D. Fayer. 2007. *Proc. Nat. Acad. Sci. (USA).* **104**:16731; Asbury, J. B., T. Steinel, C. Stromberg, S. A. Corcelli, C. P. Lawrence, J. L. Skinner, and M. D. Fayer. 2004. *J. Phys. Chem. A.* **108**:1107.
14. Schwarzer, D., J. Lindner, and P. Voehringer. 2005. *J. Chem. Phys.* **123**:161105; 2006. *J. Phys. Chem. A.* **110**:2858.
15. Vela-Arevalo, L.V., and S. Wiggins. 2001. *Int. J. Bifur. Chaos.* **11**:1359.
16. Mallik, B. S., A. Semparithi, and A. Chandra. 2008. *J. Phys. Chem. A.* **112**:5104.
17. Mallik, B. S., and A. Chandra. 2008. *J. Phys. Chem. A.* **112**:13518.
18. Kohn, W., and L. Sham. 1965. *J. Phys. Rev. A.* **140**:1133.
19. Hutter, J., A. Alavi, T. Deutsch, M. Bernasconi, S. Goedecker, D. Marx, M. Tuckerman, and M. Parrinello. *CPMD Program*, MPI füFestköerschung and IBM Zurich Research Laboratory.
20. Semparithi, A., and S. Keshavamurthy. 2003. *Phys. Chem. Chem. Phys.* **5**:5051.
21. Carmona, R., W. L. Hwang, and B. Torresani. 1998. *Practical Time–Frequency Analysis: Gabor and Wavelet Transforms with Implementations.* New York: Academic.
22. Rey, R., K. B. Moller, and J. T. Hynes. 2002. *J. Phys. Chem. A* **106**:11993; Moller, K. B., R. Rey, and J. T. Hynes. 2004. *J. Phys. Chem. A.* **108**:1275.
23. Lawrence C. P., and J. L. Skinner. 2003. *J. Chem. Phys.* **118**:264; Corcelli, S. A., C. P. Lawrence, and J. L. Skinner. 2004. *J. Chem. Phys.* **120**:8107.
24. Mallik, B. S., A. Semparithi, and A. Chandra. 2008. *J. Chem. Phys.* **129**:194512.
25. Gupta, R., and A. Chandra. 2012. *J. Mol. Liq.* **165**:1.
26. Chakraborty, D., and A. Chandra. 2012. *Chem. Phys.* **392**:96; 2011. *J. Chem. Phys.* **135**:114510.

Index

Page numbers followed by *f* and *t* indicate figures and tables, respectively.

N

O

P

T

U

V